Noise Control, Reduction and Processing in Engineering Systems

Noise Control, Reduction and Processing in Engineering Systems

Edited by **Matt Goodwin**

LANRYE INTERNATIONAL

New Jersey

Published by Clanrye International,
55 Van Reypen Street,
Jersey City, NJ 07306, USA
www.clanryeinternational.com

Noise Control, Reduction and Processing in Engineering Systems
Edited by Matt Goodwin

International Standard Book Number: 978-1-63240-386-5 (Hardback)

Contents

Preface

This book aims to highlight the current researches and provides a platform to further the scope of innovations in this area. This book is a product of the combined efforts of many researchers and scientists, after going through thorough studies and analysis from different parts of the world. The objective of this book is to provide the readers with the latest information of the field.

The book discusses a variety of practical applications of various noise control strategies in diverse engineering situations. Noise control, reduction as well as processing in engineering systems have been elucidated in this all-inclusive up-to-date book. Noise has various adverse effects on comfort, performance, and human health. Hence, noise control plays an increasingly central role in the advancement of modern industrial and engineering operations. Issues regarding noise control have attracted the attention of various scientists in a variety of fields. Noise control has a vast variety of applications in manufacturing and industrial activities. The objective of this book is to provide an extensive overview of current developments in noise control and its applications in several research areas. This book will be beneficial for researchers and engineers studying this field.

I would like to express my sincere thanks to the authors for their dedicated efforts in the completion of this book. I acknowledge the efforts of the publisher for providing constant support. Lastly, I would like to thank my family for their support in all academic endeavors.

Editor

Part 1

Background of Active Noise Control

Noise Sources in the City:
Characterization and Management Trends

Alice Elizabeth González

Environmental Engineering Department, IMFIA, Faculty of Engineering
Universidad de la República (UdelaR)
Uruguay

1. Introduction

Since man began to live sedentary, the noise in the city has been considered to be a more or less important source of annoyance. It is possible to find precedents about this subject in Greece and Rome, even before the beginning of the Christian Age. The proofs of this statement can be found in the Ancient Rome in Marcial's poetry texts about S. I b.C or in the regulations of Síbaris's city, Calabria -that was then included in the Magna Graecia- about year 600 b.C., that stated that no one could have cocks at home, and the "hammer workers" had to live outside of the city. Perhaps this is one of the oldest provisions that explicitly determine land use to noise pollution. It is also well known that, in the Middle Ages, it was forbidden to punish women at night time in London, to avoid disturbing noises (González, 2000).

Towards 1700 Bernardino Ramazzini, the "father" of the Occupational Medicine, stated for the first time the cause and effect relationship between occupational noise exposure and auditive damage. In his book "Of Morbis Artificum Diatribe" (1713), he explained this issue related to grain millers and bronze workers. About the "neighborhood of the coppersmiths" in Venice, Ramazzini wrote: *"There are bronzesmiths in all the towns and in Venice they gather around only one neighborhood. They hammer there the entire day to give ductility to the bronze and to make metal jars with it. They also have their taverns and homes there, and they cause such an intense clatter that everyone runs away from such an annoying place"* (González, 2000). It is interesting to note that noise appears for another time as a restriction for land use.

From the Middle Ages till the second half of the 19th century, the design and structure of the cities had suffered almost no changes. At the beginning of the Industrial Revolution, a process of disorderly and uncontrolled growth of urban centers emerged. People began to migrate "en masse" from countryside to the city, searching for new horizons and better job. This caused not only a fast growth of the urban population and of the built-up area, but also of the lots of problems derived from the lack of town planning and the overcrowded "around modernity".

The noise of the factories and the hustle derived from the people in the immediate areas deeply deteriorated the acoustic landscape of the cities. However, the noise and the black smoke from the chimneys of the factories were emblematic symbols that remained strongly associated with progress and economic improvement for a long time, even if these benefits were only for a little privileged part of the society.

Technological development began to generate increasingly powerful –and, perhaps, progressively also cheaper- sources of noise that people at first thought to be related to job, then to progress and some years later, to comfort and social status. However, if technological and sociocultural development are taken into account, the main noise sources in the cities have remained the same throughout history: at s. I a.C., Séneca mentions that the main noise sources in the ancient Rome were the noise of the urban traffic ("the cars of the street"), the occupational noise and the leisure noise; at s. XXI b.C. these are *still* the main sources at our-days cities, as stated for many researchers (Beristáin, 2010; Madariaga Coaquira, 2008; González & Echeverri, 2008). Face to this, it is not easy to share Orozco-Medina's affirmation: *"The noise is an unwished by-product of the modern way of life"* (Orozco-Medina & Figueroa Montaño, 2010). Though her expression is harder, this idea is the same as Lizana's when referring *"the acoustic costs associated with the progress"* (perhaps in a wider historical sense). He also states: *"The noise seems to be the most inoffensive of the pollutant agents"* (Lizana, 2010).

Regarding the main noise sources in two Latin-American our-days cities, Madariaga Coaquira indicates that nowadays the main sources of urban noise in Arequipa's Peruvian city are the traffic (road, rail and air traffic), the announcements of fruit sale and other peddlers, the police whistle, the informal shops (bakeries, welding shops, shoe repair shops, carpentries), the social events and the artistic festivals (Madariaga Coaquira, 2008). In two papers of Beristáin about noise in Mexico City, one of them published twelve years later than the other, it is interesting to notice that main sources remain the same: road traffic and transportation, industrial and commercial places, formal and informal peddlers, public and private civil works, recreation places (discotheques, rock concerts, public and private meetings and parties, local village fairs, domestic household and neighborhood noise (Beristáin, 1998, 2010).

In general, three types of noise exposure can be clearly recognized in today society (CONICYT-IMFIA, 1998):

-   Occupational exposure, that is motivated by work circumstances.
-   Social exposure, which is voluntary and involves attendance to noisy places or the voluntary "consumption" (in a wide sense) of high levels of sound -for example, the use of personal music devices at high volume, the listening to loud music, the practice of sports like shooting, as well as other activities-.
-   Environmental exposure, that is not voluntary in the sense of not being pursued by the person, but at the same time it is often inevitable. It is related to sound sources that are present in the environment where the person is -for example, traffic noise, music from stores, or neighborhood noises-.

2. Noise sources in the city

In order to describe them, the noise sources are often classified in three groups: stationary, mobile and collective sources (Miyara, 2008).

Stationary sources are those which, at the working scale, may be assumed as located at one fixed point of the space, for example a factory, a machine, a leisure place, a maneuvers zone. The civil works in urban zones can often be considered to be fixed sources.

Mobile sources instead, may also be assumed as point sources at the working scale, but they are allowed to move in the space. Is the case of road traffic, railroads and aircrafts.

The collective sources are the result of the accumulation of sources in public spaces, for example, the people in the street, a village fair, a public meeting, a city square or a popular celebration.

From another point of view, noise sources in the cities may be related to traffic and transportation, leisure, work, community services and neighborhood.

2.1 Traffic

There is no doubt that the main source of noise in the city is road traffic. Orozco-Medina reports studies from Costabal and Fagúndez showing that traffic is responsible for about 70 % of the urban noise. The major part of it comes from motor vehicles and the most relevant sources are the heavy motor vehicles, including public transport (Orozco-Medina & Figueroa Montaño, 2010).

The maritime or fluvial transport only has incidence in port cities and only near the port facilities. The air transport is the main noise source in the zones near the airports or close to the air routes. Neither the airports might be built extremely close to the cities nor the air routes be designed passing over urban areas.

In the cities where the railway is a popular transport system –for example, the metro or the subway- the whole systems had to be modernized to satisfy current environmental standards. As a consequence, most of the urban rail systems were enhanced to fulfill lower noise standards; then, they turned to reasonable noise levels. However, some very noisy trains are still running, especially in subway lines. An emblematic case is the extremely noisy subway service of Buenos Aires city in Argentine.

As it is consented worldwide, the expression "traffic noise" refers to engine and tyre noise from vehicles. But other elements should be considered to occur, as loud sound from brakes, noisy exhausts and claxons (horns), whose incidence depends on the fleet age, the driving style and the existence, fulfillment and control of a suitable regulation (González, 2000; Orozco-Medina & Figueroa-Montaño, 2010; Sanz Sa, 2008). Nowadays, researchers work on the development of vehicles powered by alternative energy sources, to reduce the dependence on fossil fuels. Some of them have another environmental advantage besides the reduction of the gases and particulate matter emissions: they are more silent. Complaints did not wait: the detractors of hybrid and electric vehicles got strongly queried the safety of a silent transport system (Observatorio de las novedades acústicas y musicales, 2010).

Regulating traffic noise is not simple, but necessary. Already in the Ancient Rome, transport regulation was needed to forbid driving during the night, to avoid disturbing people during their rest hours (Orozco-Medina & Figueroa Montaño, 2010). Currently, the standards tend to set the admissible noise emission levels for each vehicle according to its gross weight, its engine and number of seats. To know the actual noise emission level, static or dynamic tests according to international standard procedures should be performed (for example, ISO 5130: 2007 for stationary test and ISO 362:2009 for dynamic tests) (González et al., 2008).

On the other hand, the control and management of traffic noise should tend to protect public health, but even if some exceptions can be found (and it is not possible to fulfill their requirements) there has no sense to limit traffic noise as a whole (González et al., 2008). Nevertheless, some management measures that are not designed to control or to diminish the traffic noise in the cities can reach good results in this sense; it is the case of the traffic regulation known as "pico y placa"[1] in Medellín city, Colombia (Ríos Valencia, 2008). Some aspects that should be taken into account when dealing with traffic noise in the cities are discussed in Section 4.

2.2 Ambient noise from industries and commerce

Among the noise stationary sources, major cases are the industrial and trade facilities and the leisure places. These are not the only ones: if they last for a long time, the civil works in the city would be considered as stationary sources (CONICYT-IMFIA, 1998; IMM-Facultad de Ingeniería, 1999). This section is referred to ambient noise that industrial and trade facilities generate close to their location, but not to the occupational exposure of workers.

The noise troubles caused by industrial and trade facilities must consider three aspects: the propagation of the noise from the operation of the company as a stationary source and aiming for the fulfillment of regulations; the noise of the induced traffic, whose spatial scale is undoubtedly larger, and that must be added to the preexisting traffic noise; and the noise associated with the parking lots and loading and unloading zones, whose extension and dynamics should be carefully analyzed case by case in order to identify their levels of activity at different times.

Contrary to intuition, the control of environmental noise associated with the operation of large and medium industries is not the major problem to manage about stationary sources, due to the fact that the own scale of the company allows it to invest to meet current regulations. This is not only the best from the legal standpoint, but also the company can reduce or even avoid troubles that could occur later because of complaints from neighbors. In the same sense, Lizana indicates that, unlike the large factories, the small and medium-sized shops and industries are distributed all around the city, so their indirect impacts caused by the transport of materials and products affect bigger areas (Lizana, 2010). Sometimes is easy to find loudspeakers out of small shops emitting music or announcing the sales to attract people that walk by (Beristáin, 2010), often making trouble with the neighbors. In fact, the hardest problems to manage derive from the small shops working in noisy activities. Their small scale, even artisanal, often put them out of the formal working system. For these companies, fulfilling noise regulations is usually very difficult, and they often break them (Beristáin, 2010).

Nevertheless, the highest levels of noise related to big malls and hypermarkets occur inside: they are designed with basis on detailed psychology and marketing studies to satisfy fun and shopping expectation of clients. The major troubles with the neighbors may be caused by the continuous operation of air conditioning equipment, compressors and others, which run 24 hours a day, so their noise emission, that often has startings and stops, becomes much more evident in the night, when the city noise levels are low.

Noise from the loading and unloading areas usually offers true challenges to manage. It involves not only heavy traffic but trucks waiting with running engines, loud voices, horselaughs, shouts, blows and other events almost inseparable from these affairs. In turn, loads management usually is not performed by the main entrance of the company but by its back entrance. This usually stimulates the spontaneity of workers, in terms of their verbal communication and how to perform their tasks.

---

[1]"Pico y placa" (peak and plate) means "rush hour and plate": according to the type of vehicle, the last number of its plate and the day of the week, one vehicle can or cannot run during morning and afternoon rush hours.

When perimeter barriers are built with the main objective of reducing the visibility towards the inland of the company or harmonizing with the surrounding landscape, many times they can also serve as acoustic barriers. This is the case of landscaped embankments when it is possible to build them, or perimeter walls of height and design compatible with those from the buildings, attending to aesthetic criteria and current urban development regulations. The acoustic attenuation provided for these barriers could be calculated.

The incorporation of plant species to improve the appearance of a non-acoustic barrier should be always a good idea, especially in urban areas. Though, a green curtain or a vegetal barrier designed to have an acoustic function should not be confused with an aesthetic one. Indeed, according to the statements of Kotzen and English, plants and trees must meet certain characteristics to make feasible their use in acoustic barriers. According to them, with basis on experimental studies developed between 1980 and 1990, the effectiveness of plant curtains between 15 m and 40 m thickness is greater at 250 Hz and above 1000 Hz, but they warn that their maximum attenuation is about 3 dB in L_{Aeq}. These researchers also state that the maximum attenuation performance of a green barrier occurs at a wavelength twice the size of the leaves of the tree species considered (Kotzen & English, 2009).

Recent experimental studies carried out in Colombia did not find statistically significant differences between sound levels measured at a distance of 10 m from the same highway in four areas with different vegetation: low height pasture (control area), trees over 5 m high, shrub (species less than 5 m height and branched from the base), and an area with both trees and shrub (Posada et al., 2009).

2.3 Leisure noise

The voluntary assistance to noisy places, or "voluntary consumption of noise", is the main face of social exposure to noise. Most of the young people are not only voluntarily exposed to noise but they also generate it: music, voice tone, lifestyle and entertainments are noisier every day. This section is addressed to the management of noise troubles at the so called "pink zones", the zones of the cities where nightlife and leisure places are the main stars.

The noise produced by different amusement activities in the night is a growing concern for city managers. While historically the main challenge was to oblige all leisure places to have a proper acoustic isolation to avoid generating high sound levels in neighbors' houses, other conflicts and problems have emerged in recent years (González & Echeverri, 2008).

The noise problem was first centered on the fact that the recreation places had to have satisfactory isolation conditions to assure the resting of the next door neighbors. The main difficulties for the management at that time were related to having good acoustical projects, building them properly, planning the needed investments in a reasonable time (sometimes they might be really expensive) and controlling their performance in service. Even when significant investments are required, these could avoid or end with neighborhood troubles. In one case that was recently discussed, the updating of a big non-isolated building chosen to installing a nightclub to a double wall place, with double ceiling and forced ventilation system, resulted in costs equal to one month ticket sales. This issue will be deeper presented in Section 3.

2.4 Construction of civil works

The construction of public and private civil works in the urban areas can become in practice "stationary sources" of noise, especially when the construction stage lasts for a long time, regardless of the causes for this to happen (CONICYT-IMFIA, 1998; IMM-Facultad de Ingeniería, 1999). But even if they do not last more than the strict necessary time, these activities are often annoying for the neighbors because of the noise of machinery or other activities inherent to the works.

To shorten the construction timing, sometimes people should work not only during the daytime. In some cases, especially when works involve high traffic streets, many labors should be performed only during nighttime to minimize traffic and safety problems (Beristáin, 2010).

Noise levels related to civil works machinery differ from one machine to another. Table 1 presents the acoustic power levels for some machines, as they are stated in the Directive 2005/88/EC relating to the noise emission to the environment by equipment used outdoors.

In some cases, construction companies choose to adopt measures to reduce community impact of the noise from works they perform. For instance, it is the case of "Empresas Públicas de Medellín E.S.P." (Public Companies of Medellín), Colombia: since the company has reviewed its Standards and Specifications for General Construction, the NEGC 1300 ("Environmental management in public services facilities"), a chapter concerning on "Community Impact" was added. It defines five main action areas: community relations; urban environmental management; road safety and signalization; social security wage and industrial safety management in the construction place (Giraldo Arango, 2008).

As Giraldo Arango refers, the experiences of the company in incorporating management guidelines defined by needs of social responsibility and environmental management have shown that, in some cases, works have not become more expensive: just the opposite, the changes have allowed saving money and runtime in some works, as well as reducing conflicts with the neighborhood. He remarks that the performed changes also have caused the reduction of noise levels in areas close to the construction site. The company has incorporated new environmental management requirements in its works specifications: the use of modern machinery and equipment with soundproofed engines; preventive and corrective maintenance measures to keep them in proper conditions; when it is not possible to isolate the sound emission of a jackhammer or the disc from a cutter, maximum lasting of shifts of two continued hours for equipment whose noise emissions exceed the permitted standards, with breaks of the same duration; coordination of schedules with the heads of near educational and health institutions, aiming at doing pavement cuts during class breaks or shift changes. Giraldo Arango states: *"While it is true that some equipment and machinery used by the civil works construction sector are noisy, in both the public and the private sector some statements mention the impossibility of controlling their noise levels. This opinion is only partially true, since it has been proved that is possible in practice to use other methods as those normally used, and which are environmentally and economically even more attractive"*. He highlights the importance of preventive maintenance of equipment and machinery, as well as the need of a greater control about environmental management. In his opinion, control activities should be assumed by the hiring entities (Giraldo Arango, 2008).

Table 2 presents some noise measurements carried up at Montevideo, Uruguay, in 2010, one meter away from two wellpoint pumps of the same characteristics, which were being used to depress the water table level. One of them was into an acoustic encapsulation and the other was not. Both equipments belong to the same company and were working at the same time in similar works at the same area of the city.

Type of equipment	Net installed power P (kW) Electric power P_{el} in kW [1] Mass of appliance m in kg Cutting width L in cm	Permissible sound power levels in dB / 1 pW	
		Stage I as from January 3, 2002	Stage II as from January 3, 2006
Compaction machines (vibrating rollers, vibratory plates, vibratory rammers)	$P \leq 8$	108	105 [2]
	$8 < P \leq 70$	109	106 [2]
	$P > 70$	$89 + 11 \lg P$	$86 + 11 \lg P$ [2]
Tracked dozers, tracked loaders, tracked excavator-loaders	$P \leq 55$	106	103 [2]
	$P > 55$	$87 + 11 \lg P$	$84 + 11 \lg P$ [2]
Wheeled dozers, wheeled loaders, wheeled excavator-loaders, dumpers, graders, loader-type landfill compactors, combustion-engine driven counterbalanced lift trucks, mobile cranes, compaction machines (non-vibrating rollers), paver-finishers, hydraulic power packs	$P \leq 55$	104	101 [2] [3]
	$P > 55$	$85 + 11 \lg P$	$82 + 11 \lg P$ [2] [3]
Excavators, builders' hoists for the transport of goods, construction winches, motor hoes	$P \leq 15$	96	93
	$P > 15$	$83 + 11 \lg P$	$80 + 11 \lg P$
Hand-held concrete-breakers and picks	$m \leq 15$	107	105
	$15 < m < 30$	$94 + 11 \lg m$	$92 + 11 \lg m$ [2]
	$m > 30$	$96 + 11 \lg m$	$94 + 11 \lg m$
Tower cranes		$98 + \lg P$	$96 + \lg P$
Welding and power generators	$P_{el} \leq 2$	$97 + \lg P_{el}$	$95 + \lg P_{el}$
	$2 < P_{el} \leq 10$	$98 + \lg P_{el}$	$96 + \lg P_{el} P_{el}$
	$10 > P_{el}$	$97 + \lg P_{el}$	$95 + \lg P_{el}$
Compressors	$P \leq 15$	99	97
	$P > 15$	$97 + 2 \lg P$	$95 + 2 \lg P$
Lawnmowers, lawn trimmers/lawn-edge trimmers	$50 < L \leq 70$	100	98
	$70 < L \leq 120$	100	98 [2]
	$L > 120$	105	103 [2]
	$L \leq 50$	96	94 [2]

[1]P_{el} for welding generators: conventional welding current multiplied by the conventional load voltage for the lowest value of the duty factor given by the manufacturer.
P_{el} for power generators: prime power according to ISO 8528-1: 1993, clause 13.3.2
[2]The figures for stage II are merely indicative for the following types of equipment:
- walk-behind vibrating rollers;
- vibratory plates (> 3k W);
- vibratory rammers;
- dozers (steel tracked);
- loaders (steel tracked > 55 kW);
- combustion-engine driven counterbalanced lift trucks;
- compacting screed paver-finishers;
- hand-held internal combustion-engine concrete-breakers and picks (15 < m < 30)
- lawnmowers, lawn trimmers/lawn-edge trimmers.
Definitive figures will depend on amendment of the Directive following the report required in Article 20(1). In the absence of any such amendment, the figures for stage I will continue to apply for stage II.
[3]For single-engine mobile cranes, the figures for stage I shall continue to apply until 3 January 2008. After that date, stage II figures shall apply.
The permissible sound power level shall be rounded up or down to the nearest integer number (less than 0,5, use lower number; greater than or equal to 0,5, use higher number)

Table 1. Admissible levels of acoustic power for machines of functioning outdoors according to Directive 2005/88/EC.

	Non encapsulated machine	Encapsulated machine	Difference	Observations
$L_{AFmáx}$	87,2	77,1	-10,1	Significant attenuation
$L_{AFmín}$	82,6	74,3	-8,3	Significant attenuation
L_{AIeq}	86,4	76,4	-10,0	Significant attenuation
L_{AFeq}	84,8	75,5	-9,3	Significant attenuation
L_{CFeq}	91,1	90,2	-0,9	As it is expected, the encapsulation is not effective at low frequencies.
L_{AF5}	85,6	76,0	-9,6	Significant attenuation
L_{AF10}	85,4	75,9	-9,5	Significant attenuation
L_{AF50}	84,8	75,5	-9,3	Significant attenuation
L_{AF90}	84,1	75,1	-9,0	Significant attenuation
L_{AF95}	83,9	75,0	-8,9	Significant attenuation
$L_{AIeq} - L_{AFeq}$	1,6	1,0	-0,6	As L_{AFeq} and L_{AIeq} have had similar decreases, then impulsivity is only slightly reduced. Not a significant variation is found.
$L_{CFeq} - L_{AFeq}$	6,3	14,8	+8,5	As the encapsulation is not effective at low frequencies, noise levels expressed in dBC differ much more with the same levels expressed in dBA when the equipment is encapsulated.
$(L_{AF10} - L_{AF90})$	1,3	0,8	-0,5	As L_{AF10} and L_{AF90} have had similar decreases, then is only slightly reduced. Not a significant variation is found.

Table 2. Comparison of the acoustic performances of two well point pumps, one of them into an acoustic encapsulation.

All the measures were done with a type 1 sound level meter, with time responses fast, slow and impulse in real time, frequency weighting scales A and C and third-octave bands analysis also in real time. The attenuation of sound levels due to the acoustic encapsulation in the immediate environment is important in most of the measured parameters: it is about 9 dB regarding the energy levels (L_{Aeq}) and also regarding the statistic ones (the permanence levels that were considered are L_{AF5}, L_{AF10}, L_{AF50}, L_{AF90}, L_{AF95}). As expected, this does not happen with the parameters measured with C frequency weighting, that have worse response to the acoustic protection (González, 2010b).

2.5 Fairs and markets

Even though the neighbourhood fairs and markets are referred as part of the agents of lower acoustic impacts on the city regarding their scope and number of affected people (Orozco-Medina & Figueroa-Montaño, 2010), they often result in troubles that sometimes should turn into neighbourhood conflicts (Defensoría del Vecino de Montevideo, 2010).

Because of their function they must be installed within residential areas, and their performance and schedule are often the main causes of noise complaints. Management measures, then, can only be derived from agreements with and within neighbours. In Montevideo, Uruguay, with the increase in complaints in this regard, the Ombudsman's Neighbour organized a seminar to discuss this issue. The recommendations of the seminar aimed to generate a fair rotation schedule between different streets of the neighbourhood they serve. In a period of few months, twelve fairs were transferred from their usual places to other ones nearby (Defensoría del Vecino de Montevideo, 2010).

2.6 Popular celebrations, artistic festivals

Day to day, the artistic festivals congregate more people. Beginners or consecrated, most of musicians use to take part in these events that are the major way for the firsts to be known or for the idols to be in close contact to their fans. Bigger and bigger places are needed to perform these festivals. As it might be suspected, the biggest ones are open places (stadiums or parks, for example), and obviously they are not soundproof places. Not only the wellknown complaints about noise and safety might accompany these events: for instance, also bizarre conflicts may take place. In December of 2000, a great music festival that was attended by 90.000 people was performed at the hippodrome of Rosario city, Argentine. No one from the organization team remembered that also about 500 race horses were in the stables at the same time. The animals virtually madden with such a high noise and they tried to escape, clashing one and another time against the walls and gates, till many of them resulted severely hurt (Laboratorio de Acústica y Electroacústica, 2011).

Many societies have traditional cultural events that in many of the cases are accompanied by high levels of noise. Historically they have been considered as activities of interest to preserve and even potentiate to strengthen the local identity (Brito, 2011a; Defensor del Pueblo de Madrid, 2005). Nowadays, the increasing number of complaints and even trials about excessive noise levels, has forced the Administration to amend its legislation as a preventive way for preempting such actions.

For instance, in 2007 the trial judges banned the celebrations of the Carnival of Tenerife by excessive noise, based on a lawsuit filed by neighbors and a socially controversial trouble was installed (Diario El País, 2007). Then Valencia authorities preemptively conquered exceptions in terms of noise levels for 13 public celebrations, including the Fallas. These are particularly noisy, because of the pyrotechnics great display that happens at the climax of the celebration. Without referring to extreme noise levels, which often largely exceed 100 dBA, in an interview carried by the Official College of Industrial Superior Engineerings in 2004, Gaja said: *"The ideal average decibel is setted at a maximum of 65 for the day and 45 for the night but the main streets of the city always vary between 70 and 75 dBA; this level not lowers than 80 during celebrations of Fallas"* (Colegio Oficial de Ingenieros Superiores Industriales de la Comunidad Valenciana, 2004). Figure 1 present daily records of L_{Aeq} at the Plaza of España, Valencia. The atypical noise levels in March are clearly shown. The peaks are given by the celebrations of Fallas (Gaja et al., 2003).

An exhaustive report about noise pollution was published by the Ombudsman of Madrid, Spain. In this detailed document, the Ombudsman discusses almost all the expected troubles about noise (and some not so expected) in today Madrid's society. He puts in the shoes of the different social actors to present an overview as complete as possible of each case, taking into account not only legal and technical issues, but also health, ethics, economics and social ones. Management of urban noise is not an easy task. Referring to popular celebrations, he says: *"the occasional noise is noise anyway"* (Defensor del Pueblo de Madrid, 2005).

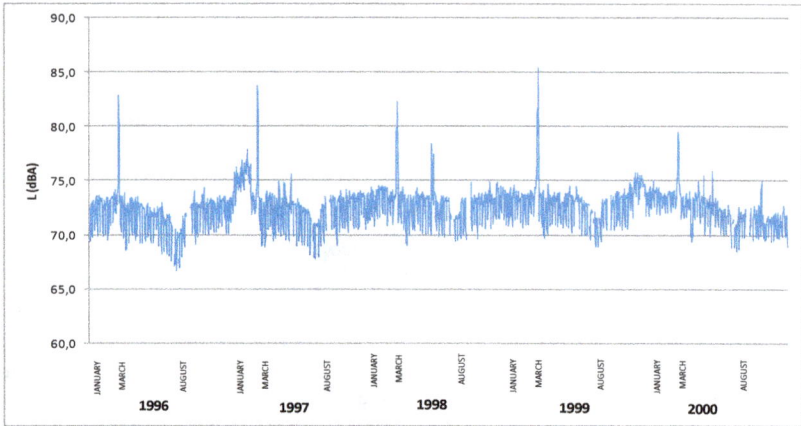

Fig. 1. $L_{Aeq,24\,h}$ at the Plaza Mayor of Valencia, Spain.

2.7 Community services: Street sweeping, solid waste collection

Some community services are inevitably associated with noises whose levels may disturb the neighborhood. Sometimes they can be minimized, but to achieve this, deep changes are required about the way the service is provided. For example, to avoid the noises caused by the mechanical sweeping of streets, a more frequent manual sweeping service could be considered.

Waste collection services, either domestic or special, should not be intended only as a noise issue about slow heavy traffic. There are different options for waste collection. The selection of one or another usually does not depend on the noise levels of them. Once the decision is made, the features of the selected option also define the sources and levels of noise that will be involved.

The elimination of noise sources from prevailing collecting system not always is possible. The proper planning and operation of the service and the preventive maintenance of all the equipment and vehicles assigned to it are the most effective ways to minimize noise impacts in the neighborhood.

2.8 Other noises in the neighborhood

All the sources of noise that affect a certain place are part of the environmental noise at it. Their interactions result in the so called "acoustic landscape", that is an expression of the "acoustic identity" of the place in a certain time.

The particular characteristics of each society can define its quotidian noise levels. These are associated with factors such as lifestyle, cultural patterns, socio-economic development, technification level, among others; these factors also have incidence on the environmental noise levels that the society could accept. Once their basic living needs are satisfied, people begin to aim for a better life quality. The acoustic comfort is one of the most appreciated aspects that people aim to conquer for improving their quality of life. Not in vane Robert Koch said: *"The day will come when people begin to fight the noise as they do so against cholera and pest"* (González, 2000).

Other agents that are included in the list of sound sources of reduced scope, that cause only minor impacts and which shape the details of the sound landscape from of each place, are for instance the crying of children, street sales, peddlers and sounds caused by pets (Orozco-Medina & Figueroa-Montaño, 2010). While these elements cause minor local impacts, they appear all the time as annoying elements for people in different cities, as Arequipa, Perú (Madariaga Coaquira, 2008) or Montevideo, Uruguay (CONICYT-IMFIA, 1998; González et al., 1997). In fact, the first acoustic map of Montevideo conducted between 1997 and 1998 by the modality of selected points that were chosen by the Municipality, has evidenced that the main source of noise annoyance in some neighbourhoods was the barking of dogs (CONICYT-IMFIA, 1998).

Referring to household appliances and increasingly powerful audio devices, Beristáin said: *"Modernity has brought some comfort to the houses of many people, but also has brought continuous and intermittent noises, and often times, high sound levels"* (Beristáin, 2010).

In one of his masterpieces, Milan Kundera says: *"Noise has one advantage. You cannot listen the words"*. This thought highlights the paradigm of isolation that is increasingly affecting the present society. About this issue, Miyara affirmed: *"The high cultural noise level threatens the ability to think, to make critical judgments about things, including the annoyance of having permanently invaded the own acoustic space. It also tends to the isolation of people. There are studies that suggest an inverse connection between sound level and interpersonal solidarity"* (González, 2000). Later, he confirms and argues about the connection between exposure to high noise levels and aggressiveness (Miyara, 2007).

3. Nightlife and noise of leisure

Conflicts with neighbors from "pink zones" become more intense and frequent day by day, not only because of the high noise levels in the environment but also by other troubles, including safety ones. In countries such as Spain, this concern began to be handled more than twenty years ago, with systematic and strong measures, such as the creation of the ZAS, the "acoustically saturated zones" (Gaja et al., 1998). In some Latin American countries including Argentina, Brazil, Colombia, Uruguay, the pink zones are still one of the main noise problems in urban areas regardless of their size, and also a very complex challenge to deal with (Beristáin, 2010; Brito, 2011a, 2011b; González et al., 2006; González & Echeverri, 2008).

Supposing that acoustic isolation requirements are fulfilled, there are still two kind of remaining problems: infrastructure problems and operation ones.

One of the main infrastructure problems related with leisure locals is related to the need for parking spaces for the assistants. This has become a "key requirement" to fulfill in order to avoid adverse impacts on the influence area and also to control the expansion of this area. Indeed, when parking lot places are scarce, the noise problem spreads from the site to their surroundings, because people not only park near the site but also far away from it. The influence area of induced traffic grows, often with significant presence of very noisy motorbikes with forbidden free escapes. While many leisure places fail to meet the requirement of the parking lot places due to its location, but perhaps it is not convenient to force them to move to another place because of their long permanence in a certain point of the city, some municipalities require hiring a private security company to quickly disperse the people who is out of the local and to avoid the occurrence of unwanted noises in the street (González et al., 2006; González &.Echeverri, 2008).

About the operational problems related to pink zones, there are two major issues: the accumulation of youths outside the nightlife clubs, and the circulating vehicles -or even parked ones- with powerful audio equipments turned on. About the accumulation of people outside the locals, there are two peak periods: before the opening, when there is a huge mass of young people waiting for accessing and at the closing time. Even though in the first case, people are usually not drunk nor under the effects of other drugs, the adrenaline rush and the anonymity of mass leads to the occurrence of shouts and songs that will undoubtedly generate annoyance to the neighborhood. While the night is going on, the need to go out to smoke because of the prohibition of smoking inside locals of public use, creates a new focus of people, sometimes encouraged by the weather conditions that make the cool night air more attractive for a while than the foul air at dancing. Then, gradually the number of people at the street increases (González &.Echeverri, 2008).

At first, noises are only derived from the communication between people. When they begin to long for high noise levels, motorcycles with noisy free escapes and music from parked or in circulation vehicles with powerful audio devices begin to be heard at high volume. In these cases, the imposition of having a hired private security service has had little effect: young people can always move to establish their "camp" further away from the nightclub, out of the effective controlled area. And so far, it has not been possible that these situations are taken as public order issues, which is the only case in which the police, as the official law enforcement agent, should proceed (González et al., 2006).

The problems associated with entertainment generate different reactions. In Spain there is a non-profit organization, PEACRAM (Plataforma Estatal de Asociaciones contra el Ruido y las Actividades Molestas – National Movement of Civil Associations against Noise and Annoying Activities), where people who feel "acoustically abused" are grouped. In the opening speech of their first congress held in Zaragoza, Spain, in 2004, it was reported that noise was the major environmental pollutant in Spain, on the basis of the number of complaints. In turn, they qualified it as "the most socially unsupportive pollutant". Sáenz Cosculluela, whose conference opened the congress, has referred to some issues related to solidarity and coexistence. He highlighted the problems of urban planning that allow the installation and even the proliferation of nightclubs in residential areas, calling the authorities to respect the "fundamental rights of the neighbors", which "take precedence over any other rights and are a non-negotiable matter". He qualified the current entertainment model as "unbearable", and he claimed for effective management measures, such as less operation hours for entertainment places, increased distances to residential areas, denial of licenses for installing new leisure locals, new urban planning regulations, or diversification of main trades of shops in commercial zones. Sáenz Cosculluela stated that "the leisure noise is the noise pollution form that attempts more directly against civic life and against the main rights of citizens, therefore the noise management approach from the environmental or from the health point of view, is not sufficient and it needs to be supplemented with approaches from ethics, civism and political points of view through the prism of human dignity, freedom and democracy" (Sáenz Cosculluela, 2004).

The management tools for leisure noise should consider the need for a solid municipal structure about licenses for the installation of leisure places and for revision and control of acoustic projects, inspection capabilities, appropriate quality and quantity equipment, regulations on land use, timetable and duty bonification program (or another kind of benefit or bonification program) for the relocation of existing facilities, all of these supported on by ongoing training of technical and semi-technical human resources (González & Echeverri, 2008).

A study about noise levels during nighttime in Salto city (Uruguay) was carried out in 2005, aiming to aware the decision-makers about the need to review the management of issues related to night leisure places. Four measurement points were selected: three of them (numered from 1 to 3) were chosen in the pink zone in the downtown -being point 2 the most impacted due to the fact that it was close of two nightclubs with great public success-; a control point was selected (point 4), in an area theoretically not affected by the night recreational activities (González et al., 2006). Environmental noise measurements lasting 30 minutes were carried out at each point in four shifts (between 9 PM and 11 PM, later designated as 10 PM; between 11 PM and 1 AM, later designated as 12 PM; between 1 AM and 3 AM, later designated as 2 AM and between 3 AM and 5 AM, later designated as 4 AM). To describe the condition without the operation of leisure locals, measurements were made on Mondays, Tuesdays and Wednesdays at all points and shifts. To describe the condition corresponding to the operation of leisure locals, measurements were carried out on Saturdays and Sundays, at all the points and shifts considered before. Measurements were made with a Type 2 integrating sound level meter owned by the municipality. In all cases the selected frequency weighting scale was "A" and the selected response time was Fast. Data from second to second were saved on an informatics device for further processing.

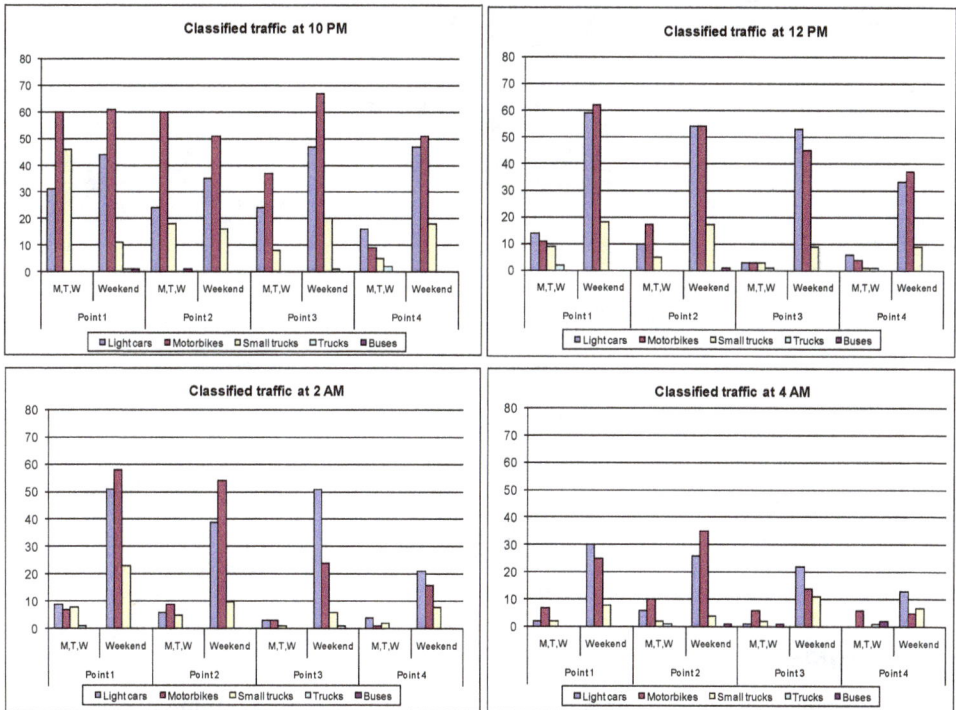

Fig. 2. Classified traffic at different points of measurement and hours, with and without nightclubs operation.

Main results are shown in figures 2 and 3, and are summarized next (González et al., 2006):
- Traffic is directly related to the activity of the nightclubs. When they are not in operation, the traffic is very low and the same happens with the noise levels.
- There is no difference between nights with and without nightclubs in operation at 10 PM.
- A great difference on L_{AF90} values is observed in points 1 to 3 between conditions with and without nightclubs in operation.
- When nightclubs are in operation, the L_{AF90} values in points 1 to 3 usually exceed 55 dBA, the noise level value stated by the municipal regulations to meet in open places during the night.
- The minimum impact caused by the nightclubs in operation is registered at 10 PM.
- The most affected hour by nightclubs in operation is 2 AM, both in terms of traffic and of noise levels.

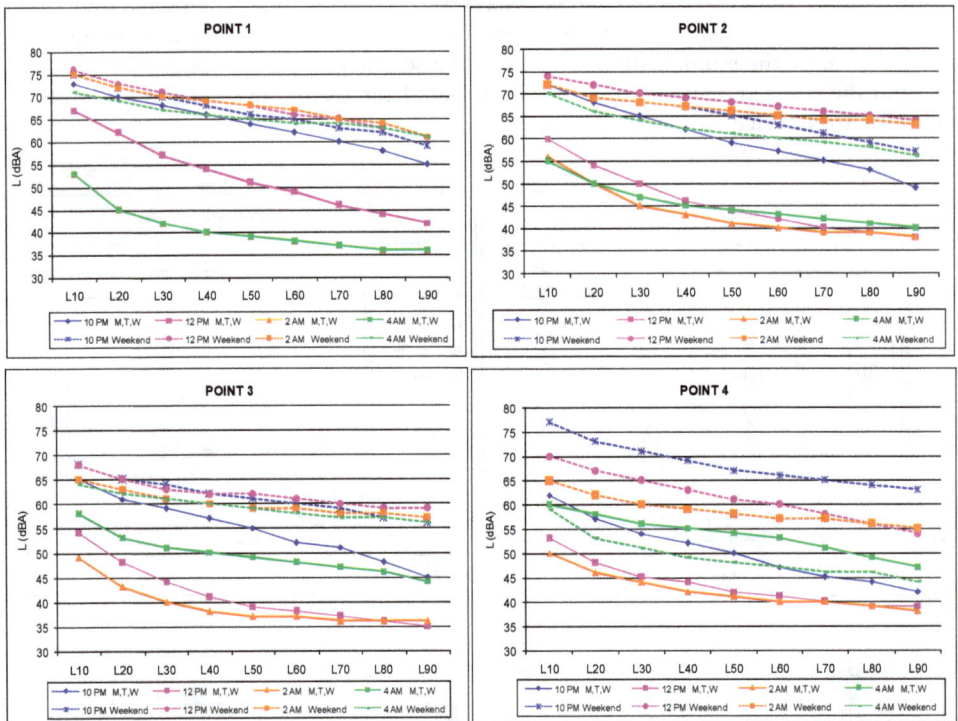

Fig. 3. Permanence curves of environmental noise levels at different points of measurement and different time, in conditions with (dotted lines) and without (full lines) nightclubs in operation.

4. Road traffic noise characterization

Traffic is usually considered the main noise source in the city. Many features of noise traffic are inherent to running vehicles, but others depend on traffic density, age and composition of the fleet, existence of interruptions such as traffic lights or crosswalks, driving style, among other factors.

4.1 Reference levels for individual vehicles

Legislation often limits the maximum emission levels for each individual vehicle, but it does not so in a generic way to traffic noise. The standardized values for different categories of vehicles may differ from one country to another. Only to exemplify, the maximum allowed levels in countries of the European Union according to Directive 2007/34/EC are presented at table 3, while those for two-wheeled vehicles in the countries of MERCOSUR according to Technical Regulation MERCOSUR/GMC/RES. N° 128/96 are presented at table 4.

Vehicle categories	Values dBA
2.1.1. Vehicles intended for the carriage of passengers, and comprising not more than nine seats including the driver's seat	74
2.1.2. Vehicles intended for the carriage of passengers and equipped with more than nine seats, including the driver's seat; and having a maximum permissible mass of more than 3,5 tonnes and:	
2.1.2.1. with an engine power of less than 150 kW	78
2.1.2.2. with an engine power of not less than 150 kW	80
2.1.3. Vehicles intended for the carriage of passengers and equipped with more than nine seats, including the driver's seat; vehicles intended for the carriage of goods:	
2.1.3.1. with a maximum permissible mass not exceeding 2 tonnes	76
2.1.3.2. with a maximum permissible mass exceeding 2 tonnes but not exceeding 3,5 tonnes	77
2.1.4. Vehicles intended for the carriage of goods and having a maximum permissible mass exceeding 3,5 tonnes:	
2.1.4.1. with an engine power of less than 75 kW	77
2.1.4.2. with an engine power of not less than 75 kW but less than 150 kW	78
2.1.4.3. with an engine power of not less than 150 kW	80

However:
- for vehicles of categories 2.1.1 and 2.1.3 the limit values are increased by 1 dBA if they are equipped with a direct injection diesel engine,
- for vehicles with a maximum permissible mass of over two tonnes designed for off-road use, the limit values are increased by 1 dB(A) if their engine power is less than 150 kW and 2 dB(A) if their engine power is 150 kW or more
- for vehicles in category 2.1.1, equipped with a manually operated gear box having more than four forward gears and with an engine developing a maximum power exceeding 140 kW/t and whose maximum power/maximum mass ratio exceeds 75 kW/t, the limit values are increased by 1 dB(A) if the speed at which the rear of the vehicle passes the line BB[2] in third gear is greater than 61 km/h.

Table 3. Current limit values for sound level of moving vehicles in the European Union

---

[2] According with ISO standard test.

MERCOSUR maximum noise for stationary and accelerating vehicle		
Category	1st phase implantation 01/01/2000	2nd phase implantation 01/01/2001
Under 80 cm³	77 dBA	75 dBA
81 cm³ to 125 cm³	80 dBA	77 dBA
126 cm³ to 175 cm³	81 dBA	
176 cm³ to 350 cm³	82 dBA	80 dBA
Over 350 cm³	83 dBA	

The noise level of stationary vehicle is the reference value at the certification process of new vehicles.

Table 4. Limit values for noise levels from moving motorcycles, scooters, tricycles, mopeds, bicycles with auxiliary motor running and similar vehicles in MERCOSUR.

4.2 Traffic noise spectrum

The spectral composition is one of the features related to the nature of traffic noise. For a continuous traffic flow, it may be assumed that the spectral composition is in correspondence with the "standardized traffic noise spectrum" that is defined both by octave bands levels and by third octave bands levels at standard EN1793-3.

Given a noise level value expressed in dBA, its spectral composition in octave bands or in third octave bands (also expressed in dBA) shall be found by adding band to band the set of tabulated values that define the standardized spectrum. Recommendation 2003/613/EC presents the standardized traffic spectrum in octave bands, based on the values given for third octave bands by the referred standard. Table 5 reproduces it.

Octave band (en Hz)	Addition values (dBA)
125	-14,5
250	-10,2
500	-7,2
1000	-3,9
2000	-6,4
4000	-11,4

Table 5. Standardized spectrum of traffic noise, according to the Recommendation 2003/613/EC

4.3 Data normality

One feature to take especially into account, as it directly affects the data processing of urban noise, is the non-normality of time series of traffic noise levels. This had already been foreseen by Don and Rees for the city of Victoria, Australia in 1985. Even if they selected the measures duration according to the recommendations of the moment, the researchers concluded that the statistic distribution of urban traffic noise level rarely fit a Gaussian. They have affirmed: "the shape is anything but Gaussian". Their explanation for the non-normality of the data was supported on the supposing that the urban noise should results from the superposition of four elements, and each one of them would fit or not a Gaussian distribution: noise levels produced by cars on the measurement side of the road; noise levels produced by trucks on the measurement side of the road; noise levels produced by all vehicles on the side of the road opposite to the measurement position; background levels that occur during the absence of vehicles (Don & Rees, 1985). Today, the shortness of their measurements (only 400 s) would be also considered as another possible way to explain the non-normality of data series.

The non-normality of traffic noise data has been reaffirmed through many years of research in various cities of different countries. It is the case of Montevideo, Rivera and Salto in Uruguay, Medellín in Colombia, Valencia and Madrid in Spain, among others (CONICYT-IMFIA, 1998; Gaja et al., 2003; Giménez Sancho, 2010; González, 2000; González et al., 2007; IMM-Facultad de Ingeniería, 1999; Jaramillo et al., 2009).

		A	B	C	D	E
Day (7:00 to 19:00)	Best fit	Log-Logistic(3P) 71,7 %	Log-Logistic(3P) 40,0 %	Cauchi 23,3 %	Log-Logistic(3P) 48,3 %	Johnson SU 75,0 %
	Gauss fit and ranking (1st – 13th)	25,0 % 12 th	3,3 % 13 th	0,0 % 13 th	0,0 % 13 th	48,3 % 6 th
Evening (19:00 to 23:00)	Best fit	Log-Logistic(3P) 90,0%	Log-Logistic(3P) 95,0 %	Laplace 70 %	Log-Logistic(3P) 45,0 %	Johnson SU 50,0 %
	Gauss fit and ranking (1st – 13th)	20,0 % 12 th	45,0 % 10 th	5,0 % 10 th	0,0 % 13 th	5,0 % 13 th
Night (23:00 to 7:00)	Best fit	Log-Logistic(3P) 52,5 % Beta 52,5 %	Log-Logistic(3P) 30,0 %	Log-Logistic(3P) 32,5 %	Log-Logistic(3P) 42,5 %	Log-Logistic(3P) 72,5 %
	Gauss fit and ranking (1st – 13th)	40,0 % 9 th	20,0 % 5 th	12,5 % 7 th	7,5 % 7 th	12,5 % 8 th

Table 6. Statistical distribution fit of noise data. Own elaboration based on Giménez Sancho's data.

Giménez Sancho determined that the adjustment of urban noise data from five consecutive years taken in the city of Madrid were *"anything but Gaussian"*, as Don and Rees said (Giménez Sancho, 2010). Table 6 presents the percentages of samples for each time of the day and for each one of the five points considered by Giménez Sancho that fit a Gaussian distribution. In each case, the statistical distribution that fits the best (among the 13 distributions he has studied) is mentioned. The ranking of the Gaussian distribution from the best fit (numbered 1st) towards the worst one (numbered 13th) is also presented.

The most immediate consequence of the non-normality of the data is provided by the statistical restrictions for their processing, as it is not possible to use parametric or Gaussian statistics. The arithmetic mean does not make sense as such, nor does the variance or standard deviation (Sachs, 1978). Then, to compare time series of noise data or to try to fix reliable values for the permanence levels based on those observed in several measurements, first of all it is needed to verify if the data sets that are statistically comparable to a confidence level selected in advance. The preferred tests to run in this data process stage are Mann-Whitney test for two samples and Kruskal-Wallis test for more than two samples (González et al., 1997; González, 2000).

4.4 Anomalous events

There are some characteristic noises that appear in different cities, specially in Latin-America, and that are the so called "anomalous events". These are noisy events that are not included in the international concept of traffic noise, that refers to engine and tyre noise. This name, as much as the concept of anomalous event, was initially developed for the city of Montevideo, while it was needed to designate these elements that were not rigorously treated in the scientific literature of the moment (CONICYT-IMFIA, 1998; González et al., 1997). Later, this concept and its name have been adopted by other researchers in different countries.

An "anomalous event" can be defined in a subjective or in an objective way. The anomalous events subjectively detected are those which the ear does not recognize as engine or tyre noise, as claxons (horns), alarms, barking, sirens, violent braking, exhaust of noisy motorcycles and other noisy vehicles.

There are also so called "evitable anomalous events", which only include noisy motorcycles, claxons (horns), and loud exhaust and braking noises. A right preventive maintenance of the vehicles should cooperate to significantly reduce their occurrence, but that is not all: a proper control system is also needed. Therefore, avoiding this kind of anomalous events could reduce noise levels in about 4 dB in many points of Montevideo city, Uruguay (González, 2000; IMM-Facultad de Ingeniería, 1999).

4.5 Acoustically anomalous events

When stationary instruments are installed to monitor environmental noise, usually there is no permanent human assistance in the place. So, in order to identify whether an anomalous event is "acoustically anomalous" in a data series, an objective criteria for its identification should be developed. In Montevideo city, the definition of an acoustic anomalous event or objective anomalous event emerges from the below detailed procedure (González, 2000):

1. Calculate the L_{Aeq} of one hour ($L_{Aeq, 1h}$) using the 60 $L_{Aeq,i}$ data from each minute of the hour.
2. List the 60 $L_{Aeq,i}$ data from each minute of the hour, from highest to lowest.
3. Cut the 12 highest values of $L_{Aeq,i}$.
4. Recalculate the hourly L_{Aeq} only using the 48 remaining $L_{Aeq,i}$. This new value will be called $L_{Aeq, 1h, corrected}$.
5. Compare each one of the 12 values of $L_{Aeq,i}$ referred in point 3 with the value of $L_{Aeq, 1h, corrected}$ obtained in point 4.
6. Every one of the with the 12 values of $L_{Aeq,i}$ referred in point 3 that exceed in 4 dB or more the level of the $L_{Aeq, 1h, corrected}$ is said to correspond to an acoustically anomalous minute, in what it may have occurred at least one acoustically anomalous event.

The relationship between the number of these events in an hour and the total traffic in that hour fit a potential curve, with a correlation coefficient $R^2 = 0,8679$. Then, the predictive equation for traffic noise levels in Montevideo city, obtained from 1361 noise samples, is (González, 2000):

$$L_{A,eq,\,1hora} = 49,4 + 10 \log (A + 2,33\ M + 9,01\ O + 6,84\ C) + 23,266\ Q^{-0.3811} - 10 \log d$$

where:

M   N° of motorbikes per hour
A   N° of light cars per hour
O   N° of buses per hour
C   N° of trucks per hour
Q   = M + A + O + C
D   distance to the axis of the street (m)

If traffic speed exceeds 60 km/h, then a linear correction should be added: (15 v - 8,67), with v expressed in km/h (González, 2000).

4.6 Measurement stabilization time

The anomalous events strongly affect the lasting of measures needed to obtain a reliable sample of urban noise. As well as a low traffic density (for example, at night) cannot be trusted to select a measurement time of less than an hour to represent the hourly L_{Aeq}, anomalous events may also enlarge the measurement time both during the day time and the night time (González, 2000).

To exemplify, if only traffic density is taken into account, the selected measurement time for traffic noise measurements in Montevideo city would be about 15 minutes. But if the occurrence of anomalous events is also considered, then the recommended measurement time should shift to 30 minutes. Indeed, in every sample of urban noise in the city of Montevideo, it is expected to register an average amount of 9 to 12 subjective anomalous events (that is, an anomalous event every 5 or 6 minutes) but only about 4 acoustic anomalous events (González, 2000).

Then, in order to avoid selecting an excessively short measurement time, which might not be representative of the real situation, or a too long time that might turn too expensive the fieldwork, the basic concept to work with is the "measurement stabilization time". The stabilization time of a noise sample of N minutes is the minimum number n of minutes after which the $L_{Aeq,\,n}$ accumulated until that moment differ from the $L_{Aeq,\,N}$ of the whole event of N minutes by less than a certain ε. The value of ε strongly affects the results, and depends on the precision required according to the objective of the measurements (González, 2000).

This expression is valid for all n ≥ T_{estab}, where N is the total number of minutes (data) in the whole sample:

$$\left| 10 \times \log \left(\frac{1}{N} \sum_{i=1}^{N} 10^{0,1L_i} \right) - 10 \times \log \left(\frac{1}{n} \sum_{i=1}^{n} 10^{0,1L_i} \right) \right| < \varepsilon$$

If ε = 1 dB is adopted for Montevideo city (that means the cumulative L_{Aeq} of the event fluctuate from that moment until the end of the measurement in the range of ± 1 dB) measurements lasting at least 30 minutes are required to reach the stabilization in a minimum of 90 % of the samples (González, 2000).

5. The acoustic maps as urban management tools

According to the definition of the Directive 2002/49/EC from the European Union, a noise map is *"the presentation of data on an existing or predicted noise situation in terms of a noise indicator, where the trespassing of any relevant regulation limit value will be indicated, also the number of people affected in a specific area or the number of households exposed to certain values of a selected noise indicator in a specific area".* It is stated that the information to the public should be given in a clear, comprehensible and easily accessible way, but the information to the authorities should be much more detailed and vast. It should include other elements such as baseline information used to construct the maps, how that information was obtained, methods employed for measurement or calculation, and –obviously- charts, which are a key part of the map.

Obtaining the information to build a noise map involves a lot of detailed field and office work. Then, if all the generated information is incorporated to the map, it will turn into a management tool that shall support a variety of uses. In fact, although the definition of "noise map" is precise and comprehensive, many diverse maps should be in practice considered under this generic designation, from diagnosis to specific exposed population maps, from acoustic conflicts maps to strategic noise maps, among others. That is due to their easy comprehension for every training level of people, which make them suitable as a tool for communicating information about noise levels, as well as other kind of information (specially about environmental issues).

Annoyance maps deserve to be treated separately, since they involve penalizing the registered noise levels according to some of their characteristics, such as its impulsiveness, the presence of pure tones or high energy contents at low frequencies, among others. These maps can help to diagnose not only annoyance but also health risks (González, 2011).

To build action plans with basis on these maps, the economic value of the environmental quality from the point of view of the citizens would be determined. Even if it is not easy, it is needed to assign monetary values to people health, annoyance, loss of intellectual performance, time lost at work, property value, among other variables. The application of surveys to detect the willingness to pay for improving the acoustic quality of the environment is a desirable way to obtain these data. Therefore, this kind of work does not only provide a monetary quantification to assess the impact of different measures to be applied in an area or, conversely, the impact that one measure might have on different areas, but it also allows for prioritization of different possible interventions. When similar reductions of environmental noise levels and similar number of benefited people are considered as consequences of different possible interventions, the most adequate one to initiate a program of actions to fight noise is that one which will be repaid sooner by the people willingness to pay for (González, 2011).

One of the most relevant applications nowadays is the acoustic mapping aimed for developing strategic plans for acoustics decontamination. As stated by the above mentioned Directive, a strategic noise map is *"a map designed for the global assessment of noise exposure in an area due to the existence of different noise sources or in order to make global predictions for the area."* As the noise maps that are built for other purposes, strategic maps must also include graphic material (charts) and detailed technical reports, in order to fulfill their original purpose. González García emphasizes, among other thematic maps that can be part of a strategic noise map: maps of noise level indicators; maps of affected areas (that show the areas where L_{den} is above 55 dBA, 65 dBA and 75 dBA); maps of exposure at facade of buildings (the noise levels are taken at a height of 4 m); maps of exposed people; maps of land use and acoustic zoning; maps of carrying noise capacity; maps of noise sensitivity (González García, 2006).

Since the enactment of Directive 2002/49/EC, strategic noise maps should be built for all cities of more than 250.000 inhabitants in the member countries. Their results have to be informed to the Council of the European Union, to implement concrete actions in order to improve the acoustic environment quality, and to ensure the review and renew of the strategic maps with a minimum frequency.

After 10 years, although there are still difficulties with its implementation and to harmonize working methods, the applicability of strategic noise maps cannot be doubted (Sanz Sa, 2010). In fact, there are many national and provincial standards that regulate the obligation to have strategic noise maps in towns that exceed a certain number of inhabitants, that may be much fewer than the 250.000 inhabitant mentioned by the Directive (to exemplify, noise Law 5/2009 of Castilla and León, Spain, states strategic noise maps for all towns with 20.000 inhabitants or more, less than 10 % of the minimum number of inhabitants considered by the European Union Directive).

Regarding regulations, the situation in Latin America is very different. Colombia is the only Latin American country that has in its current national legislation (Resolution 0627, 2006), a requirement about building noise maps, but for another urban scale: the maps are asked for for cities of more than 100.000 inhabitants. No other Latin-American country has currently noise regulations with similar requirements.

In Argentina, the Research Team on Acoustics from the Laboratorio de Acústica y Luminotecnia CIC-LAL, Buenos Aires, has been elaborating regulations proposals since 2002. In the latest version, the necessity to carry out an acoustic plan for those urban areas that exceed 25.000 inhabitants is mentioned and the minimum contents of these plans are referenced. The requirement of building acoustic maps has been removed in this version while it was present in previous ones. Although it may be discussed whether this is the best way to build a noise decontamination plan, this decision was needed because of the controversies aroused in the Parliament to force small towns to do an investment to have their acoustic maps (Velis et al., 2009).

The situation in Uruguay is even farther: nowadays there is no noise regulation of national scope, and none of the departmental ordinances even mention acoustic maps (González et al., 2008; González, 2010a).

Regarding the strategic noise maps, Bañuelos has stated: *"The matter is not only offering more or less colorful images"*. To be useful, the information presented in a strategic noise map must be realistic and representative. The high initial cost for the Administration related to a strategic noise map is not due to building the map, but to the costs of implementing the corrective measures that may be needed in consequence (Bañuelos Irusta, 2008).

It should be always taken into account that the charts of an acoustic map, regardless of the target that it has aimed to fulfill, should always be based on solid and comprehensive technical documentation, to give credibility to the exposed information. Not only its reliability becomes widely increased, but also its possible applications get greatly expanded (González, 2011).

6. Conclusion

Noise management involves great challenges: if one problem has seemed easy to solve, certainly it has not been raised properly.

To influence on politics about noise and to improve environmental acoustic quality, academic research must focus on the main concerns of the society, trying to understand these issues and to build better sustainable management proposals.

7. References

Bañuelos Irusta, A. (2008) Mapa de ruido: herramienta para la evaluación y gestión del ruido. *Proceedings of International Seminar on Environmental Noise.* ISBN: 978-958-44-3029-8, Medellín, Colombia, March 2008.

Beristáin, S. (1998) El ruido es un serio contaminante, *Proceedings of 1st Iberoamerican Congress on Acoustics,* Florianópolis, Brazil, March 1998.

Beristáin, S. (2010) Noise in the largest Mexican city, *2nd Pan-American and Iberian Meeting on Acoustics, 160th ASA meeting, 7° Congress FIA, 17° Congress IMA,* ISSN: 0001-4966, Cancún, México, November 2010.

Brito, A. (2011a) O triângulo dos eventos públicos em Fortaleza, *Poluição Sonora em Fortaleza,* 2011, available on line at aureliobrito.blogspot.com/

Brito, A. (2011b) 24 horas na vida de quem sofre com poluiçao sonora, In: *Poluição Sonora em Fortaleza,* 2011, available on line at aureliobrito.blogspot.com/

Colegio Oficial de Ingenieros Superiores Industriales de la Comunidad Valenciana (2004) Contra los ruidos. *InfoIndustrial,* No 31, (abril 2004), pp. 6-7.

CONICYT-IMFIA González, A.E., Perona, D.H., Martínez Luaces, V., Barbieri, A., Gerardo, R., Guida, M., López, J., Maneiro, M. (1998) *Contaminación sonora en ambiente urbano, Informe Final del Proyecto de Investigación CONICYT – Clemente Estable 2040,* Montevideo, Uruguay, 1998.

Defensor del Pueblo de Madrid (2005) *Informes, estudios y documentos, Contaminación Acústica,* ISBN: 84-87182-48-8, Madrid, Spain.

Defensoría del Vecino de Montevideo (2010) *Cuarto Informe Anual Año 2010,* Montevideo, Uruguay.

Diario El País, CADENASER.COM 09-02-2007. Available on line (2011) at www.elpais.com

Don, C.G. & Rees, I.G. (1985) Road traffic sound level distributions, *Journal of Sound and Vibration,* 100(1), (1985), pp. 41-53. ISSN : 0022-460X.

Gaja Díaz, E., Reig, A. Sancho, M., González, E. (1998) Evolución del nivel de Ruido Ambiental en la ciudad de Valencia. Acciones de control. *Proceedings of 1st Iberoamerican Congress on Acoustics,* Florianópolis, Brazil, March 1998.

Gaja, E., Giménez, A., Sancho, S., Reig, A. (2003) Sampling techniques for the estimation of the annual equivalent noise level under urban traffic conditions, *Applied Acoustics,* 64 (2003) pp.43–53. ISSN : 0003-682X.

Giménez Sancho, A. (2010) *Contribución al estudio de los índices europeos de valoración del ruido ambiental en ambiente urbano,* Doctoral Thesis, Universidad Politécnica de Valencia, Valencia, Spain, June 2010.

Giraldo Arango, J.M. (2008) Control del ruido en la construcción de obras públicas, *Proceedings of International Seminar on Environmental Noise.* ISBN: 978-958-44-3029-8, Medellín, Colombia, March 2008.

González, A.E., Gaja, E., Martínez Luaces, V., Gerardo, R., Reig Fabado, A. (1997) Niveles de Contaminación Sonora en la ciudad de Montevideo, *Proceedings of the Congress of the Acoustics Society of Spain Tecniacústica '97,* Oviedo, Spain, November 1997.

González, A.E. (2000) *Monitoreo de ruido urbano en la ciudad de Montevideo: determinación del tiempo óptimo de muestreo y desarrollo de un modelo predictivo en un entorno atípico.* Thesis for the Degree of Doctor in Environmental Engineering, UdelaR, Montevideo, Uruguay, March 2000.

González, A.E., Paulino, D., Tironi, M. (2006) Incidencia de actividades recreativas nocturnas sobre la calidad acústica del entorno en la ciudad de Salto (Uruguay), *XXX Iberoamerican Congress of AIDIS Internacional*, Punta del Este, Uruguay, November 2006.

González, A.E.; Gavirondo, M., Pérez Rocamora, E., Bracho, A. (2007) Urban noise: measurement time and modelling of noise levels in three different cities. *Noise Control Engineering Journal*, 55 (3), (May-June 2007), pp. 367-372. ISSN 0736–2501.

González, A.E., Echeverri Londoño, C.A. (2008) Locales de diversión nocturna y contaminación sonora. In: Actas FIA 2008, *VI Iberoamerican Congress on Acoustics FIA 2008*, compiled by Federico Miyara, ISBN 978-987-24713-1-6, Buenos Aires, Argentine, November 2008.

González, A.E., Indarte E. & Lisboa, M. (2008) *Acústica Urbana. Memorias de las Jornadas de Convergencia en Normativa de Contaminación Acústica*, Convenio MVOTMA – UdelaR (DINAMA – Facultad de Ingeniería), ISBN 978-9974-0-0541-9, Montevideo, Uruguay.

González, A.E. (2010a) Which are the main management issues about noise in Uruguay 2010? *2nd Pan-American and Iberian Meeting on Acoustics, 160th ASA meeting, 7° Congress FIA, 17° Congress IMA*, Cancún, México, November 2010. ISSN: 0001-4966

González, A.E. (2010b) *Mediciones de niveles sonoros ambientales y ocupacionales en obras civiles*, DIA-IMFIA Facultad de Ingeniería, UdelaR.

González, A.E. (2011) Mapas acústicos: Mucho más que una cartografía coloreada. *Latinoamerican Conference AES 2011*, Montevideo, Uruguay, August 2011.

González García, M.A. (2006) ¿Qué es y cómo se hace un mapa estratégico de ruido en carreteras? *Proceedings of Tecniacústica 2006*, Gandía, Spain, October 2006.

IMM – Facultad de Ingeniería (1999) *Mapa Acústico de Montevideo, Informe Final del Convenio*, Montevideo, Uruguay.

Jaramillo, A., González, A., Betancur, C., Correa, M. (2009) Estudio comparativo entre las mediciones de ruido ambiental urbano a 1,5 m y 4 m de altura sobre el nivel del piso en la ciudad de Medellín, Antioquia – Colombia. *Revista Dyna*, 157 pp. 71-79. ISSN 0012-7353.

Kotzen, B. & English, C. (2009) *Environmental noise barriers: a guide to their acoustic and visual design*, 2nd ed. ISBN 10: 0–203–93138–6.

Laboratorio de Acústica y Electroacústica, Escuela de Ingeniería Electrónica, Universidad Nacional de Rosario (2011) Available on line at the oficial Web site http://www.fceia.unr.edu.ar/acustica/

Lizana, P. (2010) Enviromental noise culture, *2nd Pan-American and Iberian Meeting on Acoustics, 160th ASA meeting, 7° Congress FIA, 17° Congress IMA*, ISSN: 0001-4966, Cancún, México, November 2010.

Madariaga Coaquira, Z. (2008) El ruido ambiental en la ciudad de Arequipa y su incidencia en la salud de la población más expuesta, *Proceedings of International Seminar on Environmental Noise*. ISBN: 978-958-44-3029-8, Medellín, Colombia, March 2008.

Miyara, F. (2007) "Ruido, juventud y derechos humanos". *I Congreso Latinoamericano de Derechos Humanos*. Rosario, Argentine, 2007.

Miyara, F. (2008) Ruido Urbano: tránsito, industria y esparcimiento, In *Acústica urbana. Módulo I. Manual de mediciones acústicas orientado a la gestión municipal*, Convenio MVOTMA – UdelaR (DINAMA – Facultad de Ingeniería), ISBN 978-9974-7610-2-5, Montevideo, Uruguay.

Observatorio de las novedades acústicas y musicales, *Coches que no hacen ruido*, 2010, available on line (2011) at
http://www.acusticaweb.com/blog/acustica-ambiental-y-ruido/

Orozco-Medina, M.G., Figueroa-Montaño, A. (2010) Urban noise and transport as a strategy of environmental quality, *2nd Pan-American and Iberian Meeting on Acoustics, 160th ASA meeting, 7° Congress FIA, 17° Congress IMA*, ISSN: 0001-4966, Cancún, México, November 2010.

Posada, M.I., Arroyave, M., Fernández, C. (2009) Influencia de la vegetación en los niveles de ruido urbanos, *Revista EIA*, 12 (December 2009), pp. 79-89, ISSN 1794-1237

Ríos Valencia, O. (2008) Niveles de ruido sobre la franja horaria del pico y placa, municipio de Medellín, *Proceedings of International Seminar on Environmental Noise*. ISBN: 978-958-44-3029-8, Medellín, Colombia, March 2008.

Sachs, L. (1978) *Estadística Aplicada*, Editorial Labor, Spain.

Sáenz Cosculluela, I. (2004) Keynote speech at *the I Congreso Nacional contra el Ruido: ruido, salud y convivencia, PEACRAM, Plataforma Estatal contra el Ruido*, Zaragoza, Spain, April 2004.

Sanz Sa, J.M. (2010) Experiencia de la 1ª fase de los MER y Perspectivas para la 2° fase. *I Jornada Acústica: Desarrollo Normativo en Acústica Ambiental*, Cádiz, Spain, November 2010.

Velis, A.G., Rizzo la Malfa, A.M., Bontti, H., Vechiatti, N., Iasi, F., Armas, A., Tomeo, D. (2009) Evolución del Proyecto de Ley de Protección Ambiental de la Calidad Acústica en la Provincia de Buenos Aires. *First Regional Conference on Acoustics from AdAA*, Rosario, Argentina, November 2009.

Part 2

Theory

2

Adaptive Fractional Fourier Domain Filtering in Active Noise Control

Sultan Aldırmaz and Lütfiye Durak–Ata

Department of Electronics and Communications Engineering, Yildiz Technical University
Turkey

1. Introduction

Acoustic noise control systems gain more importance as more and more industrial equipments, i.e., engines, fans, ventilators, and exhausters are in use (1–6). Passive acoustic noise control techniques benefit enclosures, barriers and silencers to attenuate ambient noise. However, if the noise has dominant low-frequency components, then passive techniques are either inefficient or expensive. In contrast, active noise control (ANC) systems are much more effective in canceling low-frequency noise. Various noise cancelation algorithms have been proposed in the literature (7–11). In a generic ANC scheme, a reference microphone is used to receive the ambient noise and the system produces an *anti-noise* signal which has equal amplitude but opposite phase with the primary noise to cancel it acoustically (1). As the primary noise may have time-varying characteristics, ANC systems should be able to adapt themselves to the noise rapidly.

In most of the ANC systems, either adaptive filters or neural network based structures are employed (2–6; 9; 12–18). In (15), fuzzy-neural networks are used to estimate the nonlinear response of the unknown primary acoustic path where primary and secondary paths are characterized by nonlinear functions. On the other hand adaptive filters are usually employed to increase the system performance and robustness. They are mostly employed with least mean squares (LMS)-based algorithms and the adaptation is usually realized in time domain (3; 9; 12; 13; 19). Whereas Fourier domain (20) and wavelet-based adaptive filter bank approaches (21; 22) are among the few transform-domain adaptation techniques that have been used in the ANC systems. Compared to time-domain adaptive filters, transform-domain adaptive filters may need fewer parameters (23; 24). When the noise source has dominant low-frequency components, wavelet transform-based adaptive filters provide higher performance rates. However, in case of linear frequency modulated (LFM) or chirp-type audio signals, as their frequency varies linearly with time, performance rates are limited for both Fourier and wavelet-transform domains.

LFM signals are among the frequently used signals in real life and they are good models for mechanical systems with accelerating internal components. A Gaussian enveloped,

---

[1] The authors are supported by the Scientific and Technological Research Council of Turkey, TUBITAK under the grant of Project No. 105E078.
[2] The material in this chapter was published in part at [33].

single-component LFM signal can be expressed as

$$x(t) = A \, e^{\pi \gamma (t-t_0)^2} \, e^{j\pi[\alpha(t-t_0)^2 + 2\beta(t-t_0)]} \tag{1}$$

where α is the chirp rate, t_0 and β represent the time and frequency shifts with respect to the time-frequency origin, A and γ are the parameters of the envelope. One of the most convenient analysis tools for LFM signals is the fractional Fourier transform (FrFT), which employs chirps as basis functions. FrFT is a generalization of the ordinary Fourier transform with a fractional order parameter. It is a mathematically powerful and efficiently computable linear transform. It has been employed in various application areas including time-frequency signal processing, filtering, and denoising (25). Recently, the authors have introduced the adaptive filtering scheme in fractional Fourier domain in (33).

In this chapter, we present a robust adaptive fractional Fourier domain filtering scheme in the presence of LFM signals and additive white Gaussian noise (AWGN). As the instantaneous frequency (IF) of LFM signals may show rapid variations in time, adaptation to a chirp signal is much more difficult compared to a sinusoidal signal in ANC systems. As a remedy to this problem, we propose to incorporate the FrFT.

Adaptive fractional Fourier domain filtering introduces significant improvements, since chirp-type signals are transformed into narrow-band sinusoidal signals and the non-stationary signal adaptation problem is converted to a stationary form. To improve the system performance, it is necessary to estimate the transformation order of FrFT successfully. This is directly related to the proper estimation of the IF of the chirp signal and the estimation should be kept up-to-date at certain time intervals. Many methods are proposed for IF estimation in the literature, such as polynomial phase-based estimators, LMS or RLS-based adaptive filters, and time-frequency distribution-based estimators with same inherent disadvantages (31; 32). Here IF is determined by exploiting the relationship between the Radon-Wigner transform (RWT) and FrFT of signals (29).

The chapter is organized as follows. Section 2 introduces the preliminaries of the chapter by introducing the FrFT giving its definition, important properties and its fast computation algorithm. Then, the IF estimation of single or multi–component LFM signals are investigated. Time and Fourier domain adaptive filtering schemes are explained in Section 3. In Section 4, the ANC system model, FrFT-based adaptation scheme and its performance analysis are given in detail. Finally the conclusions are drawn in Section 5.

Keywords
Active Noise Control, Adaptive filtering, Fractional Fourier transform, Fractional Fourier domains, Instantaneous frequency estimation.

2. Preliminaries

2.1 Fractional Fourier transform
FrFT is a generalization of the ordinary Fourier transform with a fractional order parameter a, which corresponds to the ath fractional power of the Fourier transform operator, $\mathfrak{F}$. The ath-order FrFT of $x(t)$ is defined as

$$x_a = \mathfrak{F}_a\{x(t)\} = \int K_a(t, t') x(t') dt' \tag{2}$$

where $0 < |a| < 2$, and the transformation kernel $K_a(t, t')$ is

$$K_a(t, t') = A_\varphi e^{-j\pi(t2\cot(\varphi) - 2tt'\csc(\varphi) + t'2\cot(\varphi))} \quad (3)$$

$$A_\varphi = e^{-j\pi sgn(\sin(\varphi))/4 + j(\varphi)/2}/|\sin(\varphi)|^{1/2}$$

with the transform angle $\varphi = a\pi/2$ (25). The first-order FrFT is the ordinary Fourier transform and the zeroth-order FrFT is the identity transformation. The ath-order FrFT interpolates between the function $x(t)$ and its Fourier transform $X(f)$. Fig. 1 shows the real part of a mono–component LFM signal $x(t)$ with a chirp rate $\alpha = 0.5$ in various fractional orders. Time domain signal, i.e. the zeroth–order FrFT, is given in Fig. 1 (a) and its Fourier transform is given in Fig. 1 (c). Moreover, Fig. 1 (b) and (d) show the signal in 0.5^{th} and 1.5^{th} order fractional domains.

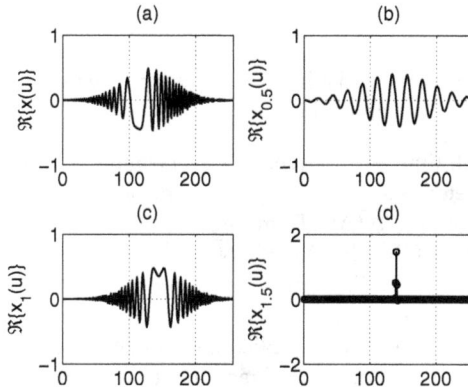

Fig. 1. The real part of the FrFT of the mono-component LFM signal in various fractional orders. (a) The signal in time domain, (b) 0.5^{th} order FrFT, (c) 1st order FrFT, i.e. the Fourier transform, and (d) 1.5^{th} order FrFT.

The inverse transform operator is simply expressed as $(\mathfrak{F}_a)^{-1} = \mathfrak{F}_{-a}$ which corresponds to $K_a^{-1}(t, t') = K_{-a}(t, t')$ as the inverse-transform kernel function. FrFT is a linear and unitary transform. One of the important properties of the FrFT is index additivity and it is expressed as

$$\mathfrak{F}_{a_1}\mathfrak{F}_{a_2} = \mathfrak{F}_{a_1 + a_2} \quad (4)$$

where a_1 and a_2 indicate the fractional transform orders.

In (26), FrFT is decomposed into a chirp multiplication followed by a chirp convolution and followed by another chirp multiplication. The chirp convolution is evaluated by using the fast Fourier transform. Thus, FrFT can be computed by $O(N\log N)$ computational complexity, where N denotes the time–bandwidth product (TBP) of the signal (26). The TBP of a signal $x(.)$ is defined as the product of time–width and bandwidth of the signal. According to the well-known uncertainty principle, signals can not be confined both in time and frequency at the same time. However, it is always possible to choose the TBP of the signal large (always greater than 1). Therefore, authors in (26) assumed that the signal is confined to the interval $[-\Delta t/2, \Delta t/2]$ in time and $[-\Delta f/2, \Delta f/2]$ in frequency domain. In order to have same length of time and frequency interval, a scaling operator must be used. When time domain scaling

$$c_1[m] := e^{j\pi\frac{1}{4}(\alpha/dx^2 - \beta/N)m^2} \qquad\qquad -N \leq m \leq N-1$$

$$c_2[m] := e^{j\pi\beta(m/2\sqrt{N})^2} \qquad\qquad -2N \leq m \leq 2N-1$$

$$c_3[m] := e^{j\pi\frac{dx^2}{4N}(\alpha/N - \beta/dx^2)m^2} \qquad -N \leq m \leq N-1$$

$$g[m] := c_1[m]x(m/2dx) \qquad\qquad\qquad -N \leq m \leq N-1$$

$$h_{a'}(m/2dx) := \frac{A_\phi}{2dx}c_3[m](c_2 * g)[m] \quad -N \leq m \leq N-1$$

where,

$$\phi'' := \frac{\pi}{2}a''$$

$$\alpha := \cot\phi''$$

$$\beta := \csc\phi''$$

$$A_\phi := \frac{exp(-j\pi\text{sgn}(\sin\phi)/4 + j\phi/2)}{|\sin\phi|^{1/2}}$$

Table 1. Table 1. Definition of the variables in Fig.2, which are used in the calculation of the fast fractional Fourier transform algorithm

is employed to the signal, time and frequency axes become $\Delta t/s$ and Δfs, respectively. By defining the scale parameter as $s = \sqrt{\Delta t/\Delta f}$, new time and frequency axis (range, interval) become same and it is $\Delta x = \sqrt{\Delta f\Delta t}$. Therefore the TBP is $N = \Delta t\Delta f$, the interval of the samples is defined in terms of the TBP, $\Delta x = \sqrt{N}$.

The fast FrFT computation block diagram is given in Fig. 2. First, the signal is interpolated by 2, then the interpolated signal is multiplied by a chirp signal c_1. After then, the obtained signal is convolved by a chirp c_2 and multiplied another chirp c_3. Finally, the obtained signal is downsampled by 2. In the algorithm, the fractional transform order, a, is assumed to be in the interval $0.5 \leq |a| \leq 1.5$. The index additivity property of the FrFT can be used to extend this range.

As mentioned in the next section, FrFT has some impacts on the Wigner distribution (WD). Roughly speaking, FrFT rotates the support of the signal on the x-y axis respect to the transform order. In order to preserve the energy of the signal in a circle with a Δx diameter, the analyzed signal must be interpolated by 2 times in the beginning of the algorithm. (26) presents the digital computation of the FrFT, moreover, discrete FrFT definitions have been developed by many researchers (27; 28).

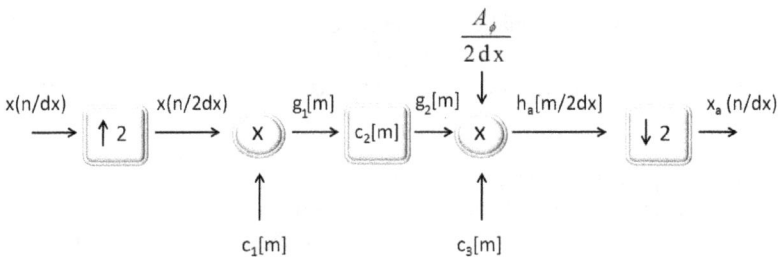

Fig. 2. The fast FrFT computation block diagram.

2.2 IF estimation of chirp-type signals

FrFT converts time-varying chirp-type signals into sinusoidals at appropriate transform orders. Thus it is crucial to estimate the instantaneous frequency (IF) value of the chirp components successfully. The IFs of the analyzed signals characterize the variation of their spectra.

The WD of a signal $x(t)$ is represented by $W_x(t, f)$ and defined as

$$W_x(t, f) = \int x(t + \tau/2)x^*(t - \tau/2)e^{-j2\pi f \tau}d\tau. \tag{5}$$

The RWT of a signal $x(t)$ is defined as the Radon transform of the WD of $x(t)$,

$$RDN[W_x](r, \varphi) = \int W_x(r\cos(\varphi) - s\sin(\varphi), r\sin(\varphi) + s\cos(\varphi))ds \tag{6}$$

where (r, φ) are the transform-domain variables in polar coordinates and the RWT gives the projection of the WD for $0 \le \varphi \le \pi$. The radial slices of the RWT, $RDN[W_x](r, \varphi)$, can be directly computed from the FrFT of the signal as,

$$RDN[W_x](r, \varphi) = |\mathfrak{F}_a\{x(r)\}|^2 = |x_a(r)|^2. \tag{7}$$

A mono-component amplitude-modulated chirp signal and its time-frequency representation by WD are shown in Fig. 3(a) and (b). FrFT rotates the WD of the signal by an angle related to the fractional transformation angle φ, as shown in Fig. 3 (c) and (d). The appropriate order of the FrFT, which is $(1/3)$ for this case, rotates the WD in the clockwise direction so that the chirp is converted to an amplitude modulated sinusoidal signal.

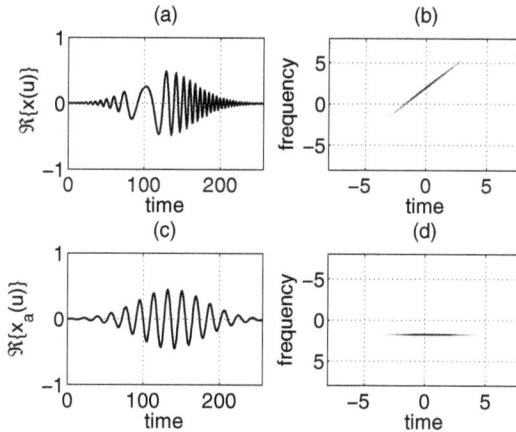

Fig. 3. (a) Real part of the LFM signal in time-domain, (b) its WD, (c) its $(1/3)$-rd order FrFT, and (d) the WD of the transformed signal.

The projections in the WD domain are related to the FrFT and we propose an efficient and a simple IF estimation technique by using the relationship between the RWT of a signal and its corresponding FrFT (29). The algorithm searches for the appropriate FrFT order a for the signal. At the appropriate order, FrFT of the signal gives the maximum peak value. By searching the peaks of $|x_a(r)|^2$ computed in $O(NlogN)$ operations at various order parameter values of $0 < |a| < 2$, the LFM rates and IF estimates can be determined robustly.

As the ath-order FrFT gives $x(t)$ for $a = 0$ and $X(f)$ for $a = 1$, binary search algorithm searches for the optimum transformation order between zero and one that maximizes the peak FrFT value. First, this algorithm calculates the FrFT of the signal for $a = 0$, $a = 0.5$, and $a = 1$ values. Secondly, it takes the maximum two peak values of the FrFT among these values. Then, FrFT calculation is repeated for two obtained peak values and their mean value. This procedure is iteratively repeated by decreasing the search region for the order parameter a. The flowchart of this algorithm is given in Fig. 4. As each FrFT computation has $O(NlogN)$ complexity, the overall complexity of the required search is of $O(3x\ L\ x\ N(logN))$, where L indicates the loop number of the search algorithm and 10 steps is usually sufficient.

Such an RWT-based IF estimation works well even the environment has AWGN besides the chirp-type noise. The peak FrFT value of the chirp signal in Fig. 5 with respect to the FrFT order $(a - 1)$ is presented for AWGN with three different SNR values.

If the signal has multi–chirp components, then each peak belonging to each different IF should be determined. For a multi–component chirp signal, Fig. 6 shows peak FrFT values of each component. Chirp rate of these components are $[\pi/18, \quad 2\pi/9, \quad 3\pi/12]$ and their time and frequency centers are $t_0 = [0, \quad 0, \quad 1]$ and $b = [-1, \quad 0, \quad 1]$, respectively. Although this technique works for the multiŰcomponent case, its performance depends on the SNR value and difference between IF values of each component of the analyzed signal. For this reason, if the ambient noise has more than one component, the minimum-essential-bandwidth-based IF estimation technique can be used as in (28).

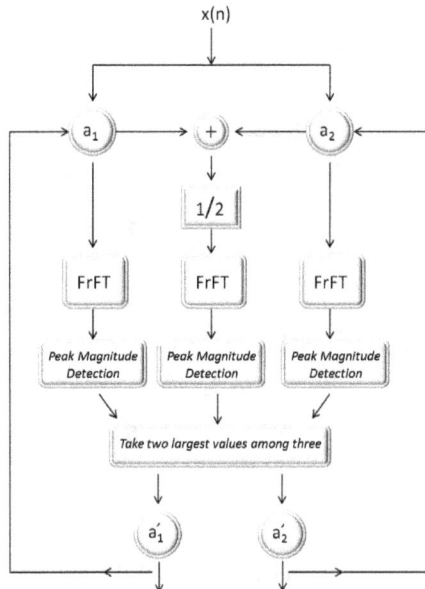

Fig. 4. A basic IF estimation scheme via RWT.

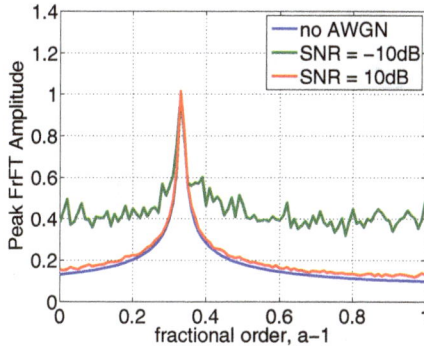

Fig. 5. Peak FrFT values as a function of the fractional order, (a-1) for a mono–component chirp signal.

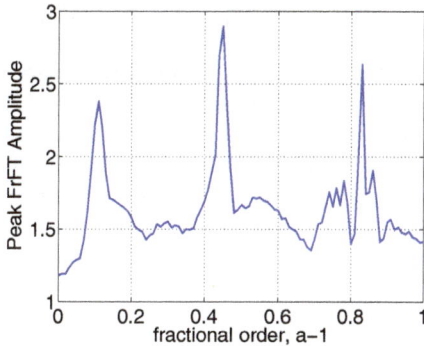

Fig. 6. Peak FrFT values as a function of the fractional order, (a-1) for multi–component chirp signal.

(28) shows that the fractional Fourier domain order corresponding to the transformed signal of minimum bandwidth gives IF estimates in sufficiently long observation periods for multiŰcomponent signals. In (28), two different IF estimation algorithms are proposed in that optimization scheme. One of them makes use of the maximum fractional time-bandwidth ratio, whereas the second one introduces a minimum essential bandwidth, which is expressed as the minimum sum of the bandwidths of the separate signal components. Genetic algorithm is employed to determine the IF of the signal components.

3. Adaptive filtering in ANC systems

Most of the ANC systems employ adaptive filters based on LMS-type algorithms, operating either in time or transform domains. In some applications, such as acoustic echo cancellation in teleconferencing, time-domain adaptive filters should have long impulse responses in order to cancel long echoes successfully. On the other hand, transform-domain adaptive filters may converge faster than time-domain adaptive filters in such cases.

We propose a fractional Fourier domain adaptive filtering scheme for ANC systems as given in Fig. 7. The effect of the environment is summarized by an unknown plant $P(z)$ from

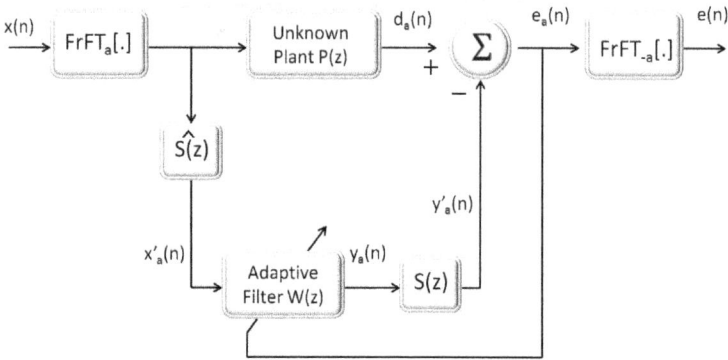

Fig. 7. The basic scheme of the FrFT-domain adaptive filter in an ANC application.

the reference microphone to the error microphone. Secondary path from the canceling loudspeaker to the error microphone is represented by $S(z)$. Secondary path effects include the effect of power amplifiers, microphones, speakers, analog-to-digital converters and digital-to-analog converters. In (19), secondary path transfer functions have been estimated, whereas in (4; 14), the secondary path is modeled as an FIR filter. (34) emphasizes on the system identification problem of ANC, thus the secondary path is not considered. Here, we employ a secondary acoustic path model as an FIR filter and assume its proper estimation as $\hat{S}(z)$, and we focus on the fractional Fourier domain adaptive filtering scheme.

3.1 Time-domain adaptation
Searching the optimum filter tap-weights to minimize the sums of squares of the cumulative error, LMS-based algorithms achieve satisfactory performance rates with low computational complexity. Assuming that $\mathbf{x}_p(n)$ is the input vector at time n, instantaneous error of the adaptive filter is

$$e(n) = d(n) - s(n)[\mathbf{w}^T(n)\mathbf{x}'_p(n)] \tag{8}$$

where $e(n)$, $d(n)$ and $\mathbf{w}(n)$ denote the error signal, reference signal, and the adaptive filter tap-weights, respectively. Error is minimized by decreasing the filter tap-weights in the direction of the gradient with a step-size μ recursively

$$\mathbf{w}(n+1) = \mathbf{w}(n) + \mu \mathbf{x}_p(n)e^*(n). \tag{9}$$

To reduce the effect of the power of the input signal on the system performance μ may be normalized by the power of the signal as in the normalized-LMS (NLMS) algorithm.

In ANC systems, filtered-X LMS (Fx-LMS) algorithm is used to reduce the secondary path effects. In the Fx-LMS algorithm, tap-weights $\mathbf{w}(n)$ are recursively adapted in the direction

---

[3] In this section, lowercase boldface italic characters generally refer to vectors and (.)* is used to denote the Hermitian conjugate operation for matrices.

of the gradient with a step-size μ by using the filtered reference signal through the secondary path model $s(n)$,

$$\mathbf{w}(n+1) = \mathbf{w}(n) + \mu \mathbf{x}'_p(n)e^*(n) \tag{10}$$

where $\mathbf{x}'_p$ is the filtered $\mathbf{x}_p$.

A summary of the Fx-LMS algorithm is given in Table 2. Time-domain adaptation to a chirp signal is presented in Fig. 11 where the reference signal is $d(n)$ and the corresponding output of the time-domain LMS-based adaptive filter is $y(n)$. The error increases as the frequency of the input signal changes rapidly. The performance of the adaptive filtering scheme is also investigated by using the NLMS algorithm. The error signal $e(n)$, of the LMS-based adaptive filter in time-domain is illustrated in Fig. 11 (a), whereas the error signal of the NLMS-based adaptive filter in the time domain is shown in Fig. 11 (b).

3.2 Fourier–domain adaptation

Fourier–domain adaptation improves the adaptive filter performance for several reasons. For example, time domain adaptation requires the convolution operation and when the impulse response gets long, the overall complexity increases. By using fast Fourier transform, the computational complexity can be reduced, which is one of the reasons in choosing Fourier domain adaptation. The other reason is that frequency domain adaptive filtering improves the convergence performance. Moreover, orthogonality properties of the discrete Fourier transform provides a more uniform convergence rate (24).

4. System model and simulations

The adaptive LMS-based ANC system in fractional Fourier domain is designed as shown in Fig 8. An adaptive filter is used to model the primary path effect $P(z)$, which is the acoustic response from the reference sensor to the error sensor. Reference signal is obtained by measuring the ambient noise and the primary signal is the output of the unknown plant. The ambient noise that shows chirp-type characteristics is modeled by an adaptive filter in fractional Fourier domain which transforms non-stationary chirp-type signals to stationary sinusoidal signals. The acoustic anti-noise signal, which is the inverse-FrFT of the adaptive filter output, is generated by the loudspeaker. Except the FrFT-order estimation at certain time intervals, adaptation scheme is the same as the Fourier-domain adaptation in practical circuits and systems applications. A chirp-type noise signal to be modeled by the fractional Fourier domain adaptive filter and its appropriately-ordered FrFT are shown in Fig. 9(a) and (b). The chirp signal is transformed to a sinusoidal signal at the estimated transformation order of $a = 1/3$ by the proposed binary search algorithm. The input and output signals of the LMS-based adaptive filter in the fractional Fourier domain are given in Fig. 9(b) and (c). By taking the inverse FrFT of (c), time-domain filter output is obtained as shown in (d). The fractional Fourier domain error signals of both LMS and NLMS-based adaptive filters are plotted in Fig. 10(a) and (b). Among the two algorithms, NLMS performed better. Moreover, the corresponding time-domain error signals of the fractional Fourier domain adaptive filters by LMS and NLMS algorithms are presented in Fig. 11(a) and (b). Compared to the error plots of time-domain adaptation in Fig. 4(b) and 5(b), the fractional Fourier domain adaptive filtering scheme achieves significantly better performance. Finally, the performance of the adaptive fractional Fourier domain LMS-based ANC is tested when the chirp-type noise signal is embedded into AWGN. Fig. 12 presents error energy with respect to SNR, e.g., when SNR is 8 dB, the error energy in fractional Fourier domain adaptation scheme is less than 10^{-1}, whereas if the adaptation is realized in time, the error energy is greater than 1. Fractional Fourier domain adaptation noticeably improves the system performance.

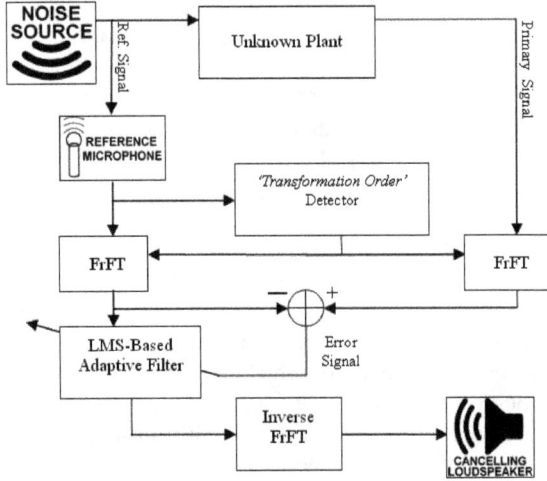

Fig. 8. The designed system model.

An LFM noise signal to be modeled by the fractional Fourier domain adaptive filter and its appropriately-ordered FrFT are shown in Fig.9(a) and (b). The chirp signal is transformed to a sinusoidal signal at the estimated transformation order of $a=1/3$. The input and output signals of the LMS-based adaptive filter in the fractional Fourier domain are given in Fig.(b) and (c). By taking the inverse FrFT of (c), time-domain filter output is obtained as shown in (d). The fractional Fourier domain error signals of both LMS and NLMS-based adaptive filters are plotted in Fig.10 (a) and (b). The adaptation step size μ can be chosen on the order of 10^{-1}. In the simulations, it is chosen as $\mu = 0.04$ in LMS and $\mu = 0.55$ in NLMS algorithm with $\alpha = 0$. In all of the simulations, adaptive filter length is chosen as 16 and among the two algorithms, NLMS performed better.

In the proposed fractional Fourier domain adaptive filtering scheme, the reference input $\mathbf{x}_p$ is transformed by the FrFT at the appropriate fractional order a so that the new input signal is

$$\mathbf{x}_p = \begin{bmatrix} x(n) \\ x(n-1) \\ \cdot \\ \cdot \\ x(n-p) \end{bmatrix} \rightarrow \mathbf{x}_{a,p} = \begin{bmatrix} x_a(n) \\ x_a(n-1) \\ \cdot \\ \cdot \\ x_a(n-p) \end{bmatrix}. \tag{11}$$

The reference and error signals are defined in the ath-order fractional Fourier domain and represented by $d_a(n)$ and $e_a(n)$, respectively. The corresponding weight-update process of the transform-domain adaptive filter becomes

$$e_a(n) = d_a(n) - \mathbf{w}_a^T(n)\mathbf{x}'_{a,p}(n). \tag{12}$$

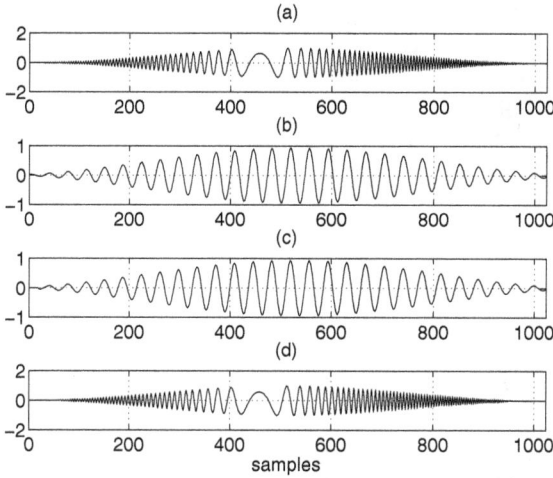

Fig. 9. An LFM signal, (b) its (1/3)rd order FrFT, (c) the output of the adaptive filter in FrFT domain and (d) the output of the adaptive filter in time domain by taking its inverse FrFT.

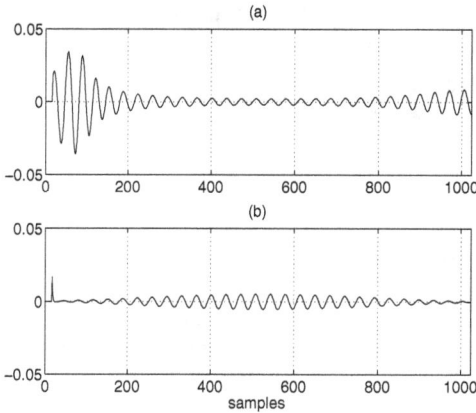

Fig. 10. Error signals in the fractional Fourier domain, by employing (a) LMS-based and (b) NLMS-based algorithms.

The tap-weights $\mathbf{w}_a(n)$ of the fractional Fourier domain adaptive filter are updated iteratively by

$$\mathbf{w}_a(n+1) = \mathbf{w}_a(n) + \mu \mathbf{x}'_{a,p}(n)e^*(n). \tag{13}$$

The FrFT parameter a should be estimated and kept updated during the adaptation process. FrFT-domain adaptation to an LFM signal is presented in Fig.12 by LMS and NLMS algorithms without introducing the secondary path effects $S(z)$. Fig.12(a) shows the chirp-type primary noise in time domain. This signal is transformed to a sinusoidal signal by taking FrFT at the appropriate order. Then, adaptation procedure is employed. Error plots of the LMS and NLMS algorithm in the fractional Fourier domain adaptation scheme are given in Fig.12(b) and (c), respectively. According to error signal of adaptive filters shown in Fig. 11 (b)-(c)

Fx-LMS Algorithm	
Input:	
Initialization vector:	$\mathbf{w}(n) = 0$
Input vector:	$\mathbf{x}(n)$
Desired output:	d(n)
Secondary path:	s(n)
Step-size parameter:	μ
Filter length:	M
Output:	
Filter output:	y(n)
Coefficient vector:	w(n+1)
Procedure:	
1) $y(n) = \mathbf{w}^H(n)\mathbf{x}(n)s(n)$	
2) $e(n) = d(n) - y(n)$	
3) $\mathbf{w}(n+1) = \mathbf{w}(n) + \mu e^*(n)\mathbf{x}(n)s(n)$	

Table 2. Fx-LMS Algorithm

and Fig. 12 (b)-(c) , it can be said that fractional Fourier domain adaptation performs well compared to time domain adaptation for chirp–type noise.

The fractional Fourier order parameter a is chosen as equal to the chirp rate, so the corresponding FrFT transforms the signal into a stationary signal. The simulation results present that fractional Fourier domain adaptive filtering is more successful at suppressing the undesired chirp-type noise compared to the time domain adaptation at the appropriate FrFT order.

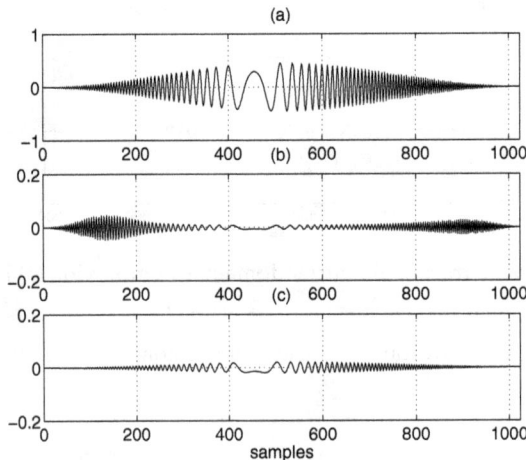

Fig. 11. (a) Primary noise in time domain, (b) LMS error, and (c) NLMS error by time domain adaptation.

The stability of the fractional Fourier domain LMS adaptation is assured by imposing limits on μ as

$$0 < \mu < 2/\lambda_m \tag{14}$$

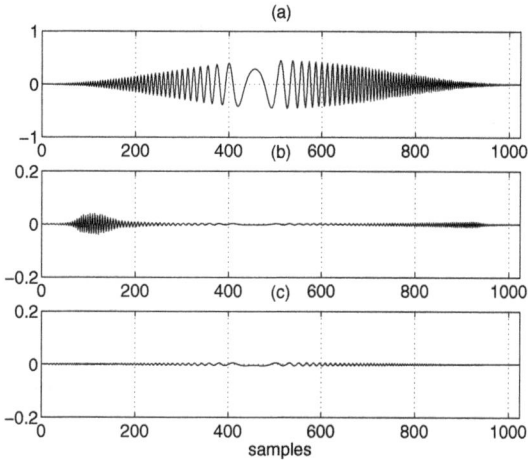

Fig. 12. (a) Primary noise in time domain, (b) LMS error, and (c) NLMS error by fractional Fourier domain adaptation.

where λ_m is the largest eigenvalue of the correlation matrix of the fractionally Fourier transformed input data. In case of the Fx-LMS algorithm, $x'_{a,p}(n)$ is used as the filtered version of $x_{a,p}(n)$ through $S(z)$. The observation period also depends on the computational power of the processor that operates the algorithm. IF estimation requires several times of FrFT algorithm giving rise to extra computational cost to the FrFT-domain adaptation.

Finally, the performance of the adaptive fractional Fourier domain LMS-based ANC is tested when the LFM noise signal is embedded into AWGN. Fig.13 presents error energy with respect to SNR, e.g., when SNR is 8 dB, the error energy in fractional Fourier domain adaptation scheme is less than 10^{-1}, whereas if the adaptation is realized in time, the error energy is greater than 1.

The fractional Fourier domain adaptation noticeably improves the system performance. The convergence analysis of the adaptive filtering schemes is realized for two alternative schemes. Fractional Fourier domain adaptive filtering scheme achieves faster adaptation compared to the time-domain adaptation algorithm as shown in Fig.14.

Fig.15 presents a real bat signal which has three different LFM components. Short–time Fourier transform (STFT) of this signal is given in Fig.16. The components of this signal are oriented along the same direction on the time-frequency plane and the FrFT transformation order is calculated as $a=0.0691$ by the IF estimation algorithm proposed in (28). Error signals of time-domain and FrFT-domain adaptation are given in Fig. 17 (a) and (b), respectively. FrFT-domain adaptive filtering scheme gives better results compared to the time-domain adaptation.

[4] The authors wish to thank Curtis Condon, Ken White, and Al Feng from Beckman Institute of the University of Illinois for permission to use the bat data it in this paper.

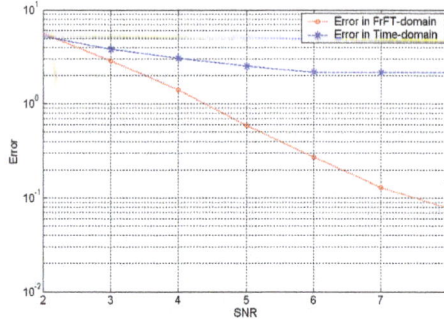

Fig. 13. Error energy in fractional Fourier and time-domain adaptation schemes with respect to SNR in AWGN.

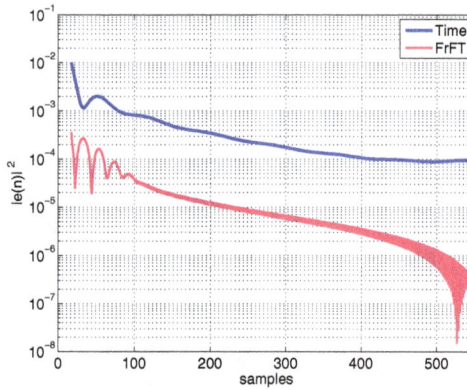

Fig. 14. Convergence analysis for both time and FrFT domain adaptive filtering.

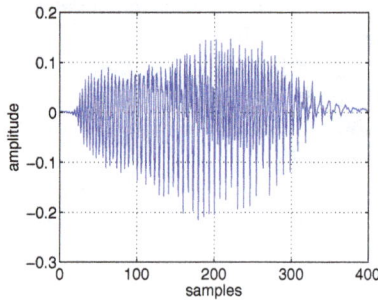

Fig. 15. Real bat echolocation signal which has multi-chirp components.

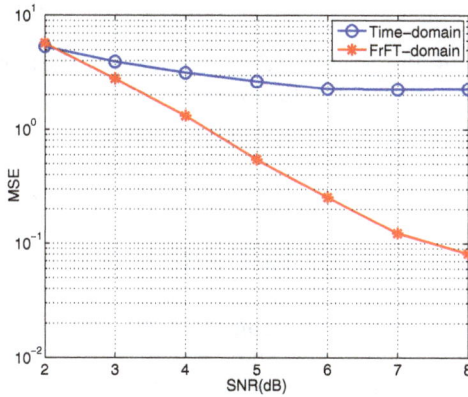

Fig. 16. STFT of the real bat signal.

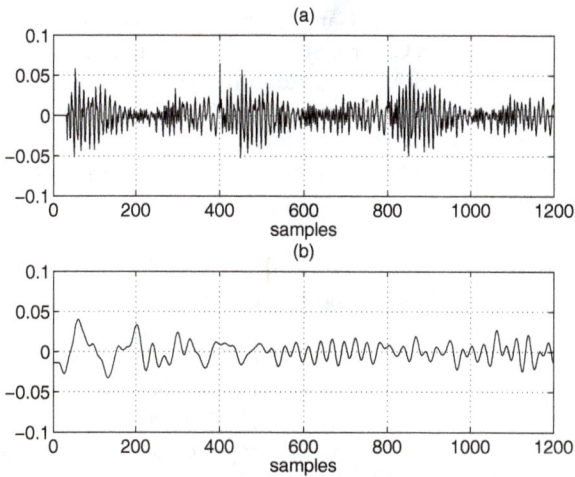

Fig. 17. (a) Error signal in Fx-NLMS algorithm by time-domain adaptation, (b) error signal in Fx-NLMS algorithm by fractional Fourier domain adaptation.

5. Conclusions

This chapter presents high-performance fractional Fourier domain adaptive filtering scheme for ANC systems. As a system parameter, the IF of the input chirp-type signal is estimated by searching the peak values of the modulus square of the FrFT. As noise signals originating from accelerating motion are chirp-type, such a fractional Fourier domain adaptive filtering approach avoids the difficulties of adaptation in a rapidly time-varying signal environment by transforming the signals to the appropriate fractional Fourier domain where the signals of interest become slowly time varying. Simulation results and error signals for mono-component chirp-type signals are compared in both time and fractional Fourier domains. It is evident that the total error quantity of adaptive filtering in fractional Fourier domain is significantly less than that of the time-domain adaptive filter.

6. Acknowledgements

This work is supported by the Scientific and Technological Research Council of Turkey, TUBITAK under the grant of Project No. 105E078.

7. References

[1] S. M. Kuo, "Active noise control: A tutorial review," Proc. IEEE, vol. 97, no. 6, (June, 1999), 943-973.

[2] C. Y. Chang and F. B. Luoh, "Enhancement of active noise control using neural-based filtered-X algorithm," Journal of Sound and Vibration, vol. 305, (2007), 348-356.

[3] G. Meng, X. Sun, S. M. Kuo, "Adaptive algorithm for active control of impulsive noise," Journal of Sound and Vibration, vol. 291, (2006),516-522.

[4] Y. L. Zhou, Q. Z. Zhang, X. D. Li, and W. S. Gan, "Analysis and DSP implementation of an ANC system using a filtered-error neural network," Journal of Sound and Vibration, vol. 285, (2005), 1-25.

[5] S. M. Kuo, S. Mitra, and W. S. Gan, "Active noise control system for headphone applications" IEEE Trans. on Control Syst. Tech., vol. 14, no. 2, (Mar., 2006), 331-335.

[6] W. S. Gan, R. G. Mitra, and S. M. Kuo,"Adaptive feedback active noise control headset: Implementation, evaluation and its extensions," IEEE Trans. on Consumer Electronics, vol. 51, no. 3, (Aug., 2005), 975-982.

[7] Z. Dayong and V. DeBrunner, "Efficient Adaptive Nonlinear Filters for Nonlinear Active Noise Control," IEEE Trans. on Circuits and Systems I, vol. 54, no. 3, (Mar., 2007), 669-681.

[8] C.Y. Chang and S.T. Li, "Active Noise Control in Headsets by Using a Low-Cost Microcontroller," IEEE Trans. on Industrial Electronics, vol. 58, no. 5, (May, 2011), 1936-1942.

[9] R. M. Reddy, I.M.S. Panahi, and R. Briggs, "Hybrid FxRLS-FxNLMS Adaptive Algorithm for Active Noise Control in fMRI Application," IEEE Trans. on Control Systems Technology, vol. 19, no. 2, (Mar., 2011), 474-480.

[10] B. J. Jalali-Farahani and M. Ismail, "Adaptive noise cancellation techniques in Sigma-Delta analog-to-digital converters," IEEE Trans. on Circuits and Systems I, vol. 54, no. 3, (Sep. 2007), 1891-1899.

[11] A. Goel, A. Vetteth, K. R. Rao, and V. Sridhar, "Active cancellation of acoustic noise using a self-tuned filter," IEEE Trans. on Circuits and Systems I, vol. 51, no. 11, (Mar., 2004),2148-2156.

[12] R.C. Selga and R.S.S. Peña, " Active Noise Hybrid Time-Varying Control for Motorcycle Helmets," IEEE Trans. on Control Systems Technology, vol. 18, no. 3, (May, 2010), 474-480.

[13] Z. Yuexian, C. Shing-Chow, and N. Tung-Sang, "Least mean M-estimate algorithms for robust adaptive filtering in impulse noise," IEEE Trans. on Circuits and Systems II vol. 47, no. 12, (Dec., 2000), 1564-1569.

[14] S. M. Kuo and D. R. Morgon, Active noise control: Algorithms and DSP implementations, New York: Wiley and Sons, 1996.

[15] W. S. Gan, Q. Z. Zhang, and Y. Zhou, "Adaptive recurrent fuzzy neural networks for active noise control," Journal of Sound and Vibration, vol. 296, (2006), 935-948.

[16] X. Liu X. Li, Q. Z. Zhang, Y. Zhou, and W. S. Gan, "A nonlinear ANC system with a SPSA-based recurrent fuzzy neural network controller," Lecture Notes in Computer Science, vol. 4491, 2007.

[17] Q. Z. Zhang and W. S. Gan, "Active noise control using a simplified fuzzy neural network," Journal of Sound and Vibration, vol. 272,(2004), 437-449.

[18] Q. Z. Zhang and W. S. Gan, A model predictive algorithm for active noise control with online secondary path modelling, Journal of Sound and Vibration, vol. 270, (2004), 1056-1066.

[19] M. Kawamata, M. T. Akhtar, M. Abe, and A. Nishihara, "Online secondary path modeling in multichannel active noise control systems using variable step size," Signal Process., vol. 88, (2008), 2019-2029.

[20] M. Joho and G. S. Moschytz, "Connecting partitioned frequency-domain filters in parallel or in cascade," IEEE Trans. on Circuits and Systems II, vol. 47, no. 8, (Aug., 2000), 685-698.

[21] S. Attallah, "The wavelet transform-domain LMS adaptive filter with partial subband coefficient updating," IEEE Trans. on Circuits and Systems II, vol. 53, no. 1, (Jan. 2006), 8-12.

[22] D. Veselinovic and D. Graupe, "A wavelet transform approach to blind adaptive filtering of speech from unknown noises," IEEE Trans. on Circuits and Systems II, vol. 50, no. 3,(Mar., 2003), 150-154.

[23] K. Mayyas and T. Aboulnasr, "Reduced-complexity transform-domain adaptive algorithm with selective coefficient update," IEEE Trans. on Circuits and Systems II, vol. 51, no. 3, (Mar., 2004), 132-142.

[24] S. Haykin, Adaptive Filter Theory, New Jersey: Prentice-Hall, 1996.

[25] H. M. Ozaktas, Z. Zalevski, and M. A. Kutay, "The Fractional Fourier Transform with Applications in Optics and Signal Processing," John Wiley and Sons, 2001.

[26] H. M. Ozaktas, O. Arıkan, M. A. Kutay, and G. Bozdagi, "Digital computation of the fractional Fourier transform," IEEE Trans. on Signal Process., 44(9), (Sept., 1996), 2141-2150.

[27] C. Candan, M. A. Kutay, and H. M. Ozaktas, "The discrete fractional Fourier transform", IEEE Trans. on Signal Process., 48(5), (May, 2000) 1329-1337,

[28] A. Serbes, L. Durak-Ata, "The discrete fractional Fourier transform based on the DFT matrix," Signal Process., 91 (3), (Mar., 2011), 571-581.

[29] A. W. Lohmann and B. H. Soffer, "Relationships between the Radon-Wigner and fractional Fourier transforms," J. Opt. Soc. Am. A, vol. 11, no. 6, (1994), 1798-1801.

[30] Q. Lin, Z. Yanhong, T. Ran, and W. Yue, "Adaptive filtering in fractional Fourier domain," IEEE Int. Symp. on Microw., Mape, (2), (Aug., 2005), 1033-1036.

[31] B. Boashash, "Estimating and interpreting the instantaneous frequency of a signal - Part II : Algorithms and applications," Proc. IEEE, vol. 80, no. 4, (Apr., 1992), 549-568.

[32] H. K. C. Kwok and D. L. Jones, "Improved instantaneous frequency estimation using an adaptive short-time Fourier transform," IEEE Trans. on Signal Process., vol. 10, no. 48, (Oct., 2000) 2964-2972.

[33] L. Durak and S. Aldırmaz, "Adaptive fractional Fourier domain filtering," Signal Processing, vol. 90, no. 4, (Apr., 2010), 1188-1196.

[34] C. F. Juang and C. T. Lin, "Noisy speech processing by recurrently adaptive fuzzy filters," IEEE Trans. on Fuzzy Systems, vol. 9 (Feb., 2001) 139-152.

3

Unsteady Flows in Turbines

Lei Qi and Zhengping Zou
Beihang University
China

1. Introduction

In order to satisfy the growing requirements of high performance aircraft, especially the civil aircraft, for increasing the economy, safety and environment protection, etc., it is imperative to understand the noise generation and control. Engine noise is one of the most important sources of aircraft noise. The main sources of noise in a high-by-pass turbofan engine, which is widely used in modern civil transport, include fan/compressor noise, combustion noise, turbine noise and jet noise. The periodic relative motion of the adjacent rows of blades is the essence of turbomachinery as used in aero engine. Fans, compressors and turbines each can generate significant tonal and broadband noise. The generation mechanisms of the noise include inlet distortion, wakes, potential interaction, tip leakage vortex, shock waves, separation flow, and so on. The interactions between rotating and stationary blade rows can cause unsteady aerodynamic force on blade surface, and thus cause the blade-passing-frequency tonal noise. And the broadband noise is generated by the interaction of the blades with random turbulence in boundary layers, wakes, vortex shedding, separation flow, etc. In turbomachinery, the interaction noise is sensitive to axial spacing between rotor and stator and choice of blade counts. Research indicates that increasing the axial spacing between rotors and stators or choosing appropriate blade number can effectively reduce the interaction noise (Crigler & Copeland, 1965; Benzakein, 1972; Tyler & Sofrin, 1962; Duncan et al., 1975). Besides, changing the phase distribution of rotor-stator interaction or using three-dimensional blade design can both reduce the interaction tonal noise in turbomachinery (Nemec, 1967; Schaub & Krishnappa, 1977; Mellin & Sovran, 1970; Suzuki & Kanemitsu, 1971).

Therefore, one of the most significant contributions to the gas turbine engine noise is due to the unsteady interactions in turbomachinery. An in-depth understanding of the unsteady flow mechanism is crucial for the effective control and reduction of the engine noise, which is especially important for the development of high performance aircraft engine. The aim of this chapter is to briefly introduce an overview of the published work about the unsteady flow in turbomachines. With a brief discussion of the basic concepts and characteristic parameters of unsteady flow, the chapter focuses on the primary unsteady flow phenomena in turbine components, including in low-pressure turbines and in high-pressure turbines. This chapter also discusses briefly the numerical methods that are applied to unsteady flow in turbomachinery. It is important to note that the contents are mainly based on the knowledge and experience of the authors. No attempt of a comprehensive overview is intended.

1.1 Turbomachinery flows

Turbomachinery flows are among the most complex flows encountered in fluid dynamic practice (Lakshminarayana, 1991). The complexity is mainly reflected in the following areas (Chen, 1989): (1) Various forms of secondary flow caused by viscosity and complex geometry, which is dominated by vortex flows: passage, leakage, corner, trailing, horseshoe and scraping vortices, etc. These form three-dimensional and rotational nature of the flow. (2) Inherent unsteadiness due to the relative motion of rotor and stator blade rows in a multi stage environment. (3) The flow pattern in the near-wall region includes: laminar, transitional and turbulent flows; besides separated flows are often exist. (4) The flow may be incompressible, subsonic, transonic or supersonic; some turbomachinery flows include all these flow regimes. (5) Due to the limitation of flow space, there are strong interactions of the solid wall surfaces with above complicated phenomena. Besides, in gas turbines, the use of cooling gas makes the flow more complex.

1.2 Unsteady flow phenomena in turbines

Flow in turbine blade rows is highly unsteady because of the periodically encountered flow distortions generated by the upstream and downstream blade rows. This unsteadiness has important consequences for the turbine stage efficiency, blade loading, mechanical fatigue due to blade flutter, heat transfer issues, thermal fatigue and noise generation. The induced unsteady flow depends on the scale of the disturbances. Usually in turbomachinery it includes two meanings (Xu, 1989): The first is the instability of the flow field. Such as the rotating stall, surge, flutter and flow distortion, etc, which must be avoided in design. The second is the inherent unsteadiness mainly due to the relative motion of rotor/stator blade rows in a multistage environment. These form unsteady characteristics with broad spectrum, as shown in Fig. 1. In this chapter the discussion will focus on the second category unsteady flows. The main generating factors of these unsteady flows can be classified based on the physical mechanisms involved as:

1. **Potential interaction.** The potential field associated with a blade row can propagate both upstream and downstream. The magnitude of this effect depends on the Mach number and the axial distance from the blade row. In high Mach number flows, potential interactions will tend to be stronger than at lower Mach numbers.
2. **Wakes.** Unlike the potential influence, a blade wake is only convected downstream. A wake profile can be characterized by a velocity deficit, and the static pressure in it does usually not vary significantly. It can influence the surface pressure, heat transfer and boundary layer nature of the downstream blades.
3. **Shock wave interaction.** When a turbine operates in the transonic regime, shock wave occurs. In addition to the losses produced by the periodic movement of the shock itself, the shock wave can cause intense unsteady effect.
4. **Streamwise vortices.** In low aspect ratio blade rows, the secondary flow and tip leakage flow, etc, in the form of streamwise vortices are significant. These vortices are convected downstream towards the next blade row where they interact with the main flow. They have an important effect on the flow distribution for the downstream blade rows.
5. **Hot streaks.** At the exit of combustor, there is a hot streak with temperature non-uniformities in both radial and azimuthal directions. It is convected downstream and will have significant effects on both the aerodynamic and the heat transfer for the downstream blade rows.

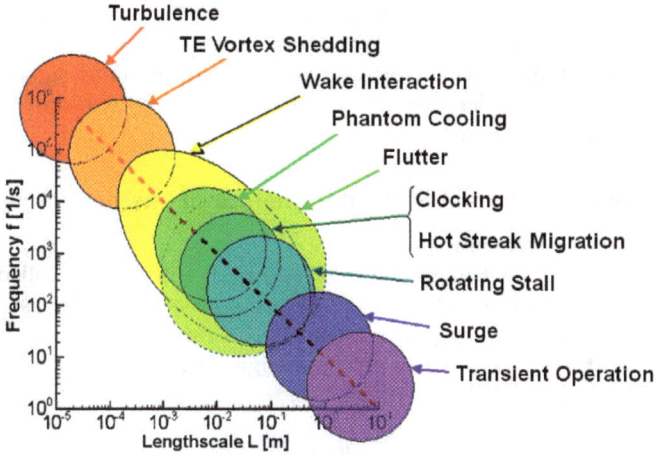

Fig. 1. Flow structures with 5 to 6 orders of magnitudes variations in length and time scales (LaGraff et al., 2006)

1.3 Unsteady flow effects on turbine performance

The impact of unsteady flow on turbomachinery performance has been extensively studied in recent decades. For the unsteady flow effects in turbomachinery, it is a source of aerodynamic noise to the acoustics experts, and it is a source of mechanical vibration to the aeroelasticity experts. For the aerodynamic designers, however, the interpretation of the unsteady flow effects in turbomachinery is still a controversial subject, which means that it can both have favorable or unfavorable influence on performance. In the past, the common opinion says that unsteadiness always brings a decrease in efficiency (Boletis & Sieverding, 1991; Sharma et al., 1992; Funazaki et al., 1997; Schulte & Hodson, 1998), such as the increase of losses, resistance and measurement errors, etc. With the deepening of understanding about the unsteady mechanisms, people gradually realize that the unsteady flow effects has a kind of latent benefit, such as the wake recovery effect, wake-boundary layer interaction (calming effect), clocking effect and hot steaks, etc.

So far, the design systems of turbomachinery are mostly based on the assumption of steady flow. Due to the lack of the realistic models for loss generation in the unsteady flow environment, designers rely on the use of experience factors for steady state loss correlations to account for these unsteady effects (Dunham, 1996). However, these factors do not necessarily reflect the true physical nature of the loss generation mechanisms in the unsteady environment. In most instances, the method based on the steady flow assumption captures the main features of the flow and we can get a high performance result. However, in high-load or off-design conditions, the performance is not satisfactory. One of the main reasons is the neglect of flow unsteadiness in actual turbomachinery. In turbine stage environment, the flow is periodic unsteady due to the relative motion of the blade rows. As modern engine design philosophy places emphasis on higher blade loading and smaller engine length, the effects of these unsteady interactions become even more important. The turbine design, up to now, has developed to a rather high level with the increasingly mature of the design technology. It is hard to make further improvement on the turbine performance under present design concept.

Therefore, it is necessary to consider the effects of unsteady flow, which have been neglected in present design and do have great effects on the performance. At present the problems we need to resolve are as follows. Which unsteady effects are the most significant ones and need to be given special consideration? How can we introduce these unsteady effects into the turbomachinery design system?

2. Basic concepts and characteristic parameters of unsteady flows

2.1 Definitions of loss coefficient

In a turbomachinery, any flow feature that reduces the efficiency will be called loss (Denton, 1993). There are many different definitions of loss coefficient in regular use for individual blade rows. Denton (1993) has given a detailed description about the loss coefficient definitions. The most useful loss coefficient for design purposes is the energy or enthalpy loss coefficient. For a turbine blade it is defined as,

$$\zeta = \frac{h_2 - h_{2s}}{h_{o2} - h_2} \tag{1}$$

where the isentropic final enthalpy, h_{2s}, is the value obtained in an isentropic expansion or compression to the same final static pressure as the actual process.

Entropy rise is one of the most commonly definitions of loss coefficient. Denton (1993) clearly illustrates that the only accurate measure of loss in a flow is entropy. Entropy is a particularly convenient measure because, unlike stagnation pressure, stagnation enthalpy or the kinetic energy, its value does not depend upon the frame of reference. It can be derived from the second law of thermodynamics that the entropy rise can be written as,

$$\Delta s = c_p \ln \frac{T}{T_{ref}} - R \ln \frac{p}{p_{ref}} \tag{2}$$

For adiabatic flow through a stationary blade row stagnation temperature is a constant and entropy rise depends only on stagnation pressure changes. So Equation 2 can be written as,

$$\Delta s = -R \ln(\frac{p_{o2}}{p_{o1}}) \tag{3}$$

For a turbine blade, another commonly definition is the stagnation pressure loss coefficient:

$$Y = \frac{p_{o1} - p_{o2}}{p_{o2} - p_2} \tag{4}$$

The reason that this definition of loss coefficient is so common is that it is easy to calculate it from cascade test data and not because it is the most convenient to use in design. However, the stagnation pressure loss coefficient can only be used in a stationary blade row in which the temperature is constant, but cannot be used in a rotational blade row. In addition, there are many other loss coefficient definitions, which are presented by Denton (1993).

It should be pointed out that in steady flow, the entropy rise and the stagnation pressure loss coefficient can both be used to estimate the loss. However, in real turbomachiney the flow is unsteady, and both the relative stagnation pressure and temperature can change. It follows that the loss coefficient should be expressed in terms of entropy, which accounts for both pressure and temperature changes. For example, experiment research by Mansour et al. (2008) showed that the overall losses and the loss distributions are misrepresented by the stagnation pressure loss coefficient. The overall losses are overestimated by more than 69% using the stagnation pressure loss coefficient. Furthermore the entropy loss coefficient identifies the tip leakage vortex as the most loss region, followed by the lower passage vortex, and then the upper passage vortex. On the other hand in terms of the stagnation pressure loss coefficient, the order of decreasing loss generation is lower passage vortex, upper passage vortex and tip leakage vortex.

The reason for this phenomenon is the isentropic rearrangement of the temperature and pressure, known as energy separation (Greitzer et al., 2004; Hodson & Dawes, 1998). It is well known that the stagnation enthalpy of a particle changes as it traverses an inviscid flow where the static pressure fluctuates. This may be written as,

$$\frac{Dh_o}{Dt} = \frac{1}{\rho}\frac{\partial p}{\partial t} \tag{5}$$

And the second law of thermodynamics relates changes in stagnation pressure and stagnation enthalpy by,

$$T_o ds = dh_o - \frac{1}{\rho_o}dp_o \tag{6}$$

From Equation (5) and (6) it can be seen that pressure changes with time not only influence the distribution of stagnation temperature, but also influence the stagnation pressure distribution by stagnation enthalpy. A detailed interpretation of the energy separation is presented by Greitzer et al. (2004). Thus, in unsteady flow field, we can only use entropy loss coefficient to represent a loss, due to the energy separation effect.

A better parameter of loss definition is the entropy generation, which is computed locally and has not to be given a reference value. The advantage of using entropy generation is to assess if and where the design could be improved. By evaluating the entropy generation in a control volume and summing many such control volumes in a blade passage, it is also possible to calculate the entropy increase for the whole blade row. Chaluvadi et al. (2003) introduced one of the methods to compute the entropy generation. According to the energy equation (Hughes & Gaylord, 1964), for a volume of "V", with a surface area "A", the rate of entropy production due to viscous dissipation can be written as,

$$\int_V \sigma dvol + \int_A \frac{k\nabla T}{T}\vec{n}dA = \frac{\partial}{\partial t}\int_V \rho s dvol + \int_A \rho s \vec{V}\vec{n}dA \tag{7}$$

where σ is the entropy production rate per unit volume due to viscous shear and k is the thermal conductivity. The unit normal vector $\vec{n}$ is positive when directed out of the volume and the velocity vector is denoted as $\vec{V}$.

2.2 Characteristic parameters of unsteady flows

The strength of flow unsteadiness can be evaluated by the Strouhal number, as

$$St = L / V_0 t_0 \tag{8}$$

where L, V_0, t_0 is the characteristic length, velocity and time, respectively. To a periodic vibration situation, for a frequency of " f ", the Strouhal number can be written as,

$$St = fL / V_0 \qquad (9)$$

In turbomachinery, the Strouhal number is equivalent to the frequency of unsteady disturbing sources, known as the reduced frequency. The reduced frequency is the ratio of time taken by the given particle for convection through the blade passage to the time taken for the rotor to sweep past one stator passage. It is expressed as (Lighthill, 1954):

$$\overline{f} = \frac{f \cdot s}{V_x} = \frac{\text{Convection time}}{\text{Disturbance time}} \qquad (10)$$

where f is the blade passing frequency, s is the blade pitch, and V_x is the axial velocity at the blade exit. The magnitude of the reduced frequency is a measure of the degree of unsteady effects compared to quasi-steady effects. If $\overline{f} \gg 1$, unsteady effects are significant and dominate the flow field, when $\overline{f} \approx 1$, unsteady and quasi-steady effects coexist. The reduced frequency $\overline{f}$ also represents the number of wakes (or other upstream unsteady features) found in a single blade passage at any instant in time.

3. Unsteady flows in low-pressure turbines

3.1 Introduction

Denton (1993) said: "The historical breakdown of loss into 'profile loss,' 'endwall loss,' and 'leakage loss' continues to be widely used although it is now clearly recognized that the loss mechanisms are seldom really independent." In low-pressure turbine blades, the profile loss is generally the largest single contributor to the total loss of efficiency, because of high aspect ratio of the blades. And the magnitude of the profile loss depends mainly on the development of the blade boundary layers, especially those on the suction surfaces.

The flow in low-pressure turbines is inherently unsteady caused by the relative motion of adjacent blade rows. There are two primary forms of periodic unsteadiness: the wakes from the upstream blade rows and the potential fields of blade rows both upstream and downstream. Potential interactions are weaker than wake interactions in most low-pressure turbines (Hodson & Howell, 2005). However, research indicates that only small changes in the static pressure field may alter the behavior of boundary layers that are close to separation or have separated on the rear of the suction surface of a blade (Opoks et al., 2006; Opoka & Hodson, 2007). Thus, care must be taken when dismissing the significance of potential interactions entirely. We will focus our discussion on the wake interactions in the present contents. This is because the wake interaction plays an important role in the development of the blade boundary layers, and with it the profile loss of the blades.

3.2 Wake transport mechanisms

The wake can be defined as any velocity deficit in the body-relative frame of reference occurring in a space much smaller than the one analyzed. There is a tendency for the wake fluid to be separated from the inviscid flow rather than mix (Casciaro, 2000). If the instantaneous velocity field is decrease of the undisturbed value, a wake looks like a facing backward jet, which is the so-called "negative jet". Many researchers have confirmed the negative jet theory in axial turbine as being the main unsteady transport mechanism. Meyer (1958) was one of the first to use the negative jet theory to explain the unsteady behavior in compressor stages. Kerrebrock & Mikolajczak (1970) observed that in a compressor the pressure side presence hinders the wake transport and consequently the stator wake fluid accumulates on the rotor pressure side, tending to decrease the load on a compressor blade. On the contrary in turbines, it looks like that the negative jet of the wake impinges on the rotor suction side, tending to increase the blade loads.

The convection of the upstream wake segment within the blade row is characterized by bowing, reorientation, elongation, and stretching (Smith, 1966; Stieger & Hodson, 2005), as Fig. 2 shows. Bowing of the wake fluid originates near the leading edge plane where the mid-passage velocities are higher than the velocities near the blade surfaces. The reorientation of the wake segment occurs due to the circulation of the blade. The velocities near the suction surface are higher than near the pressure surface, and therefore, fluid near the suction surface convects through the passage more rapidly, resulting in a reorientation of the wake segment. The difference in convection velocities also causes the wake segment to elongate, and this, in turn, decreases the wake width to conserve the vorticity of the wake fluid. Stretching occurs as the first part of the wake reaches the leading edge.

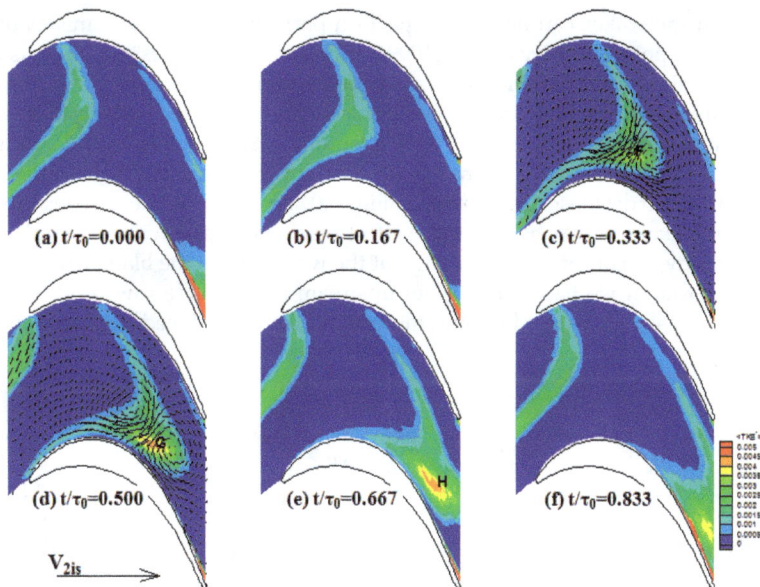

Fig. 2. Unsteady wakes convecting in blade passage (Stieger & Hodson, 2005)

The concept of the "wake avenue" was first introduced by Binder et al. (1989). Fig. 3 shows the progress of wakes through downstream blade rows. It can be seen from the plot that in the relative frame, the wake segments from the same rotor are arranged along a fixed path towards the downstream rotor. Differences in the number of rotor blades in the upstream and downstream blade rows cause the downstream flow field to be dependent upon the relative position of the upstream blade. This phenomenon is quasi-steady in the relative frame. It has also been observed in the tests carried out by other researchers (Miller et al., 2003).

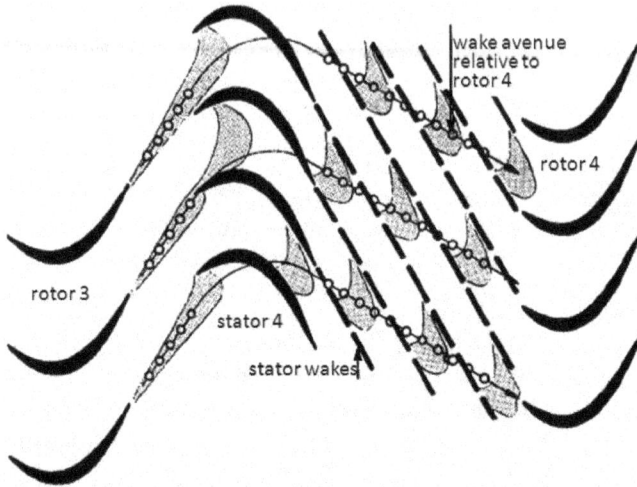

Fig. 3. Schematic diagram of the wake avenue (Binder et al., 1989)

The transport of the wakes can have an impact on performance by mechanisms other than boundary layer response. Valkov & Tan (1999a) indicates, in compressors, there is a generic mechanism with significant influence of wake transport on performance: reversible recovery of the energy in the wakes (beneficial). The loss created during the mixing of a wake with velocity deficit ΔV is proportional to ΔV^2 (Denton, 1993).The blade wake is stretched inviscidly as it is convected downstream so that, by Kelvin's circulation theorem, the velocity deficit must decrease (as shown in Fig. 4a), and with it the viscous mixing loss. Besides, the tip leakage vortex has the same effect mechanism as a wake (see Valkov & Tan, 1999b). That is why, in compressors, mixing out the wake within the blade row will generate less loss overall than if we had mixed it out upstream of the blade row. And the impact of interaction with upstream wakes becomes more significant at reduced axial spacing.

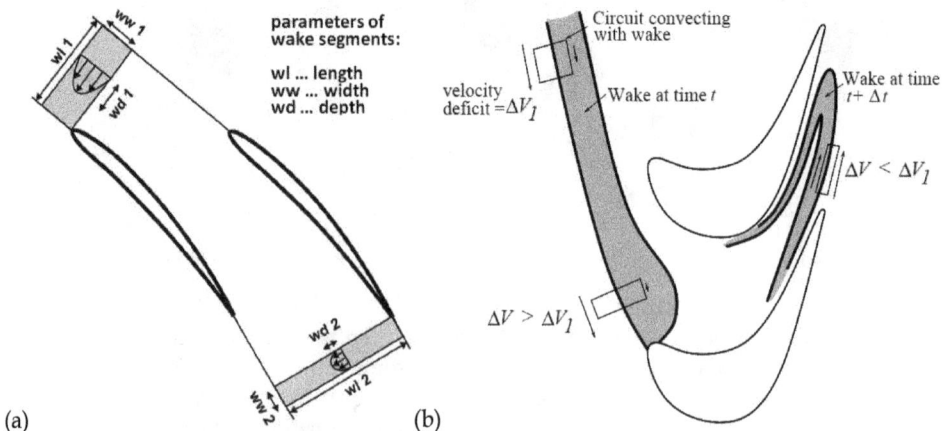

Fig. 4. Schematic diagram of wake transport: (a) in compressors (Mailach & Vogeler, 2006); (b) in turbines (Pullan, 2004)

The influence of the wake transport on performance is different in turbines. As shown in Fig. 4b, although the mixing losses of the wake increase as it is compressed near the leading and reduce as it is stretched through the passage, the converse occurs when streamwise vorticity (include the passage vortex and tip leakage vortex, etc.) enters the next blade row. Thus, in turbines, with some factors tending to increase entropy production and some reduce it, we cannot be certain whether mixing out the inlet flow within the rotor will generate more loss overall than if we had mixed it out upstream of the blade row.

3.3 Wake-boundary layer interaction

In low-pressure turbines, the wakes from upstream blade rows provide the dominant source of unsteadiness. Under low Reynolds number conditions, the boundary-layer transition and separation play important roles in determining engine performance. An in-depth understanding of blade boundary layer spatio-temporal evolution is crucial for the effective management and control of boundary layer transition or separation, especially the open separation, which is a key technology for the design of low-pressure turbines with low Reynolds number. Thus it is very important to research the wake-boundary layer interaction.

A comprehensive review of transition in turbomachinery components is given by Mayle (1991). He lists the four modes of transition and describes the mechanism: (a) Natural transition: the amplification of Tollmien-Schlichting instability waves in low free stream turbulence. (b) Bypass transition: caused by large disturbances, such as high free stream turbulence. (c) Separated-flow transition: caused by the laminar separation bubbles. (d) Periodic-unsteady transition: caused by the impingement of upstream periodic wakes. The periodic-unsteady transition is the characteristic mode in turbomachinery.

In low-pressure turbines with low Reynolds number, boundary layer separation may occur as the blade load increases. Rational use of the upstream periodic wakes can effectively inhibit the separation by inducing boundary layer transition before laminar separation can occur, so as to control loss generation. A comprehensive and in-depth research of wake-boundary layer interactions in low-pressure turbines is given by Hodson & Howell (2005). They summarized the processes of wake-induced boundary-layer transition and loss generation in low-pressure turbines. The schematic time-space diagram of wake-induced transition before separation is shown in Fig. 5a. The deep blue region denotes the turbulent wedge that results from the wake-induced strip, and followed immediately by the calmed region (light blue region) (The schematic plot of a turbulent spot is shown in Fig. 5b). The red region denotes the separation region. The periodic wake-boundary layer interaction process is as follows: When the wake passes, the wake-induced turbulent spots form within attached flows in front of the separation point. The turbulent spots continue to grow and enter into the separation zone, and consequently inhibit the formation of separation bubble. The calmed region trails behind the turbulent spots. It is a laminar-like region, but it has a very full velocity profile. The flow of the calmed region is unreceptive to disturbances. Consequently, it remains laminar for much longer than the surrounding fluid and can resist transition and separation. It is the combination of the calming effect and the more robust velocity profile within the calmed region that makes this aspect of the flow so important. After the interaction of the wake, boundary layer separation occurs in the interval between the two wakes.

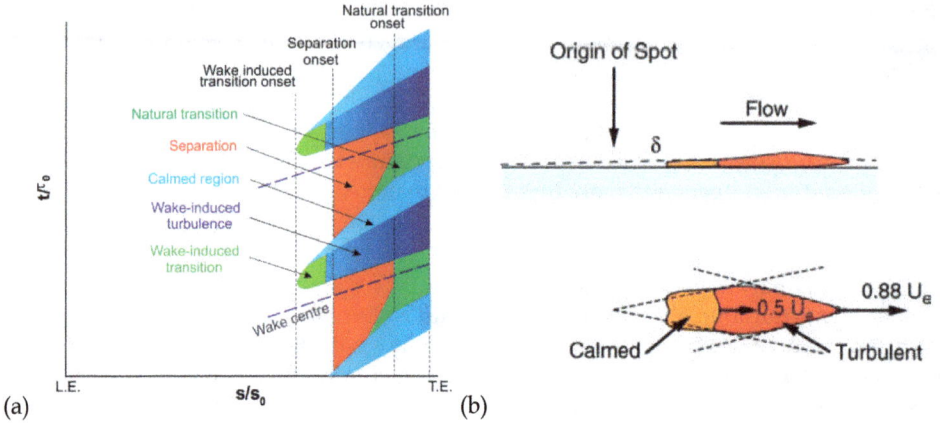

Fig. 5. Schematic diagram of wake-boundary layer interaction model : (a) the time-space diagram (Zhang & Hodson, 2004); (b) the turbulent spot (Hodson & Howell, 2005)

It can be concluded from Fig. 5 that the upstream wake-passing frequency is an important factor, which can influent the development of the boundary layers. It can effectively inhibit the boundary layer separation with appropriate wake-passing frequency, i.e., before the calmed region induced by the former wake disappears, the next wake has arrived in the area to be separated. At present, many researchers have basically understood the application of this technology in turbine design (e.g. Haselbach et al., 2002; Uliza & González, 2001).

3.4 Clocking effects

The real flow in multistage turbomachines is inherently unsteady because of the relative blade row motion. This causes unsteady interactions of pressure fields, shock waves, and wakes between stators and rotors. In recent years clocking of stator/rotor blade rows in multistage axial compressors and turbines has become an important scope of scientific investigations aiming to reduce aerodynamic losses and increase aerodynamic load. Clocking effect is to influence the unsteady flow field in multistage turbomachinery by changing the relative circumferential positions of either adjacent stator/rotor rows preferably with the same blade count.

The influence of clocking effects on aerodynamic performance is: the greatest benefits are achieved when the wake of upstream stator/rotor just impinges on the 2nd stator/rotor leading edge, while an efficiency drop was observed if the wake path entered the mid-channel (e.g. Arnone et al., 2003). Therefore, the aim of clocking is to find a relative circumferential position of successive rows of stators/rotors, so that the low momentum region in the wake of the upstream stator/rotor rows impinges the leading edge of the stator/rotor rows of the following stage.

The physical mechanism of clocking effects is as follows: when the relative circumferential positions of adjacent stator/rotor rows change that the upstream wake entering the passage also changes. It can have different impacts on the boundary layer and separation flow of the downstream blade rows, and result in the change of the transition point, separation point, and separation size in the boundary layers of blades, which will influence the efficiency of the stage. So it is important to study the effect of wake-boundary layer interactions on the clocking effects.

In the application of clocking effects, the axial and circumferential relative position of the rows, together with the blade count ratio between consecutive fixed and rotating rows, impact on the flow field unsteadiness, and consequently on the performance. Larger efficiency benefits can be achieved if the blade count ratio of consecutive stator and rotor rows is near 1:1, while practically no effect can be detected if it is far from unity (e.g. Arnone et al., 2000). In addition, larger efficiency gains can be achieved in compressors and turbines with high aspect ratio blades, conversely smaller.

Besides the total efficiency, clocking effects can influence the unsteady blade row pressure distribution. Many authors agree on the fact that larger amplitudes of unsteady pressure on the blades correspond to higher efficiency configurations (e.g. for turbine blades: Dorney & Sharma, 1996; Cizmas & Dorney, 1999), however, contrasting behavior has also been detected (Griffin et al., 1996; Dorney et al., 2001).

3.5 Shrouded tip leakage flow interaction

The necessary clearance between the rotating and the stationary components within the turbine gives rise to a clearance flow and hence loss of efficiency. Blade sealing configurations fall into two main categories: unshrouded and shrouded blades. The most obvious effect of flow leakage over the tips of both shrouded and unshrouded blades is a change in the mass flow through the blade passage, which would lead to a reduction in work for turbines.

Due to the extensive application of the shrouded blades in low-pressure turbines, here we will focus our discussion on shrouded leakage flows. The flow over the shrouded turbine blade with a single tip seal is illustrated in Fig. 6. For shrouded turbine blades the leakage will be from upstream to downstream of the blade row and so, for a fixed total volume flow, both the blade work and the pressure drop will be reduced. The pressure difference over the shroud provides the driving force for the fluid to pass into a shroud cavity and contract into a jet. The jet mixes out in the clearance space and this mixing process creates entropy. However, it was found that entropy creation due to tip leakage flows is determined almost entirely by the mixing processes that take place between the leakage flow and the mainstream; the flow processes over the shroud mainly affect the leakage flow rate (Denton, 1993). Denton (1993) provided a simple prediction model for the tip leakage loss of a shrouded blade.

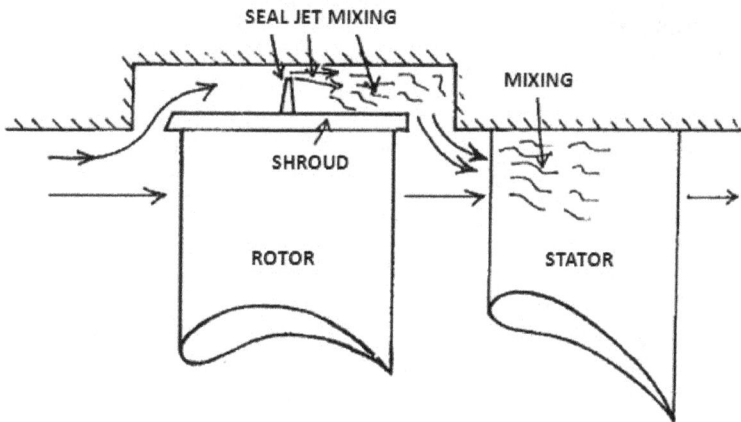

Fig. 6. Flow over a shrouded tip seal (Denton, 1993)

Little work is published on the time-resolved leakage flow-main flow interaction in shrouded axial turbines. Labyrinth leakages directly impact the boundary layers in the endwall regions before entering into the downstream row modifying the incidence angle and the secondary vortex structures (Adami et al., 2007). Denton (1993) suggested that the difference of the swirl velocity between the leakage and main flow dominates the mixing losses. The conclusion is confirmed by the research of Hunter & Manwaring (2000). However, the research by Wallis et al. (2000) has shown that although the relative swirl velocity of the leakage flow was reduced the turbine losses were increased. This indicates that a more complete understanding of unsteady interaction of the leakage flow with the main flow is required. The CFD simulations of Anker & Mayer (2002) also underlined the importance of considering unsteady measurement and simulations to obtain a more realistic representation of these effects. Hunter & Manwaring (2000) and Peters et al. (2000) showed that the size and location of the secondary flows are significantly affected by the interaction of labyrinth's leakage flow in turbines.

4. Unsteady flows in high-pressure turbines

4.1 Introduction
The blade of low aspect ratio was in popular use in recent years due to the increasing blade loading in turbomachinery (Pullan et al., 2006), especially in high-pressure and medium pressure turbines. It is well known that the loss of the flow field in the endwall regions is inversely proportional to the aspect ratio of blade. For turbines with low aspect ratio "endwall loss" (including the loss of secondary and tip leakage flows, etc.) is a major source of lost efficiency contributing even more than 60% of the total loss (Denton, 1993). The flow in high-pressure turbines is also inherently unsteady, so it is very important to consider the endwall unsteady effect in design, such as the interactions of the secondary flow vortex/tip leakage vortex with the periodic wake or potential field. Moreover in a high-loaded transonic turbine, the unsteady interactions between the shock wave and the secondary flow vortex /tip leakage vortex are also significant to the endwall loss. In order to improve the performance of high-pressure turbines, it is necessary to understand well the unsteady loss mechanisms in the endwall regions.

4.2 Shock wave interaction
It is well known that shock waves are irreversible and hence are sources of entropy. As the text books of Shapiro (1953) derive, the entropy increase across a normal or oblique shock wave varies roughly as the cube of (M^2-1), where M is the component of Mach number perpendicular to the shock front. The pressure rise across a weak shock wave is also proportional to (M^2-1). Oblique shocks will always produce less entropy than a normal one with the same upstream Mach number. The high-pressure turbine often operates with high pressure ratios and high Reynolds number and so shock waves do occur. Although local Mach numbers may be high, the shocks within the blade passage are usually oblique so that $\Delta P/P$ is small and they generate little direct loss (Denton, 1993).

In a transonic turbine stage with high loads, the trailing edge shocks from upstream blade rows are one of the most important sources of unsteadiness (Denton, 1993). There are indirect sources of loss associated with shock waves in transonic high-pressure turbines because of the interaction of the shock wave with the boundary layer. A boundary layer separation bubble will usually be formed at the foot of a weak shock and extra dissipation is likely to occur within and downstream of the bubble. If the boundary layer was laminar the bubble will almost certainly cause transition. An oblique shock impingement can produce reflected shock but a normal one can't (see Fig. 7).

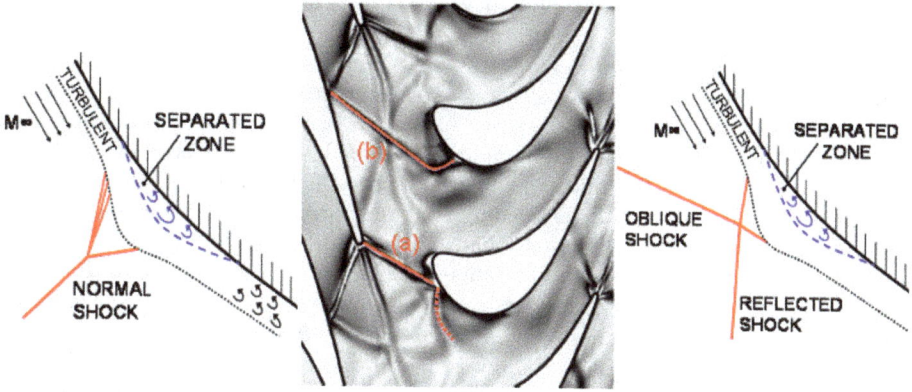

Fig. 7. Different shock-boundary layer interactions (Paniagua et al., 2007)

The motion of the upstream periodic shock waves causes the surface of the downstream blade to be subjected to a significant unsteady pressure field (see Fig. 8). When the flow unsteadiness experienced by a blade occurs at frequencies close to one of the blades' natural frequencies a significant vibration can occur. Over time this can cause high cycle fatigue and eventually, if not identified, to a catastrophic blade failure. As the trailing edge shock moves around the suction surface the vane trailing edge and the rotor leading edge form an effective throat, like Fig. 8 shows. This acts as a convergent divergent nozzle and raises the vane exit Mach number. As a rotor blade moves past a vane the divergent part of the effective nozzle gets shorter and the vane trailing edge Mach number is observed to drop.

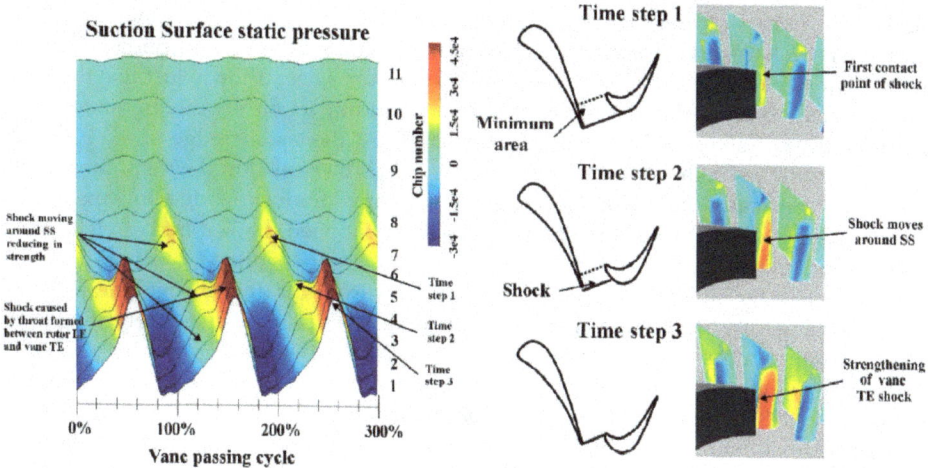

Fig. 8. The effect of upstream periodic shocks on the blade surface pressure distribution (Miller et al., 2002)

In addition, Denton (1993) said that any periodic motion of the shock itself will generate increased loss. Larger shock amplitudes cause consequently larger increases in loss. Effectively the increase in entropy generation when the shock is moving forward will be greater than the reduction when it is moving backward.

4.3 Secondary flow interaction

For high-pressure turbine blade rows with low aspect ratio the endwall losses (usually termed secondary flow losses or secondary losses) are a major source of lost efficiency contributing as high as 30-50% of the total aerodynamic losses in a blade row (Sharma & Butler, 1987). Langston (2006) said, an important problem that arises in the design and performance of axial flow turbines is the understanding, analysis, prediction and control of secondary flows. This is especially true for high-pressure turbines.

The secondary flow in a blade row can be defined as any flow, which is not in the direction of the primary or streamwise flow. The classical theories of secondary flow, as developed by Squire & Winter (1951), Hawthorne (1955) and Smith (1955) described the mechanism of the streamwise vorticity formation at blade row exit. Until now, several physical models have been developed to describe the secondary flow vortices in turbine cascade (e.g. Klein, 1966; Langston et al., 1977; Langston, 1980; Yamamoto, 1987; Wang et al., 1997), which help people have more comprehensive and vivid understanding of the secondary flow vortices structure. Review papers by Sieverding (1985), Wang et al. (1997) and Langston (2006) provide comprehensive summaries of the research on secondary flow structure and outline the most significant developments. The physical model of Wang et al. (1997) is depicted in Fig. 9. It can be seen that the passage vortex has the most significant effect on the endwall losses in turbine cascade. The classical secondary flow theories show that the flow overturning in the endwall regions and underturning in the midspan regions at the exit of the blade row.

A significant amount of research activity has recently been directed towards understanding the effect of secondary vortex unsteadiness on turbine performance. One of the major sources of unsteadiness was found to be the interaction between the streamwise vortices with the downstream blade row (e.g. Binder, 1985; Binder et al., 1986; Sharma et al., 1988). These streamwise vortices are shown to have a major influence on the secondary flow and viscous flow behavior of the downstream blade row. It has been shown, just like the wake, that the transport of streamwise vorticity can have an impact on the mixing losses in the next blade rows (Valkov & Tan, 1999b; Pullan, 2004). At the leading edge of the blade, the streamwise vortex is compressed so that, also by Kelvin's circulation theorem, the velocity deficit must decrease, and with it the rate of entropy production due to mixing. The converse occurs when the streamwise vortex is then chopped, bowed and stretched as it passes through the blade row.

In some research for multistage turbines, the blade row exit flow underturning near the hub and overturning towards the mid-span, contrary to the classical model of overturning in the endwall regions and underturning in the mid-span regions, indicating that the secondary flow is strongly influenced by the incoming secondary vortices of upstream blade row (e.g. Sharma et al., 1988; Hobson & Johnson, 1990; Chaluvadi et al., 2004). This unsteady effect raised substantial interest, as it had not been observed in the tests carried out by the other researchers (Boletis & Sieverdin, 1991; Hodson et al., 1993). It shows that the unsteady interactions on the steady performance must be related to the strength of the incoming secondary flow of the turbines.

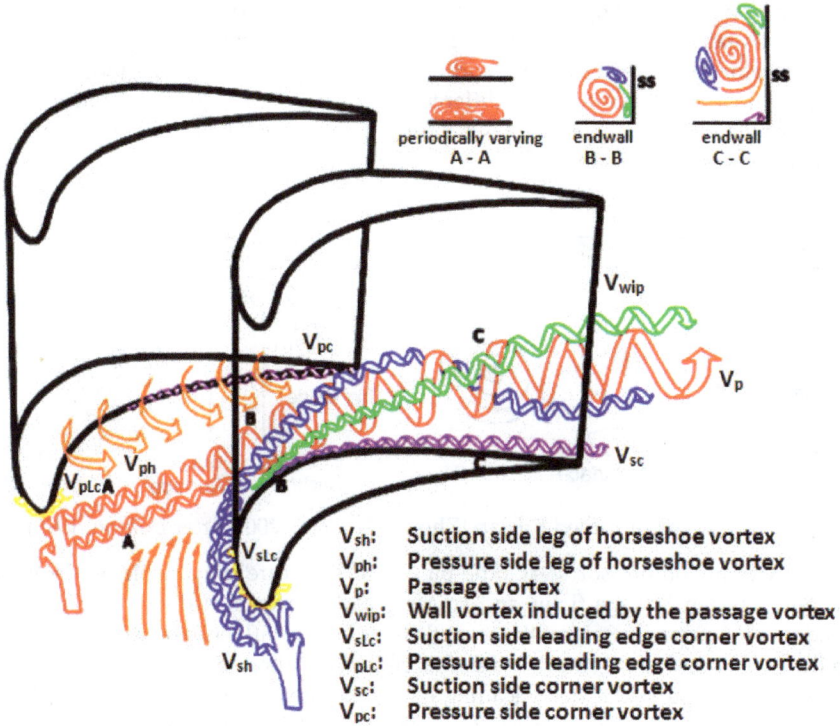

Fig. 9. Secondary flow model by Wang et al. (1997)

4.4 Unshrouded tip leakage flow interaction

The tip leakage flow is important in most turbomachinery, where a tip clearance with a height of about 1-2% blade span exists between the stationary endwall and the rotating blades. An unshrouded tip design is widely employed for a low stress and/or a better cooling in modern high-pressure turbines. Pictorial representation of the tip leakage flow in unshrouded blades is given in Fig. 10. The leakage flow over unshrouded blades occurs as a result of the pressure difference between the pressure and suction surfaces and is dominated by the vortex shed near the blade tip.

The tip leakage flow has significant effects on turbomachinery in loss production, aerodynamic efficiency, turbulence generation, heat protection, vibration and noise. As a consequence of the viscous effects, significant losses are generated by the tip leakage flow in regions inside and outside the tip gap. And the entropy creation is primarily due to the mixing processes that take place between the leakage flow and the mainstream flow. Denton (1993) gave a simple prediction model for the tip leakage loss of unshrouded blades.

So far, there are many researches about the leakage flow unsteady interactions in compressor. For example, Sirakov & Tan (2003) investigated the effect of upstream unsteady wakes on compressor rotor tip leakage flow. It was found that strong interaction between upstream wake and rotor tip leakage vortex could lead to a performance benefit in the rotor tip region during the whole operability range of interest. The experimental result of Mailach et al. (2008) revealed a strong periodical interaction of the incoming stator wakes and the compressor rotor blade tip clearance vortices. As a result of the wake influence, the tip clearance vortices are separated into different segments with higher and lower velocities and flow turning or subsequent counter-rotating vortex pairs. The rotor performance in the tip region periodically varies in time.

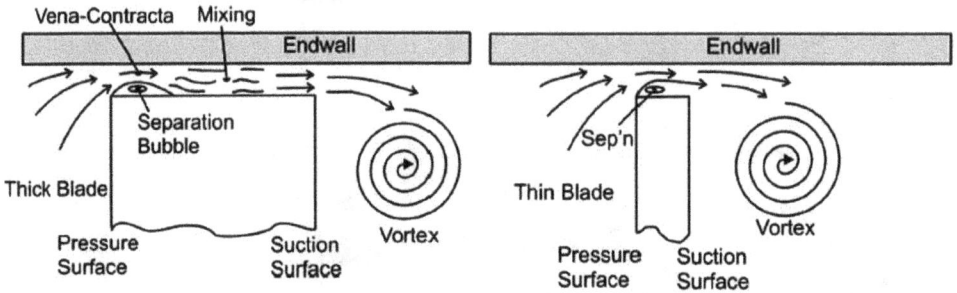

Fig. 10. Flow over an unshrouded tip gap (Zhou & Hodson, 2009, after Denton, 1993)

Compared with in compressor, very little published literature is available on the unsteady interactions between leakage flows and adjacent blade rows in turbine. Behr et al. (2006) indicated that the pressure field of the second stator has an influence on the development of the tip leakage vortex of the rotor. The vortex shows variation in size and relative position when it stretches around the stator leading edge. The present author (Qi, 2010) performed an experimental and numerical investigation of unsteady flow in the tip region of turbine with the effect of upstream wakes. The study objects included a linear turbine cascade and a turbine rotor, respectively. It shows that the presence of the upstream periodic wakes can reduce the strength and the loss of the tip leakage vortex, which is favorable to the turbine performance. In addition, due to the unsteady effect of the wakes, counter-rotating vortex pairs appear within the tip leakage vortices. These vortex pairs move downstream along the mean tip leakage vortex trajectory, which cause a significant pressure fluctuation and unsteady force on the rotor suction side.

4.5 Coolant injection and rim seal flow interactions

High-pressure gas turbines present considerable challenges to the designer because of the high aeromechanical loads at elevated temperatures. In order to cool the rotor disk and to avoid hot gas injection into the wheel space interface, cold flow is usually ejected from the cavity between the stator rim and the rotor disk. Fig. 11 illustrates a typical high-pressure gas turbine stage showing the rim seal and the wheel-space between the stator and the rotating turbine disc. Cooling air is supplied through the inner seal, and air leaves the wheel-space through the rim seal.

It was mentioned above that for high-pressure turbines, the secondary flow losses are a major source of the total aerodynamic losses in a blade row. The cooling air is injected into the mainstream and thereafter interacts with the secondary flow; consequently the turbine aerodynamic performance may be changed. So it is important to consider the coolant injection effect to the endwall region flow in unsteady environment.

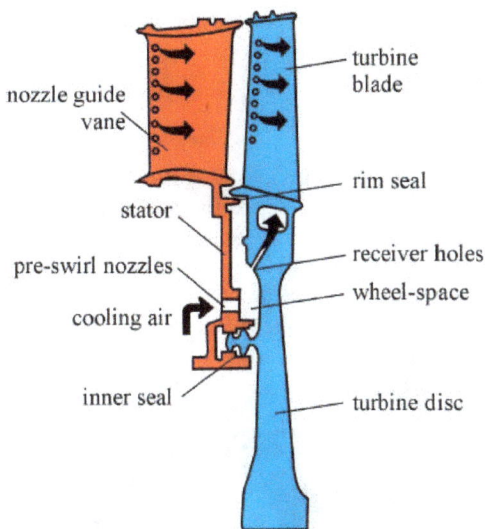

Fig. 11. Typical high-pressure turbine stage showing rim seal and wheel-space (Owen, 2009)

For subsonic turbines, the effects of rim seal purge ejection have been extensively studied in literature. It was found that even a small quantity (1 percent) of cooling air can have significant effects on the performance character and exit conditions of the high-pressure stage (McLean et al., 2001). Leakage flows from upstream cavities on engines generally emerge with lower momentum and less swirl than the mainstream flow, thus reducing the incidence at the inlet of the downstream blade row. The leakage flows may strengthen the endwall secondary flow on the downstream blade row (Anker & Mayer, 2002; Hunter & Manwaring, 2000; Paniagua et al., 2004; Pau et al., 2008). In addition, the ejection swirl angle can have a considerable effect to the efficiency of turbine stage, but the gain was restricted to the rotor due to a reduction in viscous dissipation and secondary losses (Ong et al., 2006).

In transonic turbines, trailing edge loss is a main source of losses. Denton (1993) indicates that trailing edge loss contributes even up to about 50 percent of the total two-dimensional loss in supersonic flow. The intensity of the trailing edge loss is related to the vane outlet Mach number, which is greatly affected by the rim seal. Hence, a complete study of the stator rim purge flow on transonic turbines requires not only the analysis on the downstream blade row (mixing losses and secondary flow), but the effect on the upstream transonic nozzle guide vane. Compared with in subsonic turbines, there is very little published research on transonic vane losses with rim seal. The effects of purge flow on a transonic nozzle guide vane have been studied numerically and experimentally by Pau and Paniagua (2009). In the research, a numerically predicted loss breakdown is presented, focusing on the relative importance of the trailing edge losses, boundary layer losses, shock losses and mixing losses, as a function of the purge rate ejected. Contrary to the experience in subsonic turbines, results in a transonic model demonstrate that ejecting purge flow improves the vane efficiency due to the shock structures modification downstream of the stator.

As stated previously, a small quantity of cooling air can have significant effects on the performance of the high-pressure stage. It means that the coolant injection effect should not be negligible. However, there is little research about the coolant injection effect to the performance of high-pressure gas turbines in real unsteady environment. It needs further studies in the future.

4.6 Hot streaks

Another significant unsteady phenomenon in high-pressure turbines is the interaction of hot streaks (also called "hot spots"), which refer to the radial and circumferential temperature gradients at the gas turbine combustor exit. These pronounced temperature non-uniformities in the combustor exit flow field are caused by circumferentially discrete fuel and dilution air injection within the combustor. The hot streaks can cause significant unsteadiness due to the relative motion of the blade rows. They interact with the different turbine parts and can locally cause increased blade heat transfer leading to reduced blade life and significant risks (see Fig. 12).

Fig. 12. Hot streaks migration (LaGraff et al., 2006)

In practical engine, combustor hot streaks can typically have stagnation temperatures twice of the free stream stagnation temperature. The hot streaks can lead to high heat loads and potentially catastrophic failure of the blades. It has been shown both experimentally and numerically that temperature gradients, in absence of total pressure non-uniformities, do not alter the flow (pressure field) within the first-stage turbine stator but do have a significant impact on the secondary flow and wall temperature of the first stage rotor (Butler et al., 1989).

An important concept is the "clocking effect of hot streaks". Adjusting the positions of hot streaks with respect to inlet guide vanes can be used to help control blade temperatures in gas turbines. When the hot streaks impinge on the leading edge of the blade, the impact to the blade is the maximum; on the contrary the impact decrease. Although a thermal barrier coating or cooling can reduce the local heat load to a certain degree, they will obviously increase the cost. For this reason, the method of using hot streak clocking effect is proposed. Takahashi et al. (1996) and Gundy-Burlet & Dorney (1997) showed that "clocking" the hot streaks so that it is positioned at the leading edge of the vane results in a diminishing of the effect of the hot streak on the downstream rotor. This was attributed to the deceleration and increased mixing of the hot streak as it interacted with wake of the vane. Shang & Epstein (1997) found the rotor-stator interactions between the rotor and stator can also generate significant non-uniformity of the time-averaged relative total temperature at the rotor inlet. And an optimum NGV-rotor blade count may exist that minimizes the influence of hot streaks on rotor blade life.

Another important concept is the "Kerrebrock-Mikolajczak effect" (Kerrebrock & Mikolajczak, 1970). It was a name associated with hot streak surface heating. These researchers were the first to describe why unsteady flow effects associated with compressors lead to increased pressure side heating. This effect is even more pronounced in turbines due to larger circumferential temperature variations. The description of the effect is based on the fact that the hot streak is moving at speeds significantly different from the surrounding fluid.

5. Numerical methods of unsteady flows in turbomachinery

5.1 Intruduction

Computational fluid dynamics (CFD) is now more and more used to assess unsteady effects in turbomachinery flow. The application has also been extended from original academic use to industrial design application. Most of the flows encountered in turbomachinery are turbulent. Based on the Navier-Stokes equations, the numerical methods to predicting turbulent flows can be traditionally divided into three categories (Pope, 2000): Reynolds-averaged Navier-Stokes (RANS) equations, large-eddy simulation (LES) and direct numerical simulation (DNS).

5.2 Reynolds-averaged Navier-Stokes (RANS) method

The solution of the (unsteady) Reynolds-averaged Navier-Stokes (RANS) equations is a tool that is most commonly applied to the solution of turbulent flow problems, especially in engineering applications. The Reynolds-averaged method was proposed by Osborne Reynolds in 1895. In RANS method, the equations are obtained by time- (in steady flow) or ensemble-averaging (in unsteady flow) the Navier-Stokes equations to yield a set of transport equations for the averaged momentum. In the averaging processes of the equations, the presence of the Reynolds stresses causes the equations not closed. To close the equations we must introduce a turbulence model. The turbulence models have been the object of much study over the last 30 years, but no model has emerged that gives accurate results in all flows without adjustments of the model constants. This may be due to the fact that the large, energy-carrying eddies are much affected by the boundary conditions, and universal models that account for their dynamics may be impossible to develop.

Turbulence model is a key issue in CFD simulations of turbomachines. According to the decision manner of the Reynolds stresses, turbulence model can be divided into two main categories (Wilcox, 1994; Chen & Jaw, 1998): eddy-viscosity model based on the Boussinesq eddy viscosity assumption and Reynolds stress model (RSM). By the number of turbulence model equations, the Boussinesq eddy-viscosity model in numerical simulation of turbomachinery can be divided again into: algebraic model (or zero-equation model) like the mixing length model or the Baldwin-Lomax model, one-equation model like Spalart-Allmaras model and two-equation model like various k–ε model and k–ω model. The RSM includes: algebraic stress model and Reynolds stress transport equation model. However, because of the complexity of turbulence model, so far there is not a universal turbulence model which can accurately simulate all types of complex flow in turbomachinery. So the formulation and proper application of the turbulence model is a key factor in the accurate prediction of the turbomachinery characteristics.

In addition, an accurate prediction of separation and transition flows is one of the important challenges to develop RANS method. Due to the complexity of separation and transition phenomenon, it is very difficult to make an accurate estimate. Many researchers have made an effort to develop and improve models for transition (Praisner & Clark, 2007; Praisner et al., 2007; Cutrone et al., 2007; Cutrone et al., 2008) in order to solve this problem. Anyhow, in the near future, practical calculations will continue to be carried out mainly by RANS methods with various turbulence models, which is still the most common and effective numerical method to solve engineering problems.

5.3 Large eddy simulation (LES)

Large eddy simulation (LES) is a technique intermediate between the solution of RANS and DNS. It is to simulate explicitly the larger-scale turbulent motions by solving the 3D time-dependent Navier-Stokes equations while model the small-scale motion that cannot be resolved on a given grid. LES can be more accurate than the (U)RANS approach and spend a lot less computational time than DNS.

The advantages of LES are as follows: firstly, compared with the large scales motion, the small scales are less affected by the boundary conditions and tend to be more isotropic and homogeneous, so are most likely to subject the universal models (Piomelli & Balaras, 2002); secondly, the large scales motion which is solved directly concludes most of the turbulent kinetic energy and can describe the main characteristics of the flow field better, so compared with RANS method, LES has some advantages (Rodi, 2000).

There are, however, many problems in actual application of LES that is the use of a subgrid-scale (SGS) model. Research indicates that the quality of the SGS models is of course an important issue in LES, which have a significant impact on the accuracy of the computational result. There are still difficulties in obtaining ideal SGS models. And also, there is the problem of huge calculated amount in LES, so it cannot become the major method in engineering practice at present.

5.4 Direct numerical simulation (DNS)

The direct numerical simulation (DNS) is a complete time-dependent solution of the Navier-Stokes equations. In DNS, all the scales of motion are resolver accurately, and no modeling is used. It is the most accurate numerical method available at present but is limited by its cost: because all scales of motion must be resolved, the number of grid points in each direction is proportional to the ratio between the largest and the smallest eddy in the flow.

Although DNS has huge calculated amount, it also has many advantages: firstly, DNS method is the most accurate method of all, because it directly solves the Navier-Stokes equations and need not introduce any models and assumptions; secondly, DNS can obtain all the details of the spatial-temporal evolution of the complex flow field, and can provide the most comprehensive numerical databases for the research of flow mechanisms; thirdly, the DNS database can be used to test other numerical simulation results, and can be of value in developing new turbulence models and subgrid-scale (SGS) models. Besides, DNS can be used to carry out some "virtual" numerical experimentation to study the influence of single parameter on the flow field.

In the last ten years, benefiting from the improvement of computer technology and numerical method, the applications of DNS present some new characteristic features. Firstly, to some "simple" flow problems, the computational Reynolds number is becoming higher and higher, such as the simulations of channel flow (Kaneda et al., 2003; Ishihara et al., 2007). Secondly, by DNS methods, important progress was made in mechanism research on some complex flow phenomena. For example, Krishnan and Sandham has performed the direct numerical simulations to the evolution of turbulent spots in supersonic boundary layers. It reveals the basic law of turbulent spots merging, spreading and the turbulent spot-separation bubble interactions (Krishnan & Sandham, 2006a; Krishnan & Sandham, 2006b). Finally, there are a great number of mechanism researches that is about complicated engineering flow at intermediate Reynolds number condition. For example, Wu and Durbin performed the direct numerical simulation of the flow field in low-pressure turbine cascade for the first time. They studied the interactions between upstream periodic wakes and cascade flow (Wu & Durbin, 2001; Kalitzin et al., 2003).

It can be expected that a study on the complicated engineering flow by DNS method will continually go on in-depth. But it is not practical to directly use DNS to study the engineering flow. On the one hand, DNS method need for too high computer performance; On the other hand, even the computer level can meet the requirement, the mass data produced by the DNS will not be needed by engineering researchers. Hence at present and also in the foreseeable future, DNS is not a tool for engineering calculations, but a tool for mechanism researches.

5.5 Conjugate heat transfer (CHT)

With the development of modern turbine engines, the turbine inlet temperature will be further increased in order to enhance the thermal efficiency. The high temperature of hot gas far exceeds the permissible material temperature of turbine blades. Currently, using a high efficient cooling system is the common and primary way to cool blades and reduce their thermal loads. Therefore, the precondition of designing a high efficient cooling system is the accurate prediction of the temperature field of turbine blades. With the fast development of CFD and computer technology, conjugate heat transfer (CHT) method has become an effective method for the temperature prediction of turbine and other hot components.

For the traditional method of predicting temperature field, the thermal boundary conditions must be specified and it requires tedious and costly iteration that involves sequentially performing numerical predictions of flow/temperature field of the hot gas and temperature field of blades. This method is a "decoupled" method and thus the results are unreliable. Instead, CHT method method is a more efficient and accurate way to predict temperature field. The flow/temperature field of the fluid (hot gas) and temperature field of solid (blades) are solved simultaneously and the temperature and heat flux are exchanged at fluid/solid interface during each iteration step. The CHT method is a "coupled" method and thus the results are more reliable.

The early studies of flow/temperature field coupled problem focused on basic studies, such as the work of Perelman (1961) and Luikov et al. (1971). With the fast development of CFD technology and computer hardware, it is possible to carry out full 3D conjugate heat transfer (CHT) calculation. At present, the relatively mature CHT methods or programs are: the

CHTflow solver developed by Bohn et al. (1995) at Aachen University, the Glenn-HT code developed by Rigby & Lepicovsky (2001) at NASA Glenn Research Center, the Glenn-HT/BEM code developed by Heidmann et al. (2003) and Kassab et al. (2003) at NASA Glenn Research Center, the HybFlow-Solid_CHT solver developed by Montomoli et al. (2004) at Florence University, the 2-D and 3-D conjugate heat transfer prediction codes from Han et al. (2000) at University of Texas the fully coupled conjugate heat transfer (CHT) code from Croce (2001) at University of Udine, and so on.

6. References

Adami, P., Martelli, F. & Cecchi, S. (2007). Analysis of the shroud leakage flow and mainflow interactions in high-pressure turbines using an unsteady computational fluid dynamics approach, *IMechE J. Power and Energy* Vol. 221: 837-848.

Anker, J. E. & Mayer, J. F. (2002). Simulation of the interaction of labyrinth seal leakage flow and main flow in an axial turbine, ASME Paper GT2002-30348.

Arnone, A., Marconcini, M. & Pacciani, R. (2000). On the use of unsteady methods in predicting stage aerodynamic performance, *Proceedings of Conference on Unsteady Aerodynamics, Aeroacustics and Aeroelasticity of Turbomachines, pp. 24-36, Lyon, France.*

Arnone, A., Marconcini, M. , Scotti Del Greco, A. & Spano, E. (2003). Numerical investigation of three-dimensional clocking effects in a low pressure turbine, ASME Paper GT2003-38414.

Behr, T., Kalfas, A. l. & Abhari, R. S. (2006). Unsteady flow physics and performance of a one-and-1/2-stage unshrouded high work turbine, ASME Paper GT2006-90959.

Benzakein, M. J. (1972). Research on fan noise generation, *J. Acoustical Society of America* Vol. 51(No. 5A): 1427-1438.

Binder, A. (1985). Turbulent production due to secondary vortex cutting in a turbine rotor, *J.Engineering for Gas Turbines Power* Vol. 107: 1039-1046.

Binder, A., Forster, W., Mach, K. & Rogge, H. (1986). Unsteady flow interaction caused by stator secondary vortices in a turbine rotor, ASME Paper 86-GT-302.

Binder, A., Schroeder, T. & Hourmuziadis, J. (1989). Turbulence measurements in a multistage low pressure turbine, *ASME J. Turbomachinery* Vol. 111: 153-161.

Bohn, D., Bonhoff, B. & Schönenborn, H. (1995). Combined aerodynamic and thermal analysis of a turbine nozzle guide vane, IGTC-paper-108.

Boletis, E. & Sieverding, C. H. (1991). Experimental study of the three-dimensional flow field in a turbine stator preceded by a full stage, *ASME J. Turbomachinery* Vol. 113: 1-9.

Borislav, T. S. & Choon, S. T. (2002). Effect of upstream unsteady flow conditions on rotor tip leakage flow, ASME Paper GT2002-30358.

Butler, T. L., Sharma, O. P., Joslyn, H. D. & Dring, R. P. (1989). Redistribution of an inlet temperature distortion in an axial flow turbine stage, *AIAA J. Propulsion and Power* Vol. 5(No. 1): 64-71.

Casciaro, C., Treiber, M. & Sell, M. (2000). Unsteady transport mechanisms in an axial turbine, *ASME J. Turbomachinery* Vol. 122: 604-612.

Chaluvadi, V. S. P., Kalfas, A. I. & Hodson, H. P. (2003). Blade row interaction in a high-pressure steam turbine, *ASME J. Turbomachinery* Vol. 125: 14-24.

Chaluvadi, V. S. P., Kalfas A. I. & Hodson H. P. (2004). Vortex transport and blade interactions in high pressure turbines, *ASME J. Turbomachinery* Vol. 126: 395-405.

Chen, C. J. & Jaw, S. Y. (1998). *Fundamentals of turbulence modeling*, Taylor & Francis.

Chen, M. Z. (1989). Aero-engine turbomachinery dynamics, *Proceedings of Aerothermodynamics Development Strategy Proseminar*, pp. 1-8, Beijing, China.

Cizmas, P. G. A. & Dorney, D. J. (1999). The influence of clocking on unsteady forces of compressor and turbine blades, ISABE Paper 99-7231.

Crigler, J. L. & Copeland, W. L. (1965). Noise studies of inlet-guide-vane: rotor interaction of a single stage axial compressor, NASA Technical Note, NASA TN D-2962.

Croce, G. A. (2001). A conjugate heat transfer procedure for gas turbine blades, *Annals of N.Y. Academy of Sciences*, Vol. 934: 273-280.

Cutrone, L., De Palma, P., Pascazio, G. & Napolitano, M. (2007). An evaluation of bypass transition models for turbomachinery flows, *International J. Heat and Fluid Flow* Vol. 28: 161-177.

Cutrone, L., De Palma, P., Pascazio, G. & Napolitano, M. (2008). Predicting transition in two- and three-dimensional separated flows, *International J. Heat and Fluid Flow* Vol. 29: 504-526.

Denton, J. D. (1993). Loss mechanisms in turbomachines, *ASME J. Turbomachinery* Vol. 115: 621-656.

Dorney, D. J. & Sharma, O. P. (1996). A study of turbine performance increases through airfoil clocking, AIAA Paper 96-2816.

Dorney, D. J., Croft, R. R., Sondak, D. L., Stang, U. E. & Twardochleb, C. Z. (2001). Computational study of clocking and embedded stage in a 4-stage industrial turbine, ASME Paper 2001-GT-509.

Duncan, P. E., Dawson, B. & Hawes, S. P. (1975). Design techniques for the reduction of interaction tonal noise from axial flow fans, *Proceedings of Conference on Vibrations and Noise in Pump, Fan, and Compressor Installations*, pp. 143-161, University of Southampton, U.K.

Dunham, J. (1996). Aerodynamic losses in turbomachinery, AGARD-CP-571.

Funazaki, K., Sasaki, Y. & Tanuma, T. (1997). Experimental studies on unsteady aerodynamic loss of a high-pressure turbine cascade, ASME Paper 97-GT-52.

Greitzer, E. M., Tan, C. S. & Graf, M. B. (2004). *Internal Flow: Concepts and Applications*, Cambridge University Press.

Griffin, L. W., Huber, F. W. & Sharma, O. P. (1996). Performance improvement through indexing of turbine airfoils: part II - numerical simulation, *ASME J. Turbomachinery* Vol. 118: 636-642.

Gundy-Burlet, K. L. & Dorney, D. J. (1997). Three-dimensional simulations of hot streak clocking in a 1-1/2 stage turbine, *International J. Turbo and Jet Engines* Vol. 14(No. 3): 123-132.

Han, Z. X., Dennis, B. & Dulikravich, G. (2000). Simultaneous prediction of external flow-field and temperature in internally cooled 3-D turbine blade material, ASME Paper 2000-GT-253.

Haselbach, F., Schiffer, H. P., Horsman, M., Dressen, S., Harvey, N. & Read, S. (2002). The application of ultra high lift blading in the BR715 LP Turbine, *ASME J. Turbomachinery* Vol. 124: 45-51.

Hawthorne, W. R. (1955). Rotational flow through cascades: part I-the components of vorticity, *J. Mechanics and Applied Mathematics* Vol. 8: 266-279.

Heidmann, J. D., Kassab, A. J., Divo, E. A., Rodriguez, F. & Steinthorsson, E. (2003). Conjugate heat transfer effects on a realistic film-cooled turbine vane, ASME Paper GT2003-38553.

Hobson, D. E. & Johnson, C. G. (1990). Two stage model turbine performance characteristics, Powergen Report RD/M/1888/RR90.

Hodson, H. P., Baneighbal, M. R. & Dailey, G. M. (1993). Three-dimensional interactions in the rotor of an axial turbine, AIAA Paper 93-2255.

Hodson, H. P. & Dawes, W. N. (1998). On the interpretation of measured profile losses in unsteady wake-turbine blade interaction studies, ASME J. Turbomachinery, Vol. 120: 276-284.

Hodson, H. P. & Howell, R. J. (2005). Bladerow interactions, transition, and high-lift aerofoils in low-pressure turbines, Annu. Rev. Fluid Mech. Vol. 37: 71-98.

Hughes, W. F. & Gaylord, E. W. (1964). Basic equations of engineering science, Schaum's Outline Series, McGraw Hill.

Hunter, S. D. & Manwaring, S. R. (2000). Endwall cavity flow effects on gaspath aerodynamics in an axial flow turbine: part 1 – experimental and numerical investigation, ASME Paper 2000-GT-651.

Ishihara, T., Kaneda, Y., Yokokawa, M., Itakura, K. & Uno, A. (2007). Small-scale statistics in high-resolution direct numerical simulation of turbulence: Reynolds number dependence of one-point velocity gradient statistics, J. Fluid Mech. Vol. 592: 335-366.

Kalitzin, G., Wu, X. & Durbin, P. A. (2003). DNS of fully turbulent flow in a LPT passage, Int. J. Heat and Fluid Flow Vol. 24: 636-644.

Kaneda, Y., Ishihara, T., Yokokawa, M., Itakura, K. & Uno, A. (2003). Energy dissipation rate and energy spectrum in high resolution direct numerical simulations of turbulence in a periodic box, Physics of Fluids Vol. 15(No. 2): L21-24.

Kassab, A. J., Divo, E. A., Heidmann, J. D., Steinthorsson, E. & Rodriguez, F. (2003). BEM/FVM Conjugate heat transfer analysis of a three-dimensional film cooled turbine blade, Int. J. Numerical Methods for Heat & Fluid Flow Vol. 13(No. 5): 581-610.

Kerrebrock, J. L. & Mikolajczak, A. A. (1970). Intra-stator transport of rotor wakes and its effect on compressor performance, ASME J. Engineering for Power Vol. 92: 359-370.

Klein, A. (1966). Investigation of the entry boundary layer on the secondary flows in the blading of axial turbines, BHRAT1004.

Krishnan, L. & Sandham, N. D. (2006a). Effect of Mach number on the structure of turbulent spots, J. Fluid Mech. Vol. 566: 225-234.

Krishnan, L. & Sandham, N. D. (2006b). On the merging of turbulent spots in a supersonic boundary-layer flow, Int. J. Heat and Fluid Flow Vol. 27: 542-550.

LaGraff, J. E., Ashpis, D. E., Oldfield, M. L. G. & Gostelow, J. P. (2006). Unsteady flows in turbomachinery, NASA/CP-2006-214484.

Lakshminarayana, B. (1991). An assessment of computational fluid dynamic techniques in the analysis and design of turbomachinery—the 1990 freeman scholar lecture, J. Fluids Engineering Vol. 113(No. 3): 315-352.

Langston, L. S., Nice, M. L. & Hooper, R. H. (1977). Three-dimensional flow within a turbine cascade passage, ASME J. Engineering Power Vol. 99: 21-28.

Langston, L. S. (1980). Crossflows in a turbine cascade passage, ASME J. Engineering Power Vol. 102: 866-874.

Langston, L. S. (2006). Secondary flows in axial turbines-a review, *Annals New York Academy of Sciences* Vol. 934: 11-26.

Lighthill, M. J. (1954). The response of laminar skin friction and heat transfer to fluctuations in the stream velocity, *Proc. of Royal Society* Vol. 224A: 1-23.

Luikov, A. V., Aleksashenko, V. A. & Aleksashenko, A. A. (1971). Analytical methods of solution of conjugated problems in convective heat transfer, *Int. J. Heat and Mass Transfer* Vol. 14(No. 8): 1047-1056.

Mailach, R. & Vogeler, K. (2006). Blade row interaction in axial compressors, part I: periodical unsteady flow field, *Lecture series 2006-05*, Von Kármán Institute.

Mailach, R., Lehmann, I. & Vogeler, K. (2008). Periodical unsteady flow within a rotor blade row of an axial compressor-part II: wake-tip clearance vortex interaction. *ASME J. Turbomachinery* Vol. 130: 1-10.

Mansour, M., Chokani, N., Kalfas, A.L. & Abhari, R.S. (2008). Impact of time-resolved entropy measurement on a one-and-1/2-stage axial turbine performance, ASME Paper GT2008-50807.

Mayle, R. E. (1991). The role of laminar-turbulent transition in gas turbine Engines, ASME Paper 91-GT-261.

McLean, C., Camci, G. & Glezer, B. (2001). Mainstream aerodynamic effects due to wheelspace coolant injection in a high pressure turbine stage, *ASME J. Turbomachinery* Vol. 123: 687–703.

Mellin, R. C. & Sovran, G. (1970). Controlling the tonal characteristics of the aerodynamic noise generated by fan rotors, *ASME J. Basic Engneering* Vol. 92: 143-154.

Meyer, R. X. (1958). The effect of wakes on the transient pressure and velocity distributions in turbomachines, *ASME J. Basic Engineering* Vol. 80: 1544-1552.

Miller, R. J., Moss, R. W. & Ainsworth, R. W. (2002). Wake, shock and potential field interactions in a 1.5 stage turbine, part I: vane-rotor and rotor-vane interaction, ASME Paper GT2002-30435.

Miller, R. J., Moss, R. W., Ainsworth, R. W. & Harvey, N. W. (2003). The development of turbine exit flow in a swan-necked inter-stage diffuser, ASME Paper GT2003-38174.

Montomoli, F., Adami, P., Della Gatta, S. & Martelli, F. (2004). Conjugate heat transfer modelling in film cooled blades, ASME Paper GT2004-53177.

Nemec, J. (1967). Noise of axial fans and compressors: Study of its radiation and reduction, *J. Sound and Vibration* Vol. 6(No. 2): 230-236.

Ong, J. H. P., Miller, R. J. & Uchida, S. (2006). The effect of coolant injection on the endwall flow of a high pressure turbine, ASME Paper GT2006-91060.

Opoka, M. M., Thomas, R. L. & Hodson, H. P. (2006). Boundary layer transition on the high lift T106A LP turbine blade with an oscillating downstream pressure field, ASME GT2006-91038.

Opoka, M. M. & Hodson, H. P. (2007). Transition on the T106 LP turbine blade in the pressure of moving upstream wakes and downstream potential fields, ASME Paper GT2007-28077.

Owen, J. M. (2009). Prediction of ingestion through turbine rim seals, part 1: rotationally-induced ingress, ASME Paper GT2009-59121.

Paniagua, G., Dénos, R. & Almeida, S. (2004). Effect of the hub endwall cavity flow on the flow-field of a transonic high-pressure turbine, *ASME J. Turbomachinery* Vol. 126: 578-586.

Paniagua, G., Yasa, T., Andres de la Loma, Castillon, L. & Coton, T. (2007). Unsteady strong shock interactions in a transonic turbine: experimental and numerical analysis, ISABE Paper ISABE-2007-1218.

Pau, M., Paniagua, G., Delhaye, D., De la Loma, A. & Ginibre, P. (2008). Aerothermal impact of stator-rim purge flow and rotor-platform film cooling on a transonic turbine stage, ASME Paper GT2008-51295.

Pau, M. & Paniagua, G. (2009). Investigation of the flow field on a transonic turbine nozzle guide vane with rim seal cavity flow ejection, ASME Paper GT2009-60155.

Perelman, T. L. (1961). On conjugated problems of heat transfer, *Int. J. Heat and Mass Transfer* Vol. 3(No. 4): 293-303.

Peters, P., Breisig, V., Giboni, A., Lerner, C. & Pfost, H. (2000). The influence of the clearance of shrouded rotor blades on the development of the flowfield and losses in the subsequent stator, ASME Paper 2000-GT-478.

Piomelli, U. & Balaras, E. (2002). Wall-layer models for large-eddy simulations, *Annual Review of Fluid Mechanics* Vol. 34: 349-374.

Pope, S. B. (2000). *Turbulent flows*, Cambridge University Press.

Praisner, T. J. & Clark, J. P. (2007). Predicting transition in turbomachinery, part I: a review and new model development, *ASME J. Turbomachinery* Vol. 129: 1-13.

Praisner, T. J., Grover, E. A., Rice, M. J. & Clark, J. P. (2007). Predicting transition in turbomachinery, part II: model validation and benchmarking, *ASME J. Turbomachinery* Vol. 129: 14-22.

Pullan, G. (2004). Secondary flows and loss caused by row interaction in a turbine stage, ASME Paper GT2004-53743.

Pullan, G., Denton, J. D. & Curtis E. (2006). Improving the performance of a turbine with low aspect ratio stators by aft-loading, *ASME J. Turbomachinery* Vol. 128: 492-499.

Qi, L. (2010). Investigations of unsteady interaction mechanism and flow control in the turbine endwall regions, *D. Phil. Thesis*, Beihang University.

Rigby, D.L. & Lepicovsky, J. (2001). Conjugate heat transfer analysis of internally cooled configurations, ASME Paper 2001- GT-0405.

Rodi, W. (2000). Simulation of turbulence in practical flow calculations, *European Congress on Computational Methods in Applied Sciences and Engineering*, Spanish Association for Numerical Methods in Engneering, Barcelona, Spain.

Schaub, U. W. & Krishnappa, G. (1977). The stepped stator concept: Aerodynamic and acoustic evaluation of a thrust fan, *Proceedings of 4th AIAA - Aeroacoustics Conference in America USA*, American Institution of Aeronautics and Astronautics.

Schulte, V. & Hodson, H. P. (1998). Unsteady Wake-induced boundary layer transition in high lift LP turbines, *ASME J. Turbomachinery* Vol. 120(No. 1): 28-35.

Shang, T. & Epstein, A. H. (1997). Analysis of hot streak effects on turbine rotor heat load, *ASME J. Turbomachinery* Vol. 119(No. 3): 544–553.

Shapiro, A. H. (1953). *The dynamics and thermodynamics of compressible fluid flow*, Wiley, New York.

Sharma, O. P. & Butler, T. L. (1987). Predictions of endwall losses and secondary flows in axial flow turbine cascades, *ASME J. Turbomachinery* Vol. 109: 229-236.

Sharma, O. P., Butler, T. L., Dring, R. P., Joslyn, H. D. & Renaud, E. (1988). Rotor-stator interaction in multistage axial flow turbines, AIAA Paper 88-3013.

Sharma, O. P., Pickett, G. F. & Ni, R. H. (1992). Assessment of unsteady flows in turbines, *ASME J. Turbomachinery* Vol. 114(No. 1): 79-90.

Sieverding, C. H. (1985). Recent progress in the understanding of basic aspects of secondary flows in turbine blade passages, *ASME J. Engineering Gas Turbines Power* Vol. 107: 248-257.

Sirakov, B. T. & Tan, C. S. (2003). Effect of unsteady stator wake-rotor double-leakage tip clearance flow interaction on time-average compressor performance, *ASME J. Turbomachinery* Vol. 125: 465-474.

Smith, L. H. (1955). Secondary flow in axial-flow turbomachinery, *ASME Transactions* Vol. 77: 1065-1076.

Smith, L. H. (1966). Wake dissipation in turbomachines, *ASME J. Basic Engineering* Vol. 88D: 688-690.

Squire, H. B. & Winter, K. G. (1951). The secondary flow in a cascade of aerofoils in a non-uniform stream, *J. the Aerospace Sciences* Vol. 18: 271-277.

Stieger, R. D. & Hodson, H. P. (2005). The unsteady development of a turbulent wake through a downstream low-pressure turbine blade passage, *ASME J. Turbomachinery* Vol. 127: 388-394.

Suzuki, S. & Kanemitsu, Y. (1971). An experimental study on noise reduction of axial flow fans, *Proceedings of 7th International Congress on Acoustics in Hungary*.

Takahashi, R. K., Ni, R. H., Sharma, O. P. & Staubach, J. B. (1996). Effect of hot streak indexing in a 1-1/2 stage turbine, AIAA Paper 96-2796.

Tyler, J. M. & Sofrin, T. G. (1962). Axial flow compressor noise studies, *Transactions of the Society of Automotive Engineers* Vol. 70: 309-332.

Ulizar, I. & González, P. (2001). Aerodynamic design of a new five stage low pressure turbine for the Rolls Royce Trent 500 turbofan, ASME Paper 2001-GT-0440.

Valkov, T. V. & Tan, C. S. (1999a). Effect of upstream rotor vortical disturbances on the time-averaged performance of axial compressor stators, part 1: framework of technical approach and wake-stator blade interactions, *ASME J. Turbomachinery* Vol. 121: 377-386.

Valkov, T. V. & Tan, C. S. (1999b). Effect of upstream rotor vortical disturbances on the time-averaged performance of axial compressor stators, part 2: rotor tip vortex/streamwise vortex-stator blade interactions, *ASME J. Turbomachinery* Vol. 121: 387-397.

Wallis, A. M., Denton, J. D. & Demargne, A. A. J. (2000). The control of shroud leakage flows to reduce aerodynamic losses in a low aspect ratio, shrouded axial flow turbine, ASME Paper 2000-GT-0475.

Wang, H. P., Olson, S. J., Goldstein, R. J. & Eckert, E. R. G. (1997). Flow visualization in a linear turbine cascade of high performance turbine blades, *ASME J. Turbomachinery* Vol. 119: 1-8.

Wilcox, D. C. (1994). *Turbulence modeling for CFD*, DCW Industries, Inc..

Wu, X. & Durbin, P. A. (2001). Evidence of longitudinal vortices evolved from distorted wakes in a turbine passage, *J. Fluid Mech.* Vol. 446: 199-228.

Xu, L. (1989). The experiment method of unsteady flow and aerodynamics in turbine, *Proceedings of Aerothermodynamics Development Strategy Proseminar, Beijing, China*.

Yamamoto, A. (1987). Production and development of secondary flows and losses in two types of straight turbine cascades: part I-a stator case, *ASME J. Turbomachinery* Vol. 109: 186-193.

Zhang, X. F. & Hodson, H. P (2004). Experimental study of unsteady wake-induced boundary layer transition in turbomachinery, Whittle Laboratory, University of Cambridge.

Zhou, C. & Hodson, H. P. (2009). The tip leakage flow of an unshrouded high pressure turbine blade with tip cooling, ASME Paper GT2009-59637.

4

Analysis of Quiet Zones in Diffuse Fields

Wen-Kung Tseng
National Changhua University of Education
Taiwan, R.O.C.

1. Introduction

Generally speaking, the aim of active noise control systems is to control noise at a dominant frequency range and at a specified region in space. Conventional approaches to active noise control are to cancel the noise at one point in space over a certain frequency range or at many points in space for a single-tone disturbance [Ross, 1980; Joseph, 1990; Nelson & Elliott, 1992]. Cancelling the noise at one point would produce a limited zone of quiet with no control over its shape. Although cancelling the noise at many points could produce larger zones of quiet, the optimal spacing between cancellation points varies with frequency [Miyoshi et al., 1994; Guo et al., 1997].

Previous work on active control of diffuse fields investigated the performance of pressure attenuation for single-tone diffuse field only which was produced by single frequency [Ross, 1980; Joseph, 1990; Tseng, 1999, 2000]. Recent work on broad-band diffuse fields only concentrated on analysis of auto-correlation and cross-correlation of sound pressure [Rafaely, 2000, 2001; Chun et al., 2003]. However there is only some work related to active control of broad-band diffuse fields [Tseng, 2009]. Therefore this chapter will analyze the quiet zones in pure tone and broad-band diffuse fields using ∞-norm pressure minimization.

Moreover a constrained minimization of acoustic pressure is introduced, to achieve a better control of acoustic pressure in space or both frequency and space. The chapter is organized as follows. First, the mathematical model of pure tone and broad-band diffuse fields is derived. Second, the theory of active control for pure tone and broad-band diffuse fields is introduced. Next, simulation results of quiet zones in pure tone and broad-band diffuse fields are presented. Then, preliminary experiments are described. Finally the conclusions are made.

2. The wave model of pure tone and broad-band diffuse sound fields

Garcia-Bonito used the wave model for a pure tone diffuse field, which is comprised of large number of propagating waves arriving from various directions [Garcia et al., 1997]. However, a complete mathematical derivation of this model, which was taken from Jacobson [Jacobsen, 1979], was not found and the wave model of broadband diffuse sound fields has not been derived. For completeness, this mathematical derivation for pure tone and broadband diffuse fields is given below.

When a source produces sound in an enclosure in a room, the sound field is composed of two fields. One is the sound field radiated directly from the source called the direct sound field. The other is reflection of sound waves from surfaces of the room, which contributes to the overall sound field, this contribution being known as the reverberant field. Therefore at any point in the room, the sound field is a function of direct and reverberant sound fields. The sound field in a reverberant space can be divided into two frequency ranges. In the low frequency range, the room response is dominated by standing waves at certain frequencies. In the high frequency range, the resonances become so numerous that they are difficult to distinguish from one another. For excitation frequencies greater than the Schroeder frequency, for which M (ω) = 3, where M(ω) is the modal overlap [Garcia et al., 1997], the resulting sound field is essentially diffuse and may be described in statistical terms or in terms of its average properties. The diffuse sound field model can be derived as follows.

In the model described below, the diffuse field is comprised of many propagating waves with random phases, arriving from uniformly distributed directions. Although the waves occupy a three-dimensional space, the quiet zone analysis is performed, for simplicity, over a two-dimensional area. Consider a single incident plane wave travelling along line r with its wave front parallel to lines A and B as shown in Figure 1. We assume that the plane wave has some phase when approaching line A, and has some phase shift due to the time delay when approaching line B both on the x-y plane. We next find the phase of the plane wave at (x_0,y_0) on line B. We now consider the plane perpendicular to lines A and B and parallel to line r, as illustrated in Figure 2. This incident plane wave has phase shift when approaching point (x_0,y_0) on line B. The pressure at this point can therefore be expressed as

$$P(x_0,y_0,k) = (a+jb) \exp(-jkd) \tag{1}$$

where $a+jb$ account for the amplitude and phase of this incident plane wave when approaching line A, k is the wave number and d is the additional distance travelled by the plane wave when approaching point (x_0,y_0) on line B as shown in Figure 2.

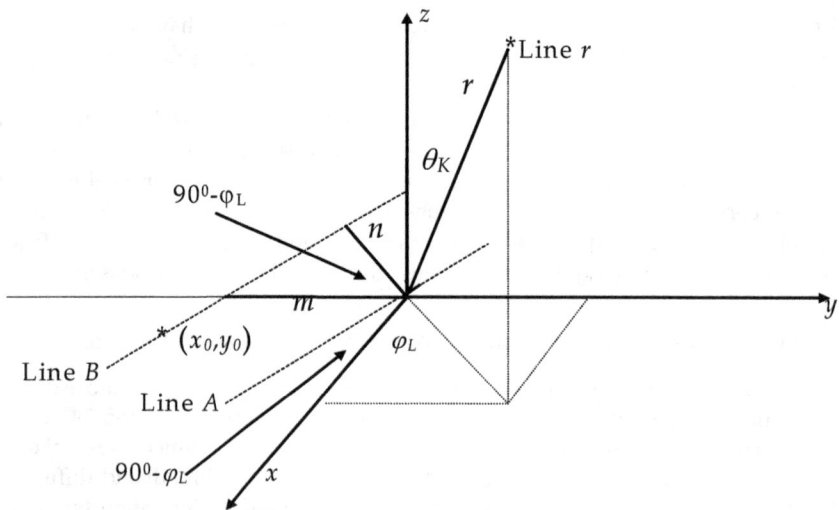

Fig. 1. Definition of spherical co-ordinates r, θ, φ for an incident plane wave travelling alone line r direction.

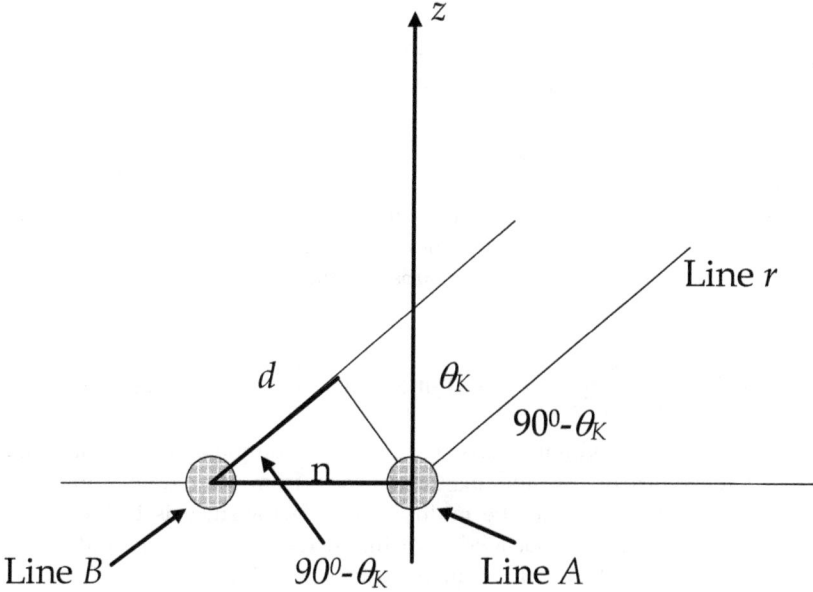

Fig. 2. The plane perpendicular to lines A and B and parallel to line r.

The equation of line A on the x-y plane can be written as

$$y = -x \tan(90^0 - \varphi_L) = -x \cot\varphi_L \tag{2}$$

The equation of line B on x-y plane can also be written as

$$y = -x \tan(90^0 - \varphi_L) + m = -x \cot\varphi_L - m \tag{3}$$

where m is the distance between lines A and B on the y-axis.
Substituting (x_0, y_0) into Equation (3) gives

$$m = -y_0 - x_0 \cot\varphi_L \tag{4}$$

The distance n between lines A and B as in Figure 1 can now be calculated as

$$n = m \cos(90^0 - \varphi_L) = m \sin\varphi_L \tag{5}$$

Substituting equation (4) into equation (5), the distance n becomes

$$n = -y_0 \sin\varphi_L - x_0 \cos\varphi_L \tag{6}$$

The distance d in figure 2 can now be calculated as

$$d = n \cos(90^0 - \theta_K) = n \sin\theta_K \tag{7}$$

Equation (6) can be substituted into equation (7) and the distance d becomes

$$d = -y_0 \sin\theta_K \sin\varphi_L - x_0 \sin\theta_K \cos\varphi_L \tag{8}$$

Therefore equation (1) can be written as

$$P(x_0, y_0, k) = (a+jb) \exp(jk(y_0 \sin\theta_K \sin\varphi_L + x_0 \sin\theta_K \cos\varphi_L)) \tag{9}$$

In our study we chose 72 such incident plane waves together with random amplitudes and phases to generate an approximation of a diffuse sound field in order to coincide with that in previous work. Thus the diffuse sound field was generated by adding together the contributions of 12 plane waves in the azimuthal directions (corresponding to azimuthal angles $\varphi_L = L \times 30^0$, $L=1,2,3, \ldots, 12$) for each of six vertical incident directions (corresponding to vertical angles $\theta_K = K \times 30^0$ for $K = 1, 2, 3, \ldots, 6$). The net pressure in the point (x_0, y_0) on the x-y plane due to the superposition of these 72 plane waves is then calculated from the expression

$$P_p(x_0, y_0, k) = \sum_{K=1}^{K\max} \sum_{L=1}^{L\max} (a_{KL} + jb_{KL}) \sin\theta_K \exp(jk(x_0 \sin\theta_K \cos\varphi_L + y_0 \sin\theta_K \sin\varphi_L)) \tag{10}$$

in which both the real and imaginary parts of the complex pressure are randomly distributed. The values of a_{KL} and b_{KL} are chosen from a random population with Gaussian distribution $N(0,1)$ and the multiplicative factor $\sin\theta_K$ is included to ensure that, on average, the energy associated with the incident waves was uniform from all directions. Each set of 12 azimuthal plane waves arriving from a different vertical direction θ_K, is distributed over a length of $2\pi r \sin\theta_K$, which is the circumference of the sphere defined by (r, φ, θ) for θ_K. This results in higher density of waves for smaller θ_K, and thus more energy associated with small θ_K. To ensure uniform energy distribution, the amplitude of the waves is multiplied by $\sin\theta_K$, thus making the waves coming from the "dense" direction, lower in amplitude. Substituting $k = \dfrac{2\pi}{c} f$ into equation (10) gives

$$P_p(x_0, y_0, f) = \sum_{K=1}^{K\max} \sum_{L=1}^{L\max} (a_{KL} + jb_{KL}) \sin\theta_K \exp(j\frac{2\pi}{c} f(x_0 \sin\theta_K \cos\varphi_L + y_0 \sin\theta_K \sin\varphi_L)) \tag{11}$$

Where f is frequency and c the speed of sound. Equation (11) is the wave model of the pure tone diffuse field since only the single frequency plane wave arriving from uniformly distributed directions is considered. If the diffuse field is broad-band within the frequency range of fl and fh, then the wave model of the broad-band diffuse field P_{pb} can be expressed as

$$P_{pb}(x_0, y_0, fl - fh) = \sum_{f=fl}^{fh} \sum_{K=1}^{K\max} \sum_{L=1}^{L\max} (a_{KL} + jb_{KL}) \sin\theta_K \exp(j\frac{2\pi}{c} f(x_0 \sin\theta_K \cos\varphi_L + y_0 \sin\theta_K \sin\varphi_L))$$

$$\tag{12}$$

Where fl-fh is the frequency range from fl to fh Hz. Equation (12) will be used for broad-band diffuse primary sound field in this work. Next we will describe the formulation of the control method, and their use in the design of quiet zones for pure tone and broad-band diffuse fields.

3. Theory of pressure minimization for pure tone and broad-band diffuse fields

In this section we present the theory of actively controlling pure tone and broad-band diffuse fields. The basic idea is to minimize acoustic pressure over an area in space for pure tone diffuse primary fields or in both space and frequency for broadband diffuse primary fields. Figure 3 illustrates the configuration of acoustic pressure minimization over space and frequency. In this work, the case of two-dimensional space is considered for pure tone diffuse fields derived in equation (11) and the case of a one-dimensional space and frequency is considered for broad-band diffuse primary fields derived in equation (12). The secondary sources are located at the (0.05m, 0) and (-0.05m, 0) point. A microphone can be placed at the desired zone of quiet or other locations close to secondary monopoles. The secondary sources are driven by feedback controllers connected to the microphone. The microphone detects the signal of the primary field, which is then filtered through the controllers to drive the secondary sources. The signals from the secondary sources are then used to attenuate the diffuse primary disturbance at the pressure minimization region.

The x-axis in figure. 3 is a one-dimensional spatial axis, which could be extended in principle, to 2 or 3D. The desired zone of quiet can be defined on this axis where a good attenuation is required. The y-axis is the frequency axis where the control bandwidth could be defined. The acoustic disturbance is assumed to be significant at the control frequency bandwidth. The shadowed region is the pressure minimization region, i.e., the desired zone of quiet over space and frequency. The region to the right of the pressure minimization region is the far field of the secondary sources, with a small control effort, and thus a small effect of the active system on the overall pressure. The region to the left of the pressure minimization region is the near field of the secondary sources, which might result in the amplification of pressure at this region. To avoid significant pressure amplification a pressure amplification constraint should be included in the design process using a constrained optimization. The region above and below the pressure minimization region represents frequency outside the bandwidth. Due to the waterbed effect (Skogestad & Postlethwaite, 1996), a decrease in the disturbance at the control bandwidth will result in amplification outside the bandwidth. Therefore, pressure amplification outside the bandwidth must be constrained in the design process.

The feedback system used in this work is shown in figure 4 and is configured using the internal model control as shown in figure 5 (Morari & Zafiriou, 1989), where P_1 is plant 1, the response between the input to the first monopole and the output of the microphone, P_{1o} is the internal model of plant 1, P_2 is plant 2, the response between the input to the second monopole and the output of the microphone, P_{2o} is the internal model of plant 2, P_{s1} and P_{s2} are the secondary fields at the field point away from the first and second monopoles respectively, d is the disturbance, the broad-band diffuse field, at the microphone location, d_s is the disturbance at the field point away from the microphone, and e is the error signal. In this work, P_{1o} is assumed to be equal to P_1 and P_{2o} is equal to P_2. Therefore the feedback system turns to a feedforward system with $x=d$, where x is the input to the control filters W_1 and W_2.

It is also assumed that the secondary and primary fields in both space and frequency, are known, and although a microphone is used for the feedback signal, pressure elsewhere is assumed to be known and this knowledge is used in the minimization formulation. Although it is not always practical to have a good estimate of pressure far from the microphone, this still can be achieved in some cases using virtual microphone techniques which provide a sufficiently accurate estimate of acoustic pressure far from the microphone (Garcia et al., 1997).

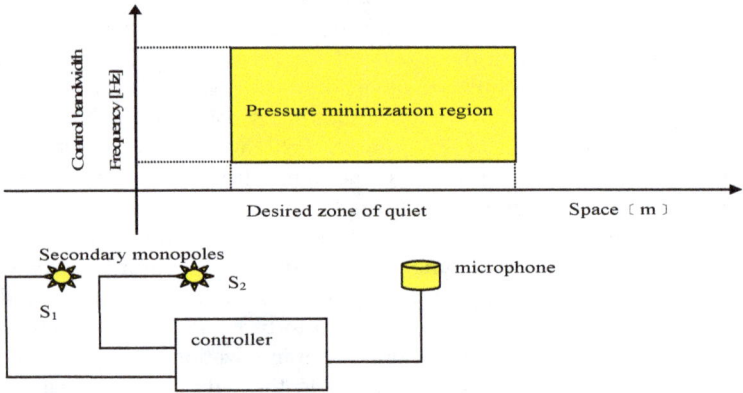

Fig. 3. Configuration of acoustic pressure minimization over space and frequency with a two-channel feedback system.

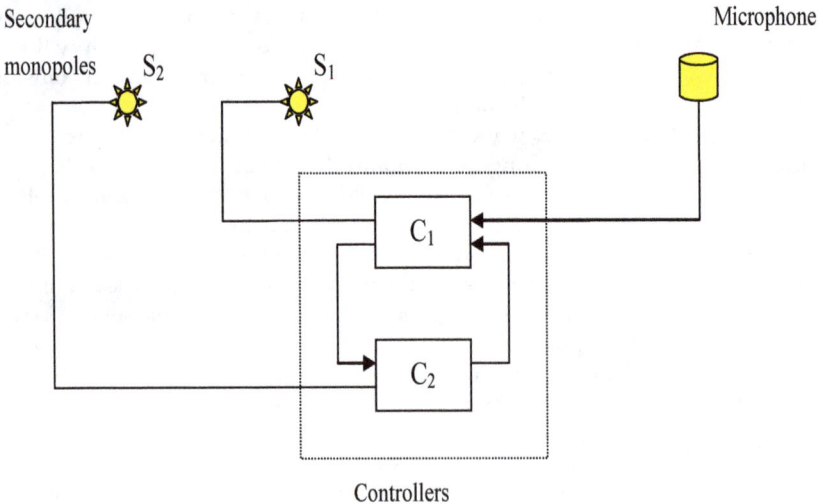

Fig. 4. Two-channel feedback control system used to control broad-band diffuse fields.

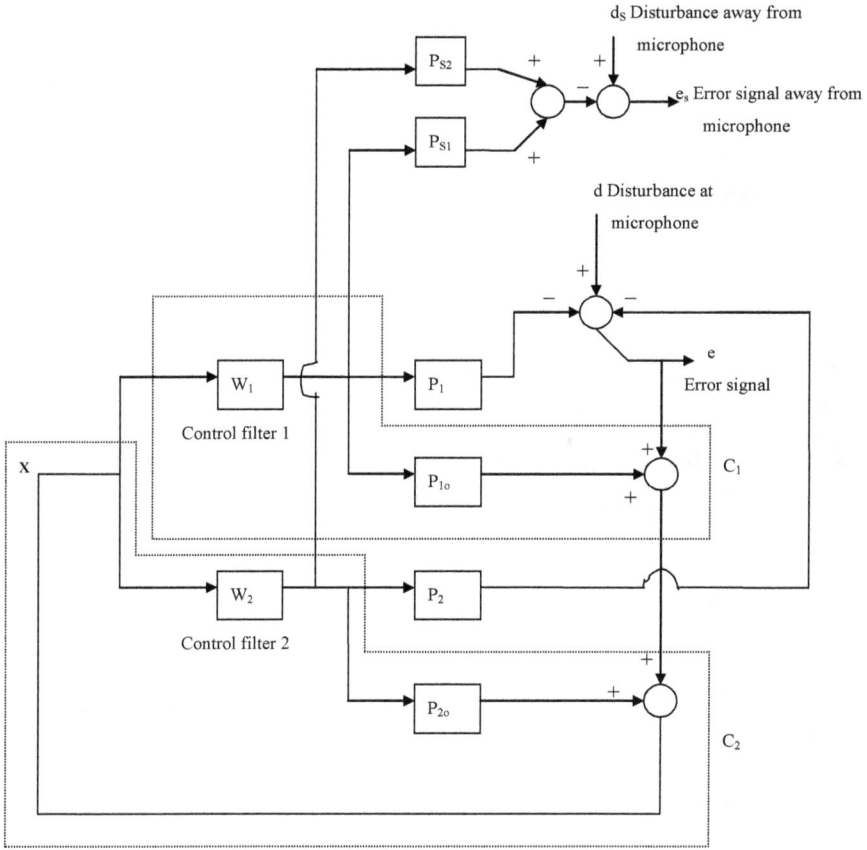

Fig. 5. Two-channel feedback control system with two internal model controllers.

The secondary fields at the field point away from the secondary monopoles could be written as (Miyoshi et al., 1994):

$$P_{s1}(r_1, f) = \frac{A_1}{r_1} e^{-j2\pi f r_1 / c} \tag{13}$$

$$P_{s2}(r_2, f) = \frac{A_2}{r_2} e^{-j2\pi f r_2 / c} \tag{14}$$

where r_1 and r_2 are the distances from the field point to the first and second monopoles, respectively, A_1 and A_2 are the amplitude constants, f is the frequency and c is the speed of sound.

The plant responses can be written as:

$$P_1(r_{1o}, f) = \frac{A_{1o}}{r_{1o}} e^{-j2\pi f r_{1o} / c} \tag{15}$$

$$P_2(r_{20}, f) = \frac{A_{20}}{r_{20}} e^{-j2\pi f r_{20}/c} \tag{16}$$

where r_{1o} and r_{2o} are the distances from the microphone to the first and second monopoles, A_{1o} and A_{2o} are the amplitude constants. The error signal could be expressed as:

$$
\begin{aligned}
e_s &= d_s - d_s W_1 P_{s1} - d_s W_2 P_{s2} \\
&= d_s (1 - W_1 P_{s1} - W_2 P_{s2}) \\
&= d_s (1 - W_1 \frac{A_1}{r_1} e^{-j2\pi f r_1/c} - W_2 \frac{A_2}{r_2} e^{-j2\pi f r_2/c})
\end{aligned}
\tag{17}
$$

The term $(1 - W_1 \frac{A_1}{r_1} e^{-j2\pi f r_1/c} - W_2 \frac{A_2}{r_2} e^{-j2\pi f r_2/c})$ is the sensitivity function [Franklin et al., 1994].

The disturbance in this work is the pure tone and broad-band diffuse fields, therefore equation (17) can also be expressed as:

$$e_s = P_p (1 - W_1 \frac{A_1}{r_1} e^{-j2\pi f r_1/c} - W_2 \frac{A_2}{r_2} e^{-j2\pi f r_2/c}) \quad \text{for pure tone diffuse fields} \tag{18}$$

$$e_s = P_{pb} (1 - W_1 \frac{A_1}{r_1} e^{-j2\pi f r_1/c} - W_2 \frac{A_2}{r_2} e^{-j2\pi f r_2/c}) \quad \text{for broad-band diffuse fields} \tag{19}$$

Where P_p is the pure tone diffuse primary field as shown in equation (11) and P_{pb} is the broad-band diffuse primary field as shown in equation (12).

The formulation of the cost function to be minimized can be written as.

$$J_\infty(r_1, r_2, f) = \left\| \sqrt{SP_p} (1 - W_1 \frac{A_1}{r_1} e^{-j2\pi f r_1/c} - W_2 \frac{A_2}{r_2} e^{-j2\pi f r_2/c}) \right\|_\infty \quad \text{for pure tone diffuse fields} \tag{20}$$

$$J_\infty(r_1, r_2, f) = \left\| \sqrt{SP_{pb}} (1 - W_1 \frac{A_1}{r_1} e^{-j2\pi f r_1/c} - W_2 \frac{A_2}{r_2} e^{-j2\pi f r_2/c}) \right\|_\infty \quad \text{for broad-band diffuse fields} \tag{21}$$

Where $\sqrt{SP_p}$ and $\sqrt{SP_{pb}}$ are the square root of the power spectral density of the pure tone and broad-band disturbance pressure at the field points respectively.

For a robust stability, the closed-loop of the feedback system must satisfy the following condition.

$$\left\| W_1 B_1 \frac{A_{1o}}{r_{1o}} e^{-j2\pi f r_{1o}/c} + W_2 B_2 \frac{A_{20}}{r_{20}} e^{-j2\pi f r_{20}/c} \right\|_\infty < 1 \tag{22}$$

where B_1 and B_2 are the multiplicative plant uncertainty bounds for plants 1 and 2 and r_{1o} and r_{2o} are the distances from the microphone to the first and second monopoles, respectively. The terms $e^{-j2\pi fr_{1o}/c}$ and $e^{-j2\pi fr_{2o}/c}$, that are the plant responses, therefore, follow the robust stability condition, $\|WPB\|_\infty < 1$. For the amplification limit, a constraint could be added to the optimization process as follows.

$$\left\|(1 - W_1 \frac{A_1}{r_1} e^{-j2\pi fr_1/c} - W_2 \frac{A_2}{r_2} e^{-j2\pi fr_2/c})D\right\|_\infty < 1 \tag{23}$$

where $1/D$ is the desired enhancement bound.
Therefore the overall design objective for the pure tone primary diffuse fields can now be written as:

$$min \quad \sigma$$

$$\left\|\sqrt{SP_p}(1 - W_1 \frac{A_1}{r_1} e^{-j2\pi fr_1/c} - W_2 \frac{A_2}{r_2} e^{-j2\pi fr_2/c})\right\|_\infty < \sigma$$

$$subject\ to \quad \left\|W_1 B_1 \frac{A_{1o}}{r_{1o}} e^{-j2\pi fr_{1o}/c} + W_2 B_2 \frac{A_{2o}}{r_{2o}} e^{-j2\pi fr_{2o}/c}\right\|_\infty < 1 \tag{24}$$

$$\left\|(1 - W_1 \frac{A_1}{r_1} e^{-j2\pi fr_1/c} - W_2 \frac{A_2}{r_2} e^{-j2\pi fr_2/c})D\right\|_\infty < 1$$

Also the overall design objective for the broad-band primary diffuse fields can now be written as:

$$min \quad \sigma$$

$$\left\|\sqrt{SP_{pb}}(1 - W_1 \frac{A_1}{r_1} e^{-j2\pi fr_1/c} - W_2 \frac{A_2}{r_2} e^{-j2\pi fr_2/c})\right\|_\infty < \sigma$$

$$subject\ to \quad \left\|W_1 B_1 \frac{A_{1o}}{r_{1o}} e^{-j2\pi fr_{1o}/c} + W_2 B_2 \frac{A_{2o}}{r_{2o}} e^{-j2\pi fr_{2o}/c}\right\|_\infty < 1 \tag{25}$$

$$\left\|(1 - W_1 \frac{A_1}{r_1} e^{-j2\pi fr_1/c} - W_2 \frac{A_2}{r_2} e^{-j2\pi fr_2/c})D\right\|_\infty < 1$$

Equation (24) can be reformulated by approximating r at discrete points only. The discrete space constrained optimization problem can now be written as:

$$min \ \sigma$$

$$subject \ to \quad \left| \sqrt{SP_p}(1 - W_1 \frac{A_1}{r_1} e^{-j2\pi fr_1/c} - W_2 \frac{A_2}{r_2} e^{-j2\pi fr_2/c}) \right| < \sigma \quad for \ all \ r_1 \ and \ r_2.$$

$$\left| W_1 B_1 \frac{A_{1o}}{r_{1o}} e^{-j2\pi fr_{1o}/c} + W_2 B_2 \frac{A_{2o}}{r_{2o}} e^{-j2\pi fr_{2o}/c} \right| < 1 \quad for \ r_{1o} \ and \ r_{2o} \tag{26}$$

$$\left| (1 - W_1 \frac{A_1}{r_1} e^{-j2\pi fr_1/c} - W_2 \frac{A_2}{r_2} e^{-j2\pi fr_2/c})D \right| < 1 \quad for \ r_1 \ and \ r_2.$$

Equation (25) can be reformulated by approximating f and r at discrete points only. The discrete frequency and space constrained optimization problem can now be written as:

$$min \ \sigma$$

$$subject \ to \quad \left| \sqrt{SP_{pb}}(1 - W_1 \frac{A_1}{r_1} e^{-j2\pi fr_1/c} - W_2 \frac{A_2}{r_2} e^{-j2\pi fr_2/c}) \right| < \sigma \quad for \ all \ f, \ r_1 \ and \ r_2.$$

$$\left| W_1 B_1 \frac{A_{1o}}{r_{1o}} e^{-j2\pi fr_{1o}/c} + W_2 B_2 \frac{A_{2o}}{r_{2o}} e^{-j2\pi fr_{2o}/c} \right| < 1 \quad for \ all \ f, \ r_{1o} \ and \ r_{2o} \tag{27}$$

$$\left| (1 - W_1 \frac{A_1}{r_1} e^{-j2\pi fr_1/c} - W_2 \frac{A_2}{r_2} e^{-j2\pi fr_2/c})D \right| < 1 \quad for \ all \ f, \ r_1 \ and \ r_2.$$

It should be noted that constraints on amplification and robust stability will be used in the simulations below. In the next section we will present the quiet zone analysis in pure tone and broad-band diffuse fields.

4. Quiet zone analysis in pure tone and broad-band diffuse fields

In this section the quiet zone analysis in pure tone and broad-band diffuse fields using two-channel and three-channel systems is investigated. The primary fields are pure tone and broad-band diffuse fields. In this work two and three monopoles are used as the secondary fields and a microphone is placed at the (0.1 m, 0) point, i.e., 10cm from the origin. The reason for choosing this configuration is because the previous study on pure tone and broad-band diffuse fields used the same configuration [Ross, 1980; Joseph, 1990; Tseng, 1999, 2000, 2009; Rafaely, 2000, 2001; Chun et al., 2003]. A series of examples are performed to analyze the quiet zones in pure tone and broadband diffuse fields. The theory described in previous sections is used for the simulations.

For the quiet zone simulations in the pure tone diffuse primary field two secondary monopoles are used to control the pure tone diffuse fields and zones of quiet for the two monopoles case are presented. Equation (26) is used to design the quiet zones. Figure 6 shows the 10 dB reduction contour line (solid curve) for two-channel system with two FIR filters having 64 coefficients with the robust constraint only using the ∞-norm strategy, minimizing the pressure over an area represented by a rectangular frame for 108Hz. The two secondary monopoles located at (0.05, 0) and (-0.05, 0) are marked by '*'. The 10 dB amplification is also shown for the ∞-norm minimization strategy (dashed line). Figure 6 shows that ∞-norm strategy minimizing the pressure over an area produces a large zone enclosed by the 10 dB reduction contour. The reason for this is because the ∞-norm is to minimize the maximum pressure within the minimization area resulting in the optimal secondary field over the area. Figure 7 shows the same results for 216Hz. As can be seen from the figure the 10 dB quiet zone becomes smaller for 216Hz than that for 108Hz. This is due to the fact that the primary diffuse field becomes more complicated when the frequency is increased. Thus the primary diffuse field is more difficult to be controlled.

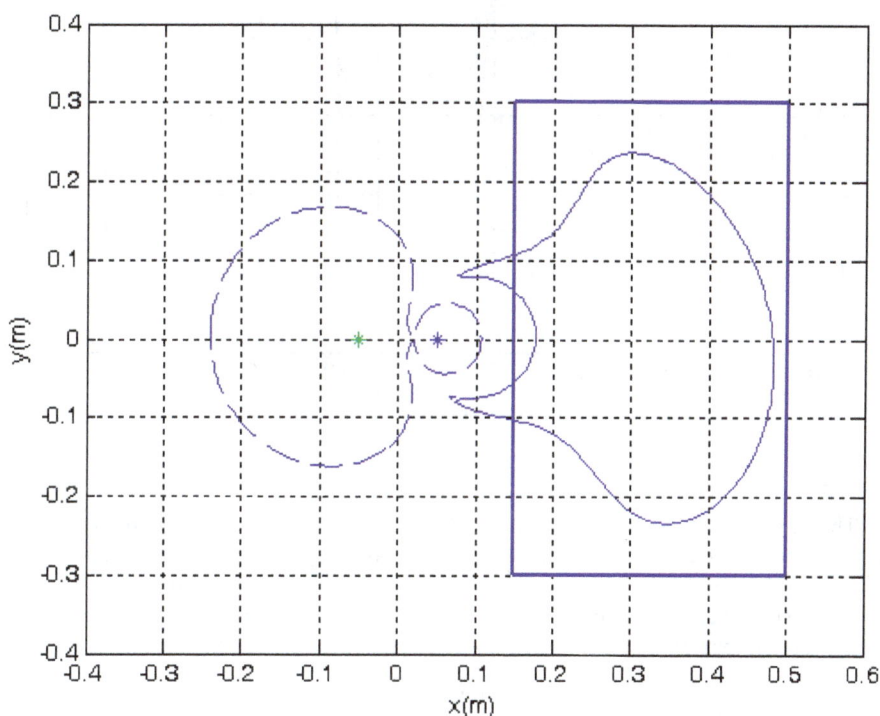

Fig. 6. The 10 dB reduction contour of the zones of quiet created by two secondary monopole sources located at positions (0.05, 0) and (-0.05, 0) for two-channel system with two FIR filters having 64 coefficients with the robust constraint only, minimizing the acoustic pressure at an area represented by a bold rectangular frame using ∞-norm minimization strategy (———), and the 10 dB increase in the primary field for 108Hz.

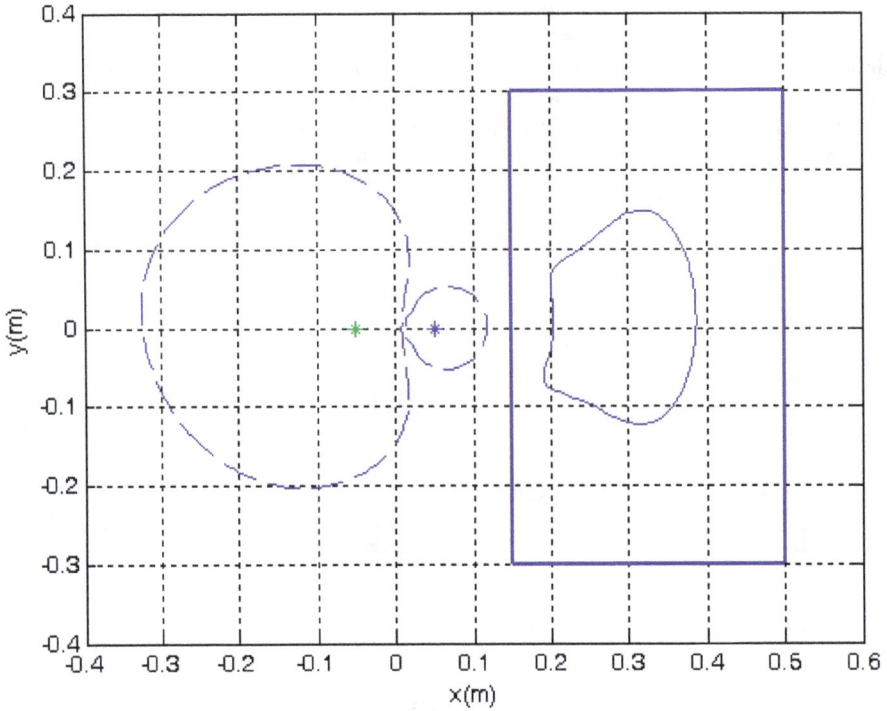

Fig. 7. The 10 dB reduction contour of the zones of quiet created by two secondary monopole sources located at positions (0.05, 0) and (-0.05, 0) for two-channel system with two FIR filters having 64 coefficients with the robust constraint only, minimizing the acoustic pressure at an area represented by a bold rectangular frame using ∞-norm minimization strategy (———), and the 10 dB increase in the primary field for 216Hz.

The zone of quiet created by introducing three secondary monopoles using ∞-norm minimization has also been explored. Figure 8 shows the 10 dB reductions in the pressure level (solid line) for ∞-norm minimization of the pressure in an area represented by the bold rectangular frame. The three secondary monopoles are located at (0, 0), (0.05, 0) and (-0.05, 0) represented by '*', and the 10 dB amplification in the acoustic pressure of the diffuse primary field is represented by a dashed line. Figure 8 shows that three secondary monopoles create a significantly larger zone of quiet than that in the two secondary monopoles case. However the size of the 10 dB amplification in the acoustic pressure away from the zone of quiet is also larger in this case. This shows that larger number of secondary sources provide better control over the secondary field, with the potential of producing larger zones of quiet at required locations.

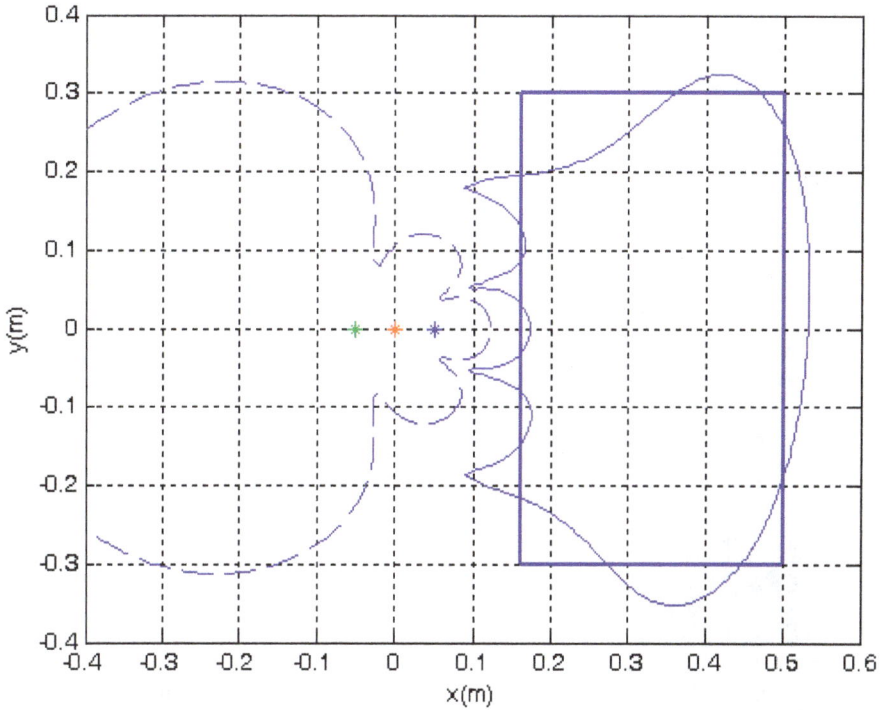

Fig. 8. The 10 dB reduction contour of the zones of quiet created by three secondary monopole sources located at positions (0.05, 0) (0, 0) and (-0.05, 0) for three-channel system with three FIR filters having 64 coefficients without constraints, minimizing the acoustic pressure at an area represented by a bold rectangular frame using ∞-norm minimization strategy (———), and the 10 dB increase in the primary field for 108Hz.

For quiet zone simulations in the broad-band diffuse primary field two secondary monopoles are used to control the broad-band diffuse fields. Equation (21) is used as the cost function to be minimized and equation (27) is used to design the quiet zones. The coefficients of the control filters with 64 coefficients were calculated using the function *fmincon()* in MATLAB. The attenuation contour over space and frequency for the two-channel system is shown in figure 9. The secondary monopoles are located at the (0.05 m, 0) and (-0.05 m, 0) points, and the minimization area is the region enclosed in the rectangle as shown in figure 9. From the figure we can observe that a high attenuation is achieved in the desired region. It can also be noted that the shape of the high-attenuation area is similar to that of the minimization region. This is because two monopoles could generate complicated secondary fields. Thus a good performance over the minimization region was obtained. A high amplification also appears at high-frequency regions and at the region close to the secondary monopoles. The attenuation contours on x-y plane at 400Hz and 600Hz are also shown in figures 10 (a) and (b). As can be seen from the figures the shape of the attenuation contour is shell-like.

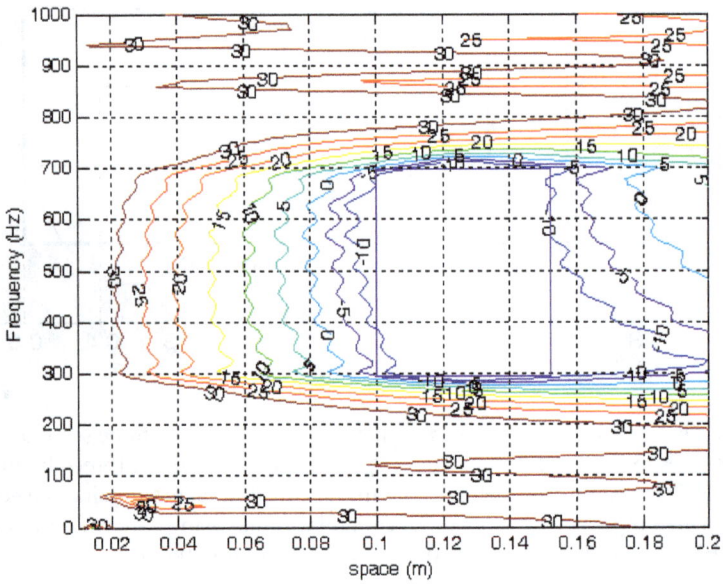

Fig. 9. Attenuation in decibels as a function of space and frequency for two-channel system with two FIR filters having 64 coefficients without constraints for broadband diffuse primary fields.

(a)

(b)

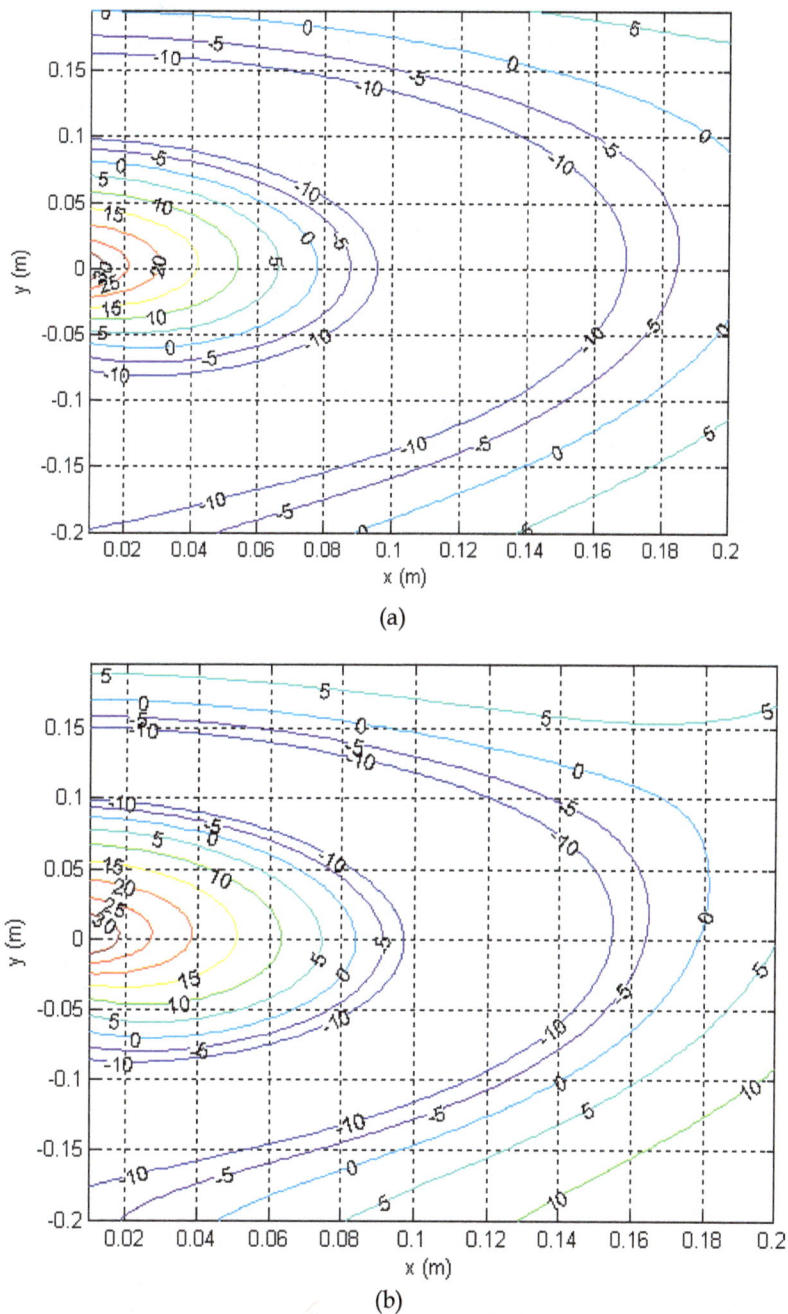

Fig. 10. Attenuation contour in decibels on x-y plane for the two-channel system with two FIR filters having 64 coefficients without constraints. (a) 400 Hz. (b) 600Hz.

In this work constraints on amplification and robust stability are added to the optimization process to prevent a high amplification and instability. Equation (27) was used in the design process. Figure 11 shows the attenuation contour over space and frequency with an amplification constraint not exceeding 20dB at the spatial axis from r=0.1m to r=0.2m for all frequencies and a constraint on robust stability with $B_1=B_2=0.3$. We can see that the attenuation area becomes smaller than that without the amplification and robust stability constraints.

In the next simulation three secondary monopoles are used to control the broadband disturbance. The attenuation contour over space and frequency for three-channel system is shown in figure 12. The secondary monopoles are located at the origin, (-0.05m, 0) and (0.05m, 0) points and the minimization region is larger than that in the two-channel system as shown in the figure. From the figure we can see that high attenuation is achieved in the desired region which is larger than that in the two secondary monopole case as shown in figure 9. It can also be seen that the shape of the high attenuation area is similar to that of the minimization region. This is because three secondary monopoles created more complicated secondary fields than those in the two secondary monopoles. Thus better performance over the minimization region was obtained as expected. High amplification also appears at high frequencies and at the region close to the secondary monopoles.

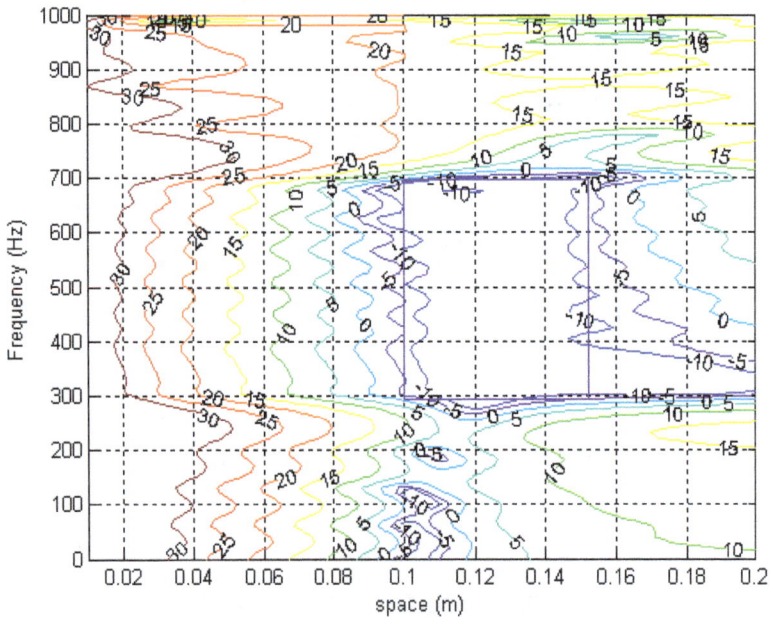

Fig. 11. Attenuation in decibels as function of space and frequency for a two-channel system with two FIR filters having 64 coefficients and with constraints on amplification not exceeding 20dB at spatial axis from r=0.1m to r=0.2m for all frequencies and constraints on robust stability with $B_1 = B_2 = 0.3$.

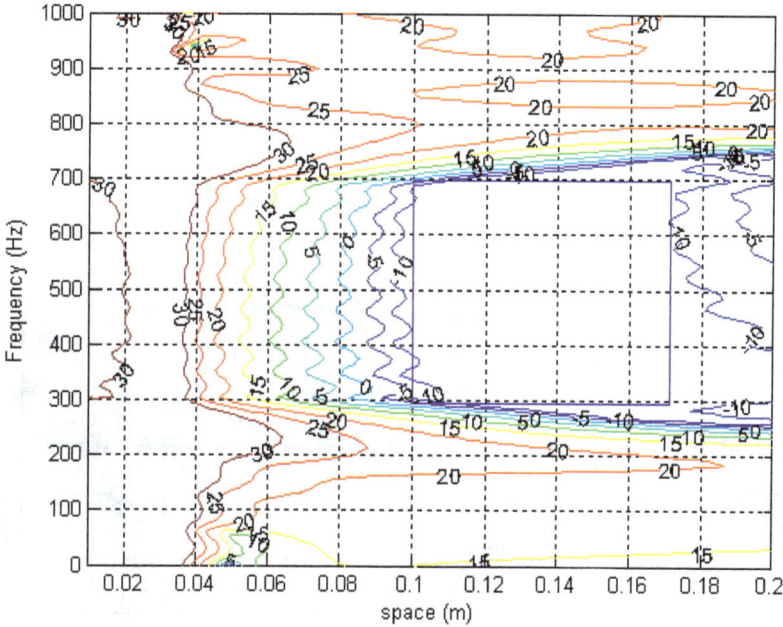

Fig. 12. Attenuation in decibels as a function of space and frequency for a three-channel system with FIR filters having 64 coefficients without constraints.

In the next simulation constraints on robust stability and amplification for the three-channel system are added in the optimization process to avoid unstable and high amplification. Figure 13 shows the attenuation contour over space and frequency for three secondary monopoles with a constraint on robust stability for $B_1=B_2=B_3=0.3$ and an amplification constraint not to exceed 20dB at the spatial axis from r=0.1m to r=0.2m for all frequencies. It can be seen that the attenuation area becomes smaller than that without constraints on robust stability and amplification for three secondary monopoles.

In the fifth example the effect of different minimization shapes on the size of the attenuation contours for three secondary monopoles has also been investigated in this study. Figures 14 (a) and (b) show the attenuation contours over space and frequency for three secondary monopoles without constraints on robust stability and amplification for different minimization shapes. It can be seen that the shape of the 10dB attenuation contour changes with the minimization shape. In figure 14 (a) the 10dB attenuation contour has a narrow shape in frequency axis and longer in space axis similar to the minimization shape. When the minimization shape changes to be narrower in space axis and longer in frequency axis, the 10dB attenuation contour tends to extend its size in the frequency axis as shown in figure 14 (b). Therefore the shape of the 10dB attenuation contour can be designed using the method presented in the work.

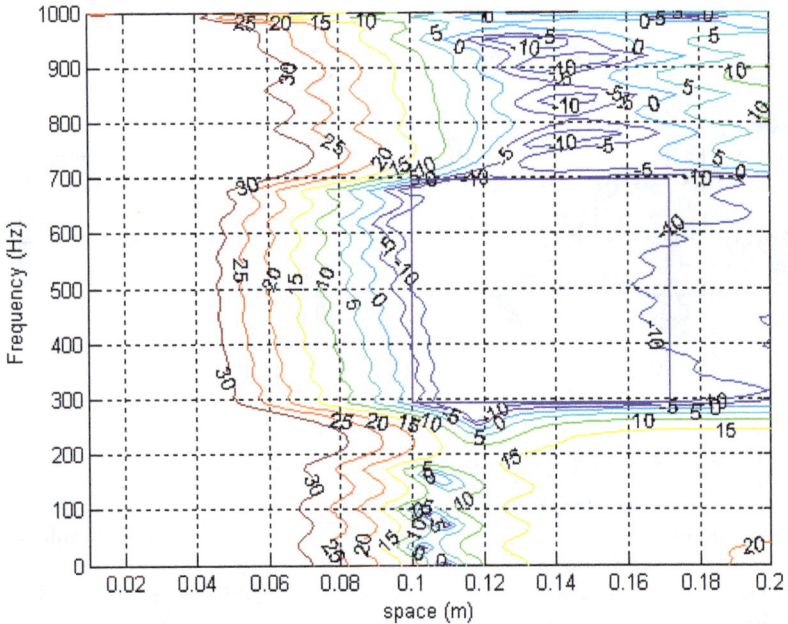

Fig. 13. Attenuation in decibels as a function of space and frequency for a three-channel system with FIR filters having 64 coefficients and constraints on robust stability for $B_1=B_2=B_3=0.3$ and amplification not to exceed 20dB at the spatial axis from r=0.1m to r=0.2m for all the frequencies.

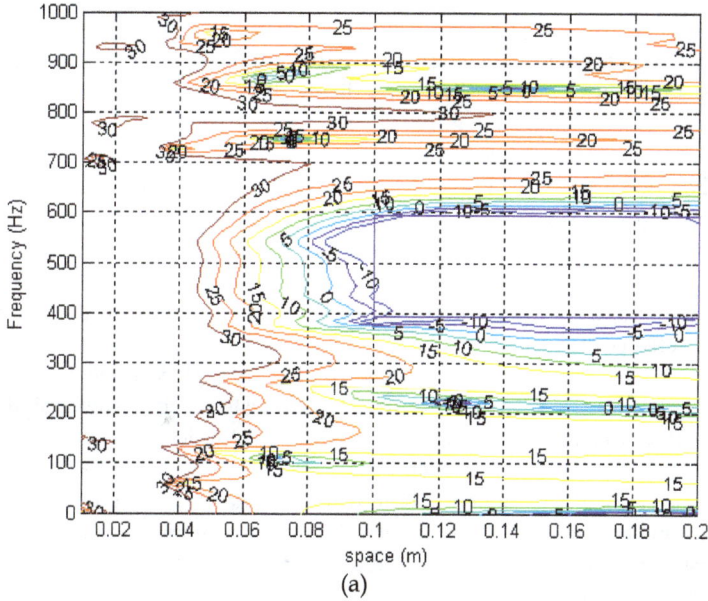

(a)

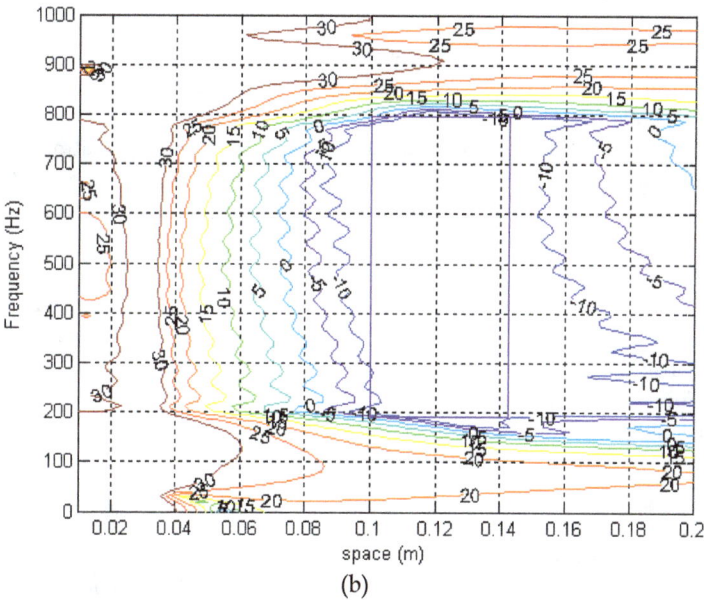

(b)

Fig. 14. Attenuation in decibels as a function of space and frequency for a three-channel system with FIR filters having 64 coefficients without constraints for the different minimization shape represented by a bold rectangular frame. (a) The rectangular frame is narrow in the frequency axis direction and longer in the space axis direction. (b) The rectangular frame is narrow in the space axis direction and longer in the frequency axis direction.

5. Experiments

In this section the experiment to validate the results of the active noise control system using ∞-norm pressure minimization has been described. The excitation frequency of 108Hz was chosen for the primary source. Figure 15 shows the experimental set-up used in the measurements. The secondary sources are two 110mm diameter loudspeakers placed separately. The grid is 30mm pitch made of 3mm diameter brass rod. The dimensions of the grid are 600×600 mm. The electret microphones of 6mm diameter are located at the corresponding nodes of the grid. The size of the room where the experiment has been performed is 10m×8m×4m and it is a normal room. The primary source was located at 4m away from the microphone grid. The primary field can be assumed to be a slightly diffuse field due to the effect of reflection.

The primary and secondary sources are connected to a dual phase oscillator that allows the amplitude and phase of the secondary sources to be adjusted. The reference signal necessary for the acquisition system to calculate the relative amplitude and phase of the complex acoustic pressure at the microphone positions is connected to a dual phase oscillator whose output can be selected with a switch that allows the signal fed to the primary source or to the secondary source to be used as a reference. An FFT analyser is connected to the reference signal to measure the frequency of excitation accurately. All the microphones are connected to an electronic multiplexer which sequentially selects three microphone signals which are filtered by the low pass filter and then acquired by the Analogue Unit Interface (AUI). The sampling frequency is 1,000 Hz and 2,000 samples are acquired for every microphone. The input signal to the AUI through channel 4 is taken as a reference to calculate the relative amplitude and phase of all the signals measured by the microphones on the grid. The calculation of the relative amplitude and phase of the pressure signals was carried out by the computer by Fourier transforming the four input signals and calculating the amplitude and phase of the microphone signals at the excitation frequency with respect to the reference signal. After a complete cycle a matrix of complex pressure values at all the grid points is therefore obtained.

At this stage, the quiet zones created by two secondary loudspeakers were investigated through experiments for one sample of primary field at 108 Hz in a room. The primary field was measured first, and the transfer functions between the secondary loudspeakers and all the microphones on the microphone grid were then measured. The primary field and transfer functions were then taken to calculate the optimal filter coefficients as in equation (26). The zone of quiet is calculated as the ratio of the total (controlled) squared pressure and the primary squared pressure. Figure 16 (a) shows the 10dB zone of quiet created by using ∞-norm pressure minimization over an area represented by the rectangular frame through computer simulations. Figure 16 (b) shows the equivalent results as in Figure 16 (a) through experiments for one sample of the primary field. It shows that the shape and size of 10dB quiet zones are similar in computer simulations and experiments. In figures 16 (a) monopole sources were used as secondary sources in simulations. In figure 16 (b), however, loudspeakers were used as secondary sources in experiments. Although monopole sources are not an accurate model of loudspeakers, it simplifies the secondary source modelling and assists comparison between simulations and experiments.

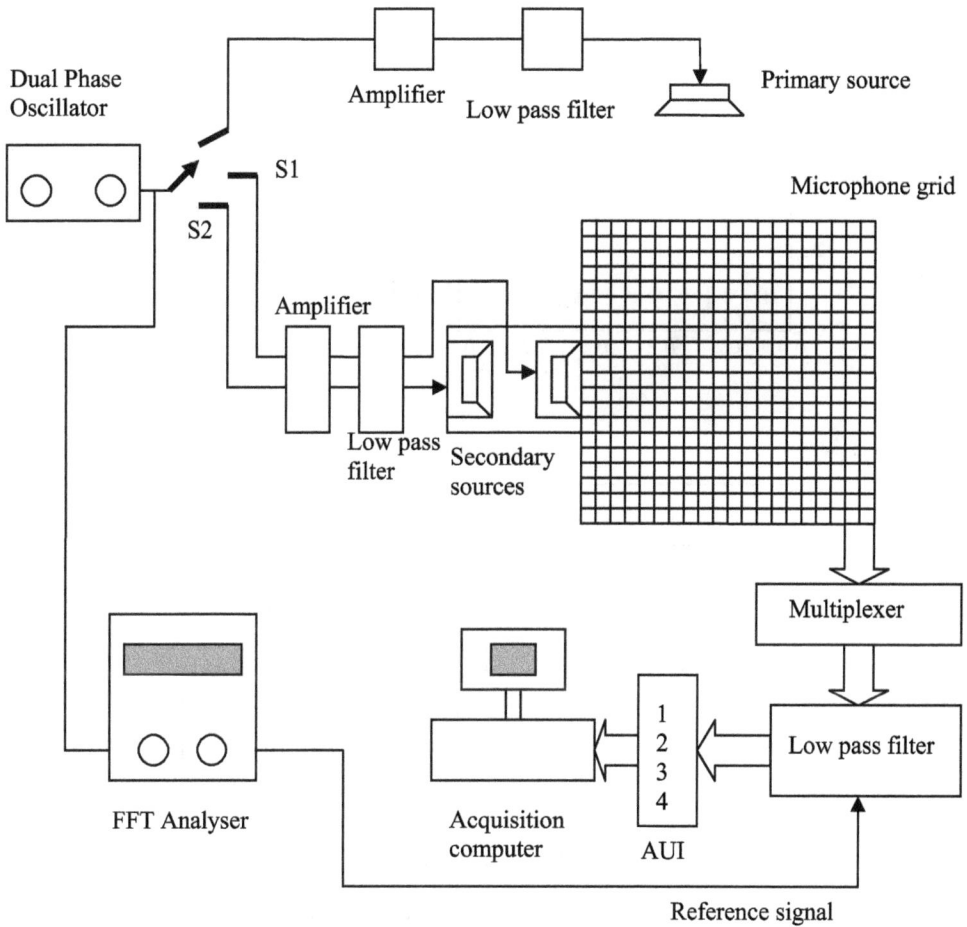

Fig. 15. Configuration of experimental set-up.

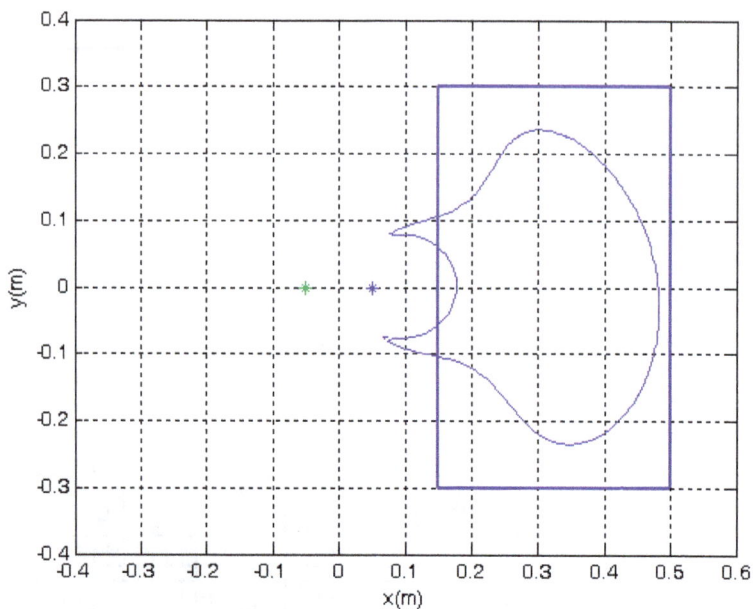

(a)

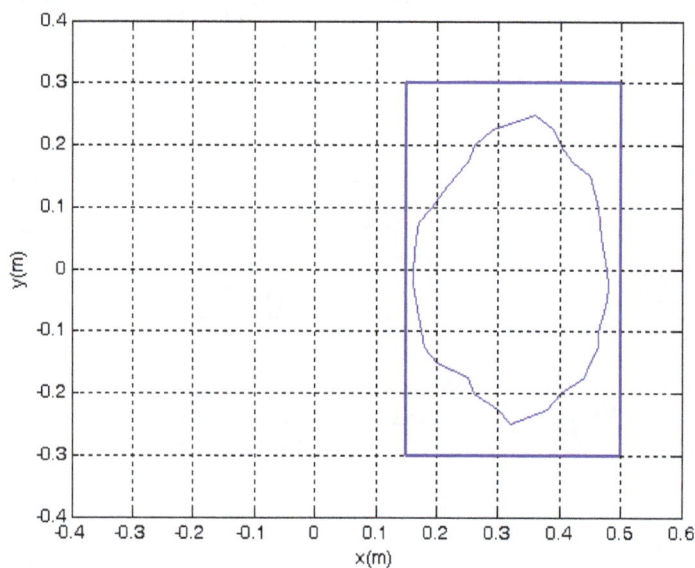

(b)

Fig. 16. The 10 dB reduction contour of the average zones of quiet created by two secondary sources located at positions (0.05, 0) and (-0.05, 0) using ∞-norm pressure minimization for 108Hz of diffuse primary fields. (a) Computer simulations. (b) Experiments.

6. Conclusions

The theory of active control for pure tone and broad-band diffuse fields using two-channel and three-channel systems has been presented and the quiet zone analysis has been investigated through computer simulations and experiments. The acoustic pressure was minimized at the specified region over space or both space and frequency. Constraints on amplification and robust stability were also included in the design process. The results showed that a good attenuation in the desired quiet zone over space or both space and frequency could be achieved using a two-channel system. However, a better performance was achieved using a three-channel system. When limits on amplification and robust stability were introduced, the performance began to degrade. It has also been shown that acoustic pressure could be minimized at a specific frequency range and at a specific location in space away from the microphone. This could be realized by virtual microphone methods. Moreover, the shape of the 10dB attenuation contour could be controlled using the proposed method.

7. Acknowledgement

The study was supported by the National Science Council of Taiwan, the Republic of China, under project number NSC-96-2622-E-018-004-CC3.

8. References

C. F. Ross, C. F. (1980). Active control of sound. *PhD Thesis*, University of Cambridge.

Chun, I.; Rafaely, B. & Joseph, P. (2003). Experimental investigation of spatial correlation of broadband diffuse sound fields. *Journal of the Acoustical Society of America*, 113(4), pp. 1995-1998.

Franklin, G. F.; Powell, J. D. & Emamni Naeini, A. (1994). *Feedback control of dynamic systems*, Addison-Wesley, MA. 3rd ed.

Garcia-Bonito, J.; Elliott, S. J. & Boucher, C. C. (1997). A novel secondary source for a local active noise control system, *ACTIVE 97*, pp.405-418.

Guo, J.; Pan, J. & Bao, C. (1997). Actively created quiet zones by multiple control sources in free space. *Journal of Acoustical Society of America*, 101, pp. 1492-1501,.

Jacobsen, F. (1979). The diffuse sound field, *The Acoustics Laboratory Report* no. 27, Technical University of Denmark.

Joseph, P. (1990). Active control of high frequency enclosed sound fields, *PhD Thesis*, University of Southampton.

Miyoshi, M.: Shimizu, J. & Koizumi, N. (1994). On arrangements of noise controlled points for producing larger quiet zones with multi-point active noise control, *Inter-noise 94*, pp. 1229-1304.

Morari, M. & Zafiriou, E. (1989). *Robust process control*, Prentice-Hall, NJ.

Nelson, P. A. & Elliott, S. J. (1992). *Active Control of Sound*, Academic, London.

Rafaely, B. (2000). Spatial-temporal correlation of a diffuse sound field. *Journal of the Acoustical Society of America*. 107(6), pp. 3254-3258.

Rafaely, B. (2001). Zones of quiet in a broadband diffuse sound field. *Journal of the Acoustical Society of America*, 110(1), pp. 296-302.

Tseng, W. K.; Rafaely, B. & Elliott, S. J. (1999). 2-norm and ∞-norm pressure minimisation for local active control of sound, *ACTIVE'99*, pp.661-672.

Tseng, W. K.; Rafaely, B. & Elliott, S. J. (2000). Local active sound control using 2-norm and infinity-norm pressure minimisation. *Journal of Sound and Vibration* 234(3) pp. 427-439.

Tseng, W. K. (2009). Quiet zone design in broadband diffuse fields. *International MultiConference of Engineers and Computer Scientists*. pp. 1280-1285.

Skogestad, S. & Postlethwaite, I. (1996). *Multivariable feedback control*, John Wiley and Sons, Chichester, UK..

Part 3

Experimental Tests

5

The Convergence Analysis of Feedforward Active Noise Control System

Guoyue Chen
Akita Prefectural University
Japan

1. Introduction

Most active noise control (ANC) systems [1-3] are based on feedforward structure with adaptive filters, which are updated with the filtered-x LMS algorithm [4, 5] or the multiple error filtered-x (MEFX) LMS algorithm [6, 7]. The convergence characteristics of these algorithms have been studied mostly in the time domain, and it was found that the convergence characteristics were subject to eigenvalue distribution of the autocorrelation matrix of the filtered reference signal [4, 7]. Analysis in the time domain, however, requires a great deal of computation, and its physical meaning is unclear.

This chapter presents a new method for evaluating the adaptive algorithm for the feedforward ANC system, which can be approximately analysed in the frequency domain at each frequency (FFT) bin separately, which can provide significant computational savings and a better understanding of the physical meaning. Some convergence characteristics in the frequency domain can be understood easily, and a preprocessing method is proposed to improve the whole performance of the adaptive algorithm, especially when the reference pathssre unknown or measured in prior. Most contents of this chapter are based on the previous works [8-11].

The chapter is orgnized as follows. In section 2, the model of adaptive algorithm for multiple noises and multiple control points system is introduced in the time domain and the frequency domain, separately. Section 3 analyzes the convergence characteristics of some adaptive algorithms in the frequency domain, like the filtered-x LMS algorithm, the Delayed-x LMS, and the MEFX LMS algorithm, and the effects of the secondary path and the reference path on the convergence performance are analysed. Some results are represented by computer simulations in section 4.

2. A multiple noise source and multiple control point ANC system

Figure 1 shows a general ANC system for multiple noise sources and multiple control points. In Figure 1, I is the number of noise sources, K is the number of the reference sensors of the adaptive digital filter (ADF) array, which are finite impulse response (FIR) filters, M is the number of secondary sources, and L is the number of the control points (error sensors). Such a system will be referred to as CASE[I, K, M, L] in this chapter. Noises are recorded by K reference sensors and the impulse response of the reference paths are modelled with the transfer matrix $\mathbf{B}(n)$. There are $M \times L$ different secondary paths (secondary path matrix) between all secondary sources and error sensors, and all secondary paths are assumed to be time invariant and are modelled as $\mathbf{C}(n)$. The outputs of the adaptive filter arrays are used to drive M secondary sensors to reduce the effect of noises at the error sensors as large as possible according to the estimated secondary paths and the recorded reference signals.

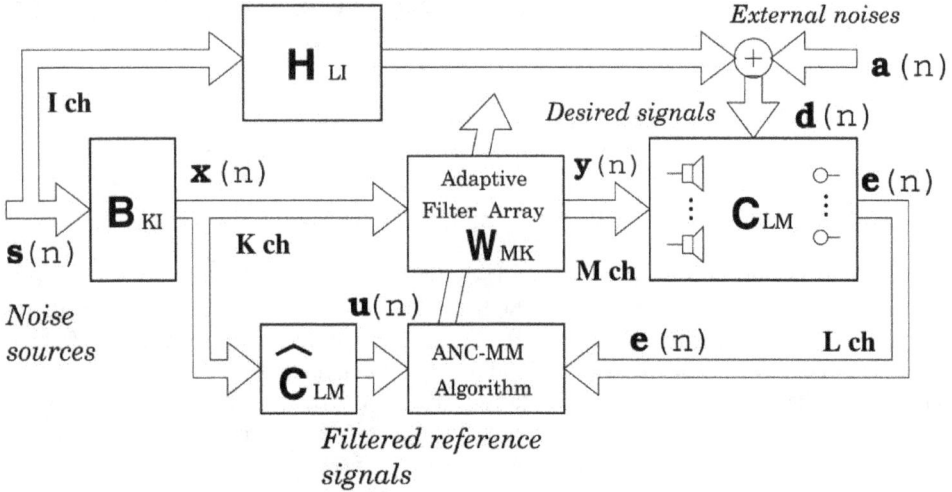

Fig. 1. Block diagram of the general active noise control system for multiple noise sources and multiple control points, CASE[I, K, M, L].

2.1 The adaptive algorithm in the time domain

As is shown in Figure 1, the adaptive controller generates the outputs to construct M secondary sources, and the squared sum of L error signals is minimized to update the coefficients of all adaptive filters. The updating mechanism of adaptive filter coefficients is controlled by the MEFX LMS algorithm, which is an extension of the filtered-x LMS algorithm for the CASE[I, K, M, L] ANC system.

The MEFX LMS algorithm can be summarized as follows [11],

$$e(n) = \mathbf{d}(n) + \mathbf{U}(n)\mathbf{w}(n) , \tag{1}$$

$$\mathbf{w}(n+1) = \mathbf{w}(n) - 2\mu \mathbf{U}^{\mathrm{T}}(n)e(n) , \tag{2}$$

where the superscript ^{T} denotes the transpose, μ is the step-size parameter,

$$e(n) = \begin{bmatrix} e_1(n) & e_2(n) & \cdots & e_L(n) \end{bmatrix}^{\mathrm{T}} , \tag{3}$$

$$\mathbf{d}(n) = \begin{bmatrix} d_1(n) & d_2(n) & \cdots & d_L(n) \end{bmatrix}^{\mathrm{T}} \tag{4}$$

and $\mathbf{w}(n)$ is the stacked column vector of all adaptive filter coefficients,

$$\mathbf{w}(n) = \begin{bmatrix} \mathbf{w}_{11}(n) & \cdots & \mathbf{w}_{M1}(n) & \cdots & \cdots & \mathbf{w}_{MK}(n) \end{bmatrix}^{\mathrm{T}}. \tag{5}$$

The element of $\mathbf{w}(n)$, $\mathbf{w}_{mk}(n)$, is the adaptive filter coefficients vector between the m-th secondary source and the k-th reference signal,

$$\mathbf{w}_{mk}(n) = \begin{bmatrix} w_{mk_1}(n) & w_{mk_2}(n) & \cdots & w_{mk_{N_w}}(n) \end{bmatrix}, \tag{6}$$

where N_w is the length of the adaptive filters.

The filtered reference signal matrix $\mathbf{U}(n)$ is an $L \times MKN_w$ matrix and is defined by

$$\mathbf{U}(n) = \begin{bmatrix} \mathbf{u}_{111}^{\mathrm{T}}(n) & \mathbf{u}_{1M1}^{\mathrm{T}}(n) & \cdots & \mathbf{u}_{1MK}^{\mathrm{T}}(n) \\ \mathbf{u}_{211}^{\mathrm{T}}(n) & \mathbf{u}_{2M1}^{\mathrm{T}}(n) & \cdots & \mathbf{u}_{2MK}^{\mathrm{T}}(n) \\ \vdots & \vdots & \vdots & \vdots \\ \mathbf{u}_{L11}^{\mathrm{T}}(n) & \mathbf{u}_{LM1}^{\mathrm{T}}(n) & \cdots & \mathbf{u}_{LMK}^{\mathrm{T}}(n) \end{bmatrix}, \tag{7}$$

where the element of $\mathbf{U}(n)$, $\mathbf{u}_{lmk}(n)$, is the N_w-vector of the filtered reference signals

$$\mathbf{u}_{lmk}(n) = \begin{bmatrix} u_{lmk}(n) & u_{lmk}(n-1) & \cdots & u_{lmk}(n-N_w+1) \end{bmatrix}^{\mathrm{T}}, \tag{8}$$

and the filtered reference signal $u_{lmk}(n)$ is obtained by

$$u_{lmk}(n) = \sum_{j=1}^{N_c} c_{lm_j}(n) x_k(n-j+1), \tag{9}$$

$$l = 1, \quad \cdots \quad L; \quad m = 1, \quad \cdots \quad M; \quad k = 1, \quad \cdots \quad K$$

where N_c is length of the estimated secondary path $\mathbf{c}_{lm} = \begin{bmatrix} c_{lm_1} & c_{lm_2} & \cdots & c_{lm_{N_c}} \end{bmatrix}^{\mathrm{T}}$, which models the response between the m-th secondary source to the l-th error sensor. The reference signal $x_k(n)$ at the k-th reference sensor is

$$x_k(n) = \sum_{i=1}^{I} b_{ki} * s_i(n) = \sum_{i=1}^{I} \sum_{j_b=1}^{N_b} b_{kij_b}(n) s_i(n - j_b + 1), \tag{10}$$

where $s_i(n)$ is the i-th noise source and $b_{ki}(n)$ is the impulse response from the i-th noise source to the k-th reference sensor, which is assumed to be a finite length of N_b.

2.2 Analysis in the time domain

It is well known that the convergence speed of the LMS algorithm is determined by the convergence time of different modes, which depend on the eigenvalues of the autocorrelation matrix, $E[\mathbf{x}(n)\mathbf{x}^{\mathrm{T}}(n)]$, of the input signals to adaptive filter [12-14]. Similar conclusion is applied to the MEFX LMS algorithm to analyse the eigenvalue spread of the autocorrelation matrix of the filtered reference signals [2], which is defined by

$$\mathbf{R}(n) = E\left[\mathbf{U}^{\mathrm{T}}(n)\mathbf{U}(n) \right], \tag{11}$$

where $E[\ ]$ is the statistical expectation operator. The reference signals are assumed to be statistically stationary and time index n of the matrix $\mathbf{R}$ has been emitted.

To analyse the convergence characteristics of the MEFX LMS algorithm in the time domain, it is necessary to calculate the autocorrelation matrix $\mathbf{R}$ of the filtered reference signal, whose size is $MKN_w \times MKN_w$. Therefore, it is difficult to calculate the eigenvalues of the matrix $\mathbf{R}$ so as to investigate the convergence characteristics of the adaptive filters in the time domain. Another disadvantage is that the physical meanings of both maximum and minimum eigenvalues of the matrix $\mathbf{R}$ are unclear since the filtered reference signals include the reference paths and secondary paths by convolution operation.

2.3 Analysis in the frequency domain

In this section, the convergence characteristics of the MEFX LMS algorithm are analysed in the frequency domain. It is known that the system using the filtered-x LMS algorithm may be unstable, even for a very small step-size parameter (resulting in a slow adaptive speed), because small estimated errors to the secondary paths will enlarge the filtered reference signals to make the whole algorithm diverge in the time domain [2]. Since the convergence speed of the MEFX LMS algorithm is slow, the adaptive filters can be considered as time invariant linear filters for a short period. Thus, the MEFX LMS algorithm described in equations (1) and (2) in the time domain can be approximately expressed in the frequency domain as [10, 11]

$$E(n,\omega) = D(n,\omega) + U(n,\omega)W(n,\omega), \tag{12}$$

$$W(n+1,\omega) = W(n,w) - 2\mu U^{\mathrm{H}}(n,\omega)E(n,\omega), \tag{13}$$

where the superscript ^{H} denotes the Hermitian transpose, and

$$E(n,\omega) = [E_1(n,\omega),\quad E_2(n,\omega),\quad \cdots \quad E_L(n,\omega)]^{\mathrm{T}}, \tag{14}$$

$$D(n,\omega) = [D_1(n,\omega),\quad D_2(n,\omega),\quad \cdots \quad D_L(n,\omega)]^{\mathrm{T}}, \tag{15}$$

$$W(n,\omega) = [W_{11}(n,\omega),\quad \cdots \quad \cdots \quad W_{MK}(n,\omega)]^{\mathrm{T}}. \tag{16}$$

The filtered reference matrix $\mathbf{U}(n,\omega)$ is an $L \times MK$ matrix defined by

$$\mathbf{U}(n,\omega) = \begin{bmatrix} U_{111}(n,\omega) & \cdots & U_{1M1}(n,\omega) & \cdots & \cdots & U_{1MK}(n,\omega) \\ U_{211}(n,\omega) & \cdots & U_{2M1}(n,\omega) & \cdots & \cdots & U_{2MK}(n,\omega) \\ \vdots & \vdots & \vdots & \vdots & \vdots & \vdots \\ U_{L11}(n,\omega) & \cdots & U_{LM1}(n,\omega) & \cdots & \cdots & U_{LMK}(n,\omega) \end{bmatrix}, \tag{17}$$

where

$$U_{lmk}(n,\omega) = C_{lm}(\omega)X_k(n,\omega), \tag{18}$$

$C_{lm}(\omega)$ is the estimated transfer function from the m-th secondary source to the l-th error sensor, $X_k(n,\omega)$ is the reference signal obtained at the k-th reference sensor,

$$X_k(n,\omega) = \sum_{i=1}^{I} B_{ki}(n,\omega)S_i(n,\omega).$$ (19)

The matrix $R(\omega)$ is defined by the matrix of the filtered reference signal $U(n,\omega)$, as follows,

$$R(\omega) = E\left[\; U^H(n,\omega)U(n,\omega) \; \right].$$ (20)

In the time domain, the matrix $R(n)$ shown in equation (11) is the autocorrelation matrix [13] . In the frequency domain, the matrix $R(\omega)$ in equation (20) is called the power spectrum matrix. Since the dimensions of the power spectrum matrix $R(\omega)$ are $MK \times MK$ at each frequency bin ω such that much less computation is required to find the eigenvalues of the matrix $R(\omega)$ in the frequency domain than those in the time domain.
As is the same with the analysis in the time domain [12-14], the upper limit of the step-size parameter $\mu(\omega)$ and the longest time constant $\tau(\omega)$ can be given at each frequency bin by

$$0 < \mu(\omega) < \frac{1}{\lambda_{max}(\omega)},$$ (21)

$$\tau(\omega) > \frac{1}{2\mu(\omega)\lambda_{min}(\omega)}.$$ (22)

Where $\lambda_{max}(\omega)$ and $\lambda_{min}(\omega)$ are the largest and smallest eigenvalues of the matrix $R(\omega)$ shown in equation (20) at each frequency bin. Over the whole frequency range of interest, the upper limit of the step-size parameter μ is given by

$$0 < \mu < \frac{1}{\max_{\omega}\{\lambda_{max}(\omega)\}}.$$ (23)

Where $\max_{\omega}\{\;\}$ denotes the maximum value over the whole frequency range. In practice, we can choose the unique step-size parameter determined by equation (23) to keep the system stable. Hence, smaller $\lambda_{max}(\omega)$ leads to a slower convergence at some frequency bins. Especially, if a large sharp dip exists over the whole frequency range, the corresponding convergence speed will be slowed down. The sharper the dip becomes, the slower the convergence speed will be.
It was also found from simulation results [16] that the smaller $\lambda_{max}(\omega)$ leads to a larger computational error and a smaller noise reduction at some frequency bins, results in a slow convergence speed over the entire frequency range, which is the same as in the time domain. Substituting equation (23) into equation (22) will give the longest time constant at each frequency bin, $\tau_{max}(\omega)$, as follows,

$$\tau_{max}(\omega) > \frac{\max_{\omega}\{\lambda_{max}(\omega)\}}{2\lambda_{min}(\omega)},$$ (24)

It is clear that the convergence speed is subject to the ratio of the maximum to the minimum eigenvalues.

It is clear from equation (24) that the convergence speed is subject to the minimum eigenvalue of the matrix $R(\omega)$ at the frequency bin ω. Comparing the convergence speed of the adaptive filters at different frequency bin ω, it is found that a smaller eigenvalue of the power spectrum matrix $R(\omega)$ results in a longer convergence time. Then, the convergence speed of the adaptive filters over the whole frequency bin (in the time domain) becomes slower.

The longest time constant, τ_{max}, over the whole frequency range can be obtained as

$$\tau_{max} > \frac{\max_{\omega}\{\lambda_{max}(\omega)\}}{2\min_{\omega}\{\lambda_{min}(\omega)\}},\tag{25}$$

where $\min_{\omega}\{\ \}$ denotes the minimum values over the whole frequency range. It is clear from equation (25) that the longest time constant is related to the ratio of the maximum to the minimum eigenvalue of the matrix $R(\omega)$ over the whole frequency range. The convergence analysis of the time domain MEFX LMS algorithm may be evaluated generally in the frequency domain such that we can obtain insight into the convergence characteristics of the MEFX LMS algorithm in the time domain.

2.4 The power spectrum matrix R

The filtered reference signal $u(n) = \sum_{i=1}^{N_c} x(n-i+1)c_i$ in the time domain is expressed by the convolution of $x(n)$ and $c(n)$, while $U(n,\omega) = C(\omega)X(n,\omega)$ in the frequency domain is expressed by a simple multiplication. Rearrange the filtered reference matrix $U(n,\omega)$ in equation(17) and combine the result of equation (18), the matrix $U(n,\omega)$ can be written as

$$U(n,\omega) = X^T(n,\omega) \otimes C(\omega),\tag{26}$$

where $\otimes$ is the Kronecker product, and

$$X(n,\omega) = B(\omega)S(n,\omega),\tag{27}$$

where the reference path $B(\omega)$ and the secondary path $C(\omega)$ are

$$C = \begin{bmatrix} C_{11} & C_{12} & \cdots & C_{1M} \\ C_{21} & C_{22} & \cdots & C_{2M} \\ \cdots & \cdots & \cdots & \cdots \\ C_{L1} & C_{L2} & \cdots & C_{LM} \end{bmatrix} \text{ and } B = \begin{bmatrix} B_{11} & B_{12} & \cdots & B_{1I} \\ B_{21} & B_{22} & \cdots & B_{2I} \\ \cdots & \cdots & \cdots & \cdots \\ B_{K1} & B_{K2} & \cdots & B_{KI} \end{bmatrix}.$$

The matrix $R(\omega)$ can be given by

$$R(\omega) = E[U^H(n,\omega)U(n,\omega)] = E\{[X^T(n,\omega) \otimes C(\omega)]^H[X^T(n,\omega) \otimes C(\omega)]\}\tag{28}$$

Utilizing the Kronecker product characteristics,

$$\mathbf{R}(\omega) = E\{[\mathbf{X}^T(n,\omega) \otimes \mathbf{C}(\omega)]^H [\mathbf{X}^T(n,\omega) \otimes \mathbf{C}(\omega)]\}$$
$$= [E\{\mathbf{X}(n,\omega)\mathbf{X}^H(n,\omega)\}]^* \otimes [\mathbf{C}^H(\omega)\mathbf{C}(\omega)]$$
$$= [\mathbf{B}(\omega)E\{\mathbf{S}(n,\omega)\mathbf{S}^H(n,\omega)\}\mathbf{B}^H(\omega)]^* \otimes [\mathbf{C}^H(\omega)\mathbf{C}(\omega)]$$
$$= |s_1|^2 \cdots |s_I|^2 [\mathbf{B}(\omega)\mathbf{B}^H(\omega)]^* \otimes [\mathbf{C}^H(\omega)\mathbf{C}(\omega)]$$

(29)

$MK \times MK$ power spectrum matrix $\mathbf{R}(\omega)$ is dependent completely on the reference path $\mathbf{B}(\omega)$ and the secondary path $\mathbf{C}(\omega)$.

It is easy to prove that $\mathbf{R}(\omega)$ is Hermitian matrix, i.e. $\mathbf{R}(\omega) = \mathbf{R}^H(\omega)$, so all eigenvalues of $\mathbf{R}(\omega)$ are nonnegative, that is to say, $\mathbf{R}(\omega)$ is a nonnegative define matrix. $\mathbf{B}(\omega)\mathbf{B}^H(\omega)$ and $\mathbf{C}^H(\omega)\mathbf{C}(\omega)$ are also Hermitian. According to the characteristics of the Kronecker product, the determinant and trace of $\mathbf{R}(\omega)$ satisfy the following equations

$$\det[\mathbf{R}(\omega)] = \det[\mathbf{C}^H(\omega)\mathbf{C}(\omega)]^K \det[\mathbf{B}(\omega)\mathbf{B}^H(\omega)]^M$$

(30)

$$Trac[\mathbf{R}(\omega)] = Trac[\mathbf{C}^H(\omega)\mathbf{C}(\omega)]Trac[\mathbf{B}(\omega)\mathbf{B}^H(\omega)]$$

(31)

The determinant of the power spectrum matrix $\mathbf{R}(\omega)$ may be expressed in terms of the input power spectra $|s_1|^2, \cdots, |s_I|^2$, K times the determinant of the matrix $\mathbf{C}^H(\omega)\mathbf{C}(\omega)$ and M times determinant of the matrix $\mathbf{B}(\omega)\mathbf{B}^H(\omega)$.

From equation (25), the longest time constant $\tau_{\max}$ is only determined by the ratio of the maximum and minimum eigenvalues of $\mathbf{R}(\omega)$, so the $\tau_{\max}$ is independent of the noise power $|S_1|^2, \cdots, |S_I|^2$ and only determined by the characteristic of the reference path $\mathbf{B}(\omega)$ and the secondary path $\mathbf{C}(\omega)$. In general, if the determinant of the matrix $\mathbf{C}^H(\omega)\mathbf{C}(\omega)$ or $\mathbf{B}(\omega)\mathbf{B}^H(\omega)$ is small, the smallest eigenvalue of the matrix $\mathbf{C}^H(\omega)\mathbf{C}(\omega)$ or $\mathbf{B}(\omega)\mathbf{B}^H(\omega)$ is small, and the smallest eigenvalue of the matrix $\mathbf{R}(\omega)$ is also small. Therefore, the convergence characteristics of the MEFX LMS algorithm can be evaluated separately by the distributions of eigenvalues of the matrix $\mathbf{C}^H(\omega)\mathbf{C}(\omega)$ or $\mathbf{B}(\omega)\mathbf{B}^H(\omega)$.

3. The behavior of adaptive algorithm in the frequency domain

3.1 The filtered-x LMS algorithm

Firstly, Let us investigate the convergence characteristics of the filtered-x algorithm influenced by the secondary path C for a simple ANC system, CASE[1,1,1,1]. For convenience, assuming that the primary noise s(n) is white noise with zero mean and unit variance ($\sigma^2 = 1$), and the primary path b(n) is 1, so

$$X(n,\omega) = B(\omega)S(n,\omega) = S(n,\omega).$$

(32)

In the CASE[1,1,1,1], a simple ANC system, the power spectrum $\mathbf{R}(\omega)$ is real at each frequency bin ω

$$R(\omega) = |S_1|^2 |C(\omega)|^2 |B(\omega)|^2$$

In this case, $R(\omega)$ can be expressed as follows:

$$R(\omega) = |S_1|^2 |C(\omega)|^2 |B(\omega)|^2 = \sigma^2 |C(\omega)|^2 |B(\omega)|^2 = |C(\omega)|^2, \tag{33}$$

and the upper limit of the step-size parameter μ from equation (21) can be rewritten as

$$\mu < \frac{1}{\max_\omega \{|C(\omega)|^2\}}, \tag{34}$$

It is clear that the step-size parameter μ is subject to the maximum value of $|C(\omega)|^2$. The time constant, $\tau(\omega)$, at a frequency bin can be obtained from equation (22), as

$$\tau(\omega) \approx \frac{1}{2\mu |C(\omega)|^2}, \tag{35}$$

$$t(\omega) > \frac{\max\{|C(\omega)|^2\}}{2|C(\omega)|^2}. \tag{36}$$

It is found from equation (35) that the smaller value of $|C(\omega)|^2$ leads to a slower convergence speed of the adaptive filter at the frequency bin ω. In other words, if $|C(\omega)|^2$ shows a large dip at a frequency bin ω, the convergence speed of the adaptive filter is slow at that frequency bin, and the total convergence speed is also slow. This means that the performance of the filtered-x LMS algorithm is not good if the power gain of the secondary path C is not flat over the whole frequency range. In practical cases, the transfer function of the secondary path C has to be measured prior to the active noise cancelation. Therefore, the convergence characteristics can be evaluated by the power gain of the measured secondary path C.

With the secondary path C measured experimentally or generated by computer, the simulations show that the smaller value of the power gain of the secondary path C leads to slower a convergence speed, a larger computation error and a smaller cancellation at a given frequency bin.

3.2 The delayed-x LMS algorithm

The filtered-x LMS algorithm is widely used in feedforward ANC systems. This adaptive algorithm is an alternate version of the LMS algorithm when the secondary path C from the adaptive filter output to the error sensor is represented by a non-unitary transfer function. The Filtered-x LMS algorithm requires a filtered reference signal, which are the convolution of the reference signal and the impulse response of a secondary path C. As a result, this algorithm has a heavy computational burden for real-time controllers.

The Delayed-x LMS algorithm [9] is a simplified form of the filtered-x LMS algorithm, where the secondary path C from the secondary source to the error sensor is represented by a pure delay of k samples (the delayed model D) to reduce computation and system complexity. This simplified version has been applied in telecommunications applications [16,17]. In the ANC system, the simplification will bring a modelling error [18,19], which causes deterioration in the ANC performance. The ANC system with the Delayed-x LMS algorithm was empirically studied in the time domain [20-22], and stability has been evaluated by using a frequency domain model of the "filtered" LMS algorithm [23]. The theoretical study of the convergence characteristics will be summarized here.

In the Delayed-x LMS algorithm, the model of the secondary path C is replaced by a delayed model D, and no convolution is required to obtain the filtered reference signal. A block diagram of the ANC system with the Delayed-x LMS algorithm is shown in Figure 2. The filtered reference signal is given by

$$u'(n) = gx(n-k) \tag{37}$$

where the gain g is usually 1 and k is the number of points from 0 to the peak of the impulse response of the secondary path C.

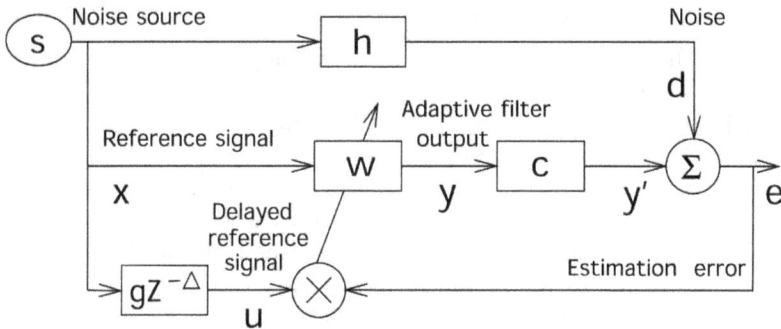

Fig. 2. Block diagram of an ANC system with the Delayed-x LMS algorithm.

The Delayed-x LMS algorithm can reduce computation loading significantly by eliminating the convolution. However, the modelling error caused by simplifying the filtered reference signal deteriorates the performance of the adaptive control system. In the frequency domain, $U(\omega,n)$ is the DFT of the filtered reference signal $u(n)$ filtered by the real secondary path, $U(\omega,n) \approx X(\omega,n)C(\omega)$. $U'(\omega,n)$ is the DFT of the filtered reference signal $u'(n)$ in equation (37), $U'(\omega,n) \approx X(\omega,n)G(\omega)$ and $\overline{U'(\omega,n)}$ is the complex conjugate of $U'(\omega,n)$. The transfer function of the delayed model D is defined as $G(\omega)$. From equation (20), the convergence characteristics of the Delayed-x LMS algorithm are defined by matrix $E[\overline{U'(\omega,n)}U(\omega,n)]$ at each frequency bin ω. Assuming that the primary noise $s(n)$ is a white noise with zero mean and unit variance, the matrix can be written as

$$E[\overline{U'(\omega,n)}U(\omega,n)] = E[\overline{G(\omega)}C(\omega)] \tag{38}$$

It is clear that the stability of the Delayed-x LMS algorithm is assured by

$$0 < \left|1 - 2\mu\overline{G(\omega)}C(\omega)\right| < 1, \tag{39}$$

where

$$\overline{G(\omega)} = |G(\omega)| \exp(-j\theta_g(\omega)),\qquad\qquad(40)$$

$$C(\omega) = |C(\omega)| \exp(j\theta_c(\omega)),\qquad\qquad(41)$$

and $\theta_g(\omega)$ and $\theta_c(\omega)$ are phases of $G(\omega)$ and $C(\omega)$, respectively. For convenience, the frequency bin ω will be omitted hereafter. From equations (40) and (41), equation (39) can be rewritten as

$$0 < \left|1 - 2\mu|G||C|\exp(j(\theta_c - \theta_g))\right| < 1\qquad\qquad(42)$$

It is clear that the change of the gain $|G|$ can be included into the adjustment of the step-size parameter μ. The stability condition of the Delayed-x LMS algorithm is determined by the phase error $\theta_s = \theta_c - \theta_g$ in the following range

$$-\pi/2 < \theta_s (\mathrm{mod}\, 2\pi) < \pi/2.\qquad\qquad(43)$$

In other words, if the phase error θ_s is out of the above range, the Delayed-x LMS algorithm will not be stable. The theoretical result in equation (43) is the same as that obtained by Feinutuch [23]. For easy understanding, the stability and convergence characteristics of the Delayed-x LMS algorithm are also discussed in the complex plane [9, 24]. It is found the Delayed-x LMS algorithm is stable when the phase error will keep in the range between $-\pi/2$ and $-\pi/2$. It is also found that the convergence speed of the adaptive filter is slower and cancellation is smaller when the phase error with the stability condition in equation (43) is large in the frequency domain.

Since the secondary path C can be measured generally prior to active cancellation, stability and convergence characteristics are easily evaluated by calculation the phase error before cancellation. A possible way to achieve good performance is to adjust the position of the loudspeaker and error microphone or to adjust the number of delayed points.

3.3 The behavior of the multichannel filtered-x algorithm

In this section, the behavior of the MEFX LMS algorithm will be evaluated in CASE[I,K,L,M] system with $M \times L$ secondary paths $\mathbf{C}$ and $K \times I$ reference paths $\mathbf{B}$. As stated in equation (29), the power spectrum matrix $\mathbf{R}(\omega)$ is determined by the secondary paths $\mathbf{C}(\omega)$ and the reference paths $\mathbf{B}(\omega)$. The effect of $\mathbf{C}(\omega)$ and $\mathbf{B}(\omega)$ on the convergence behavior of the MEFX LMS algorithm will be discussed separately, and a new preprocessing method to the reference path is proposed to improve the whole performance of the adaptive algorithm.

3.3.1 The matrix C

In general, if the smallest eigenvalue of the matrix $\mathbf{R}$ is small, the smallest eigenvalue of the matrix $\mathbf{C}^H\mathbf{C}$ is small, so the determinant of the matrix $\mathbf{C}^H\mathbf{C}$ is also small. An approximate method to avoid computing the eigen-decomposition of $\mathbf{C}^H\mathbf{C}$ is to replace the ratio of the maximum to minimum eigenvalue with the ratio of determinant. Define the ratio $\rho(\omega)$ instead of equation (25) as follows,

$$\rho_C(\omega) = \frac{\max\limits_{\omega}\{|C^H(\omega)C(\omega)|\}}{|C^H(\omega)C(\omega)|}.$$ (44)

The ratio $\rho_C(\omega)$ of the maximum to the minimum value of the determinant of matrix $C^H C$ over the whole frequency range is defined by

$$\rho_C = \frac{\max\limits_{\omega}\{|C^H(\omega)C(\omega)|\}}{\min\limits_{\omega}\{|C^H(\omega)C(\omega)|\}}.$$ (45)

Thus, we can evaluate the convergence speed approximately by using the ratio ρ_C, instead of calculating the eigenvalues of the matrix R or $C^H C$. This is to say, the frequency domain analysis requires much less computation compared with the time domain analysis. The physical meaning of the matrix $C^H C$ will be further discussed in the next section.

3.3.2 The physical meaning

The physical meaning of the matrix $C^H C$ will be discussed in detail in this section. For simplicity, an $M = L = 2$ system will be considered, of course equation $|C^H C| = |C|^2$ is valid in this case. At each frequency bin ω, the matrix C can be expressed by

$$C = \begin{bmatrix} C_{11} & C_{12} \\ C_{21} & C_{22} \end{bmatrix}.$$ (46)

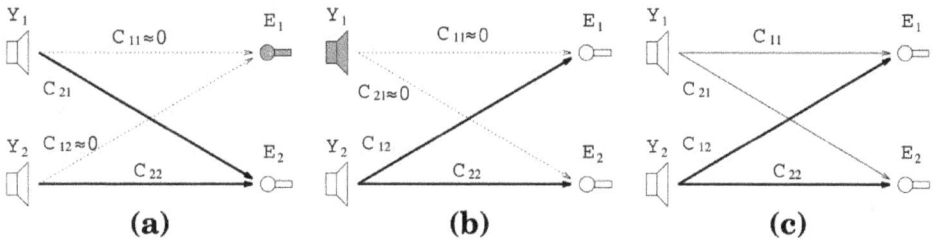

Fig. 3. The condition for $|C| \approx 0$, in the case of an $M = L = 2$.

Referring to Figure 3, at a frequency bin ω, the determinant of the matrix is small ($|C| \approx 0$) in the following three cases: (a) All elements of a row of the matrix C are small (e.g., $C_{11} \approx 0$ and $C_{12} \approx 0$), which implies the existence of a common zero in the transfer function of secondary paths from two secondary sources to the error sensor 1; (b) All elements of a column of the matrix C are small (e.g., $C_{11} \approx 0$ and $C_{12} \approx 0$), which means the secondary source 1 does not work at this frequency. Hence, the effective number of secondary sources is decreased. (c) The transfer paths from the secondary sources to the error sensors are similar. This can be expressed mathematically as

$$C_{11} / C_{21} \approx C_{12} / C_{22}.$$ (47)

It follows from equation (47) that two secondary sources act as one secondary source. This means that the effective number of the secondary sources is decreased. Therefore, the multiple channel ANC system does not work properly. It is well known that characteristics of the secondary paths are dependent upon the arrangement of secondary sources / error sensors, as well as frequency responses. Therefore, if the frequency responses of the secondary sources to the error sensors are proportional as equation (47), which implies that either the secondary sources or the error sensors will be located close to each other, which reduces the effective number of secondary sources. On the other hand, in practical applications, frequency responses of the secondary sources to the error sensors are frequently different at each frequency bin. Under this condition, even if their arrangement is not close together as y mentioned above, equation (47) may be valid, which results in a divergence of the adaptive process. In summary, either arrangement of the secondary sources to the error sensors, or their frequency responses can make equation (47) valid, the determination of matrix C may be equal approximately to zero and the eigenvalue spread of the power spectrum matrix R will become large.

For the three cases mentioned above, since $|c|$ or $|c^H c|$ is small, the smallest eigenvalue of the power spectrum matrix R is correspondingly small, which results in a low convergence speed. If the value $|c^H c|$ varies significantly over the whole frequency range, the convergence speed of the MEFX LMS algorithm is also slow. From the previous works on the single-channel ANC system, conditions (a) and (b) can easily be considered. However, such conditions are very rare in practical applications. Condition (c), which occurs more frequently in practical applications, should be emphasized in a practical ANC system for multiple control points. The physical meaning of the general multiple channel ANC system can be discussed by using the rank and the linear independence theory of the matrix C.

In practical applications, since the transfer function for the secondary paths, which is time invariant in most cases, can be measured prior to the ANC processing, the influence of multiple secondary paths on the convergence speed should be evaluated prior to the ANC processing.

3.4 The matrix B

In equation (29), the transfer function matrix B of the reference paths is similar to the transfer matrix C of the secondary paths, so that the characteristics of the matrix C are also similar to those of matrix B. But the physical meaning of smaller $|B|$ is different that of $|C|$.

Here, the physical meaning of the case that $|BB^H| \approx 0$ is discussed for an $I = K = 2$ system for simplicity. The equation $|BB^H| = |B|^2$ is valid in this case.

For each frequency bin ω, B is of the form

$$B = \begin{bmatrix} B_{11} & B_{12} \\ B_{21} & B_{22} \end{bmatrix}$$

$$(48)$$

Referring to Figure 4, at a frequency bin ω, the determinant of the matrix is small ($|B| \approx 0$) in the following three cases:

a. All elements of a row of the matrix $\mathbf{B}$ are small (e.g., $B_{11} \approx 0$ and $B_{12} \approx 0$). This implies the existence of a common zero in the transfer function from two noise sources to the reference sensor #1.

b. All elements of a column of the matrix $\mathbf{B}$ are small (e.g., $B_{11} \approx 0$ and $B_{21} \approx 0$). In this case, noise source #1 doesn't exist at this frequency. Hence, the reference sensor cannot receive the correct reference signal from noise source #1.

c. The transfer rates from the noise sources to the reference sensors are similar $(B_{11} / B_{21} \approx B_{12} / B_{22})$. In this case, $X_2 \approx (B_{21} / B_{11}) X_1$, and this is equivalent to using only one reference sensor at this frequency, even though two sensors are used. This means that the effective number of the reference sensors is decreased, and perfect noise cancellation is impossible. But minimization of the noise level is achievable in the sense of the least mean square error by the adaptive processing of the ANC system. Therefore, the multiple channel ANC system does not work properly.

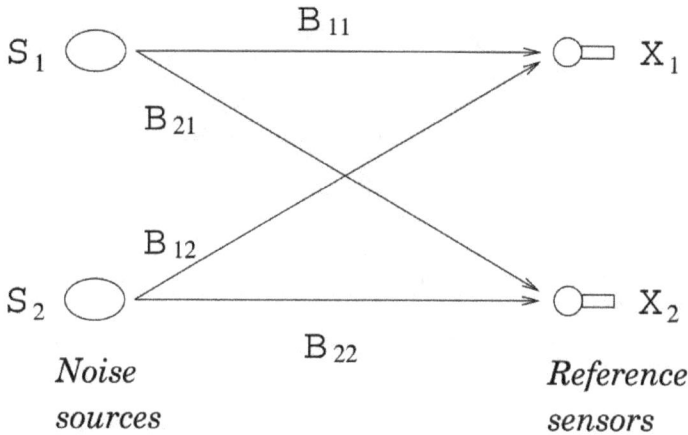

Fig. 4. Block diagram for 2 noise sources and 2 reference sensors.

For the three cases mentioned above, the value of the determinant of the matrix $|\mathbf{B}|$ or $|\mathbf{BB}^H|$ become small, and the physical meaning is the correlation between the reference signals X_1 and X_2 become large [24]. In this case, the determinant or the smallest eigenvalue of the matrix $|\mathbf{B}|$ is small, the smallest eigenvalue of the power spectrum matrix $\mathbf{R}$ is also small, and the convergence speed of the adaptive filters is slow at that frequency bin. Then, the convergence speed of the mean square error at the error sensors is also slow. If the value $|\mathbf{BB}^H|$ varies over the whole frequency range, the convergence speed of the MEFX LMS algorithm is slower.

It is analysed from equation (29) that exchanging the roles of the transfer matrices $\mathbf{B}$ and $\mathbf{C}$, the same conclusion can be obtained as in section 3.3.

An adaptive blind method for reducing the eigenvalue spread of the correlation matrix of reference signals is discussed in next section. The preprocessed outputs used as input of an ANC system are approximately uncorrelated noises and the power spectrums are flat approximately. The MEFX LMS algorithm converges rapidly and a small MSE is obtained [25, 26].

3.5 Blind preprocessing method for multichannel freedforward ANC system

Sometimes, the effects of the reference path $\mathbf{B}$ can't be ignored, especially when the reference sensors cannot be located near the noise sources. As stated in equation (30), the value $\left|\mathbf{B}\mathbf{B}^H\right|$ at each frequency bin ω affects the eigenvalue spread of the autocorrelation matrices $\mathbf{R}$, and the step size μ and the longest time constant of the adaptive filter. In order to improve the whole performance of the MEFX LMS algorithm, some necessary preprocessing methods are proposed to reduce the effects of the reference path. However, noise signals are often unknown and time-varying in practice, and the accurate transfer function of the reference path is difficult to be measured in prior, so it seems impossible to cancel this effects. A blind preprocessing method is proposed to deal with this case, where noises are assumed to be independent or uncorrelated each other and the channel impulse responses are unknown.

An arbitrary linear system can be factored into the product of an all-pass system and a minimum phase system

$$\mathbf{B}(\omega) = \mathbf{B}_{min}(\omega)\mathbf{B}_{all}(\omega), \tag{49}$$

where the all-pass system satisfy $\mathbf{B}_{all}(\omega)\mathbf{B}_{all}^H(\omega) = \mathbf{I}$, and the minimum phase component $\mathbf{B}_{min}(\omega)$ has a stable inverse. $\mathbf{B}(\omega)\mathbf{B}^H(\omega)$ can be simplified as

$$\mathbf{B}(\omega)\mathbf{B}^H(\omega) = \mathbf{B}_{min}(\omega)\mathbf{B}_{min}^H(\omega), \tag{50}$$

In order to eliminate or reduce the eigenvalue spread of $\mathbf{B}(\omega)\mathbf{B}^H(\omega)$, a natural choice is to find an inverse system matrix $\mathbf{V}(\omega)$ to filter the reference signals, and the new transfer system matrix from the noise sources to the inputs of the adaptive filter array is $\mathbf{G}(\omega) = \mathbf{V}(\omega)\mathbf{B}(\omega)$, which has a smaller eigenvalue spread. The correlation matrix of the new system is

$$\mathbf{V}(\omega)\mathbf{B}(\omega)\mathbf{B}^H(\omega)\mathbf{V}^H(\omega) = \mathbf{V}(\omega)\mathbf{B}_{min}(\omega)\mathbf{B}_{min}^H(\omega)\mathbf{V}^H(\omega), \tag{51}$$

and the optimal inverse system matrix is $\mathbf{V}(\omega) = \mathbf{B}_{min}^{-1}(\omega)$.

If the reference path $\mathbf{B}(\omega)$ can be measured in prior, the optimal inverse system $\mathbf{V}(\omega)$ can be computed easily and fixed into the application to cancel the effect of the reference path. However, $\mathbf{B}(\omega)$ cannot be obtained in most applications in prior and may be time-varying in complicated application, an adaptive algorithm to find the optimal inverse system is expected. A blind spatial-temporal decorrelation algorithm is proposed in [25], which is based on maximization of entropy function in the time domain. More details and the final performance evaluation of the adaptive algorithm can be referred in [25, 26]. Computer simulations show that blind preprocessing algorithm can obtain lower MSE for multichannel feedforward ANC system.

4. Computer simulations

Numerical simulations are carried out to demonstrate the convergence characteristics discussed above by using a simple CASE[2, 2, 2, 2] system, as shown in Figure 5. All simulations are performed in the time domain, but their evaluation is carried out both in the time and frequency domains.

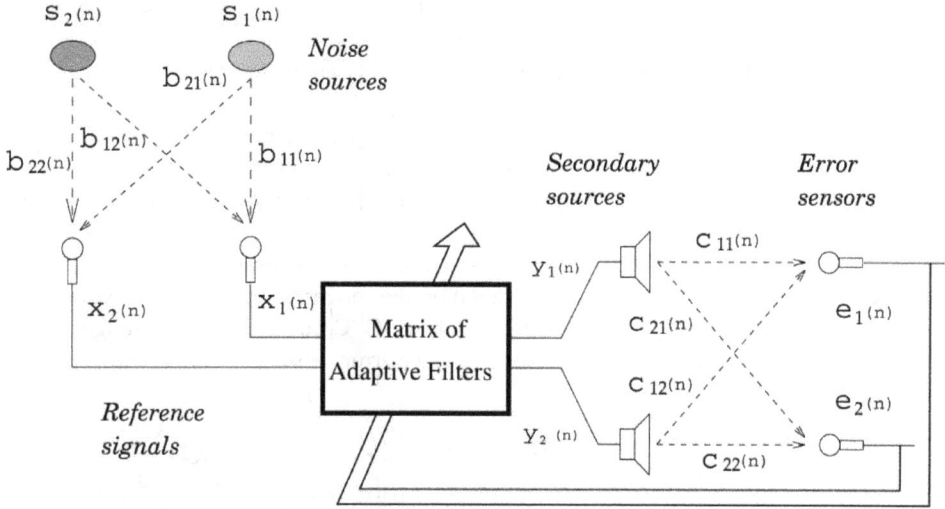

Fig. 5. Block diagram for CASE[2,2,2,2] ANC system.

For convenience, the experimental conditions are assumed as follows: (1) Noise sources $s_1(n)$ and $s_2(n)$ are uncorrelated white noise with zero mean and unit variance. (2) The responses of the primary paths H_{11}, H_{12}, H_{21}, and H_{22}, and the secondary paths C_{11}, C_{12}, C_{21}, and C_{22} are experimentally obtained in an ordinary room. (3) In order to evaluate the influence of the matrix C, the filtered reference signal $x_1(n) = s_1(n)$ and $x_2(n) = s_2(n)$ are selected. In this case, $B = I$, where I is the identity matrix.

The simulation is carried out by using four secondary paths as shown in Figure 6. The step-size parameter μ is set to 0.00001 to keep the system stable and achieve a better convergence.

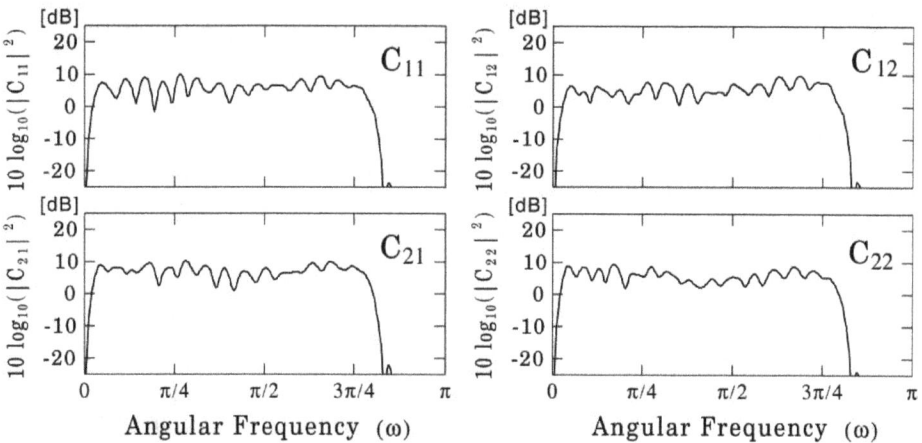

Fig. 6. The four transfer functions from the secondary sources to the error sensors.

There is no common zero in the four secondary paths C_{11}, C_{12}, C_{21}, and C_{22}, as shown in Figure 6. However, it can be seen from Figure 7(a) that there are some dips in the determinant of the matrix $\mathbf{C}^H\mathbf{C}$. The two eigenvalues of the matrix $\mathbf{C}^H\mathbf{C}$ are plotted in Figure 7(b). As can be seen from Figure 7, when the $|\mathbf{C}^H\mathbf{C}|$ is small, the eigenvalues of the matrix $\mathbf{C}^H\mathbf{C}$ are also small. Spectra of the residual signal at different iterations at the error sensor #1 are plotted in Figure 8. Nearly the same results are obtained at the error sensor #2, but these Figures are omitted in this chapter. Comparing Figures 7 and 8, if the $|\mathbf{C}^H\mathbf{C}|$ at some frequencies is small, the residual power at the corresponding frequencies is high. In actual application, since the secondary paths, which is time invariant in most cases, should be measured prior to the ANC processing, the influence of the secondary path(s) on the convergence speed can be evaluated prior to the ANC processing. It is possible to make $|\mathbf{C}^H\mathbf{C}|$ over the whole range flat by adjusting the locations of the secondary sensors and the error sensors.

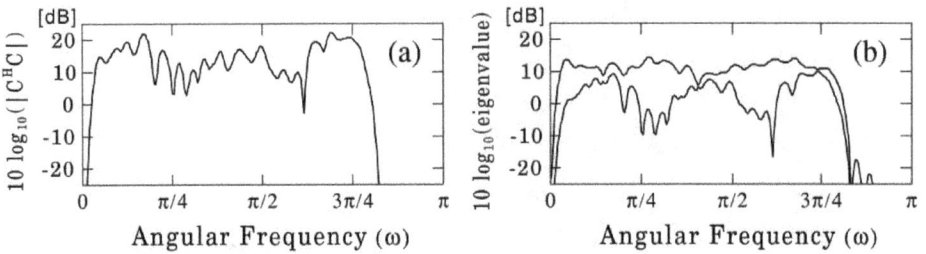

Fig. 7. The matrix $\mathbf{C}$ is composed of four secondary paths shown in Figure 6. (a): the determinant of the matrix $\mathbf{C}^H\mathbf{C}$, $\left[10\log_{10}\left(|\mathbf{C}^H\mathbf{C}|\right)\right]$ (b): two eigenvalues of the matrix $\mathbf{C}^H\mathbf{C}$, $\left[10\log_{10}\left(eigenvalue\right)\right]$.

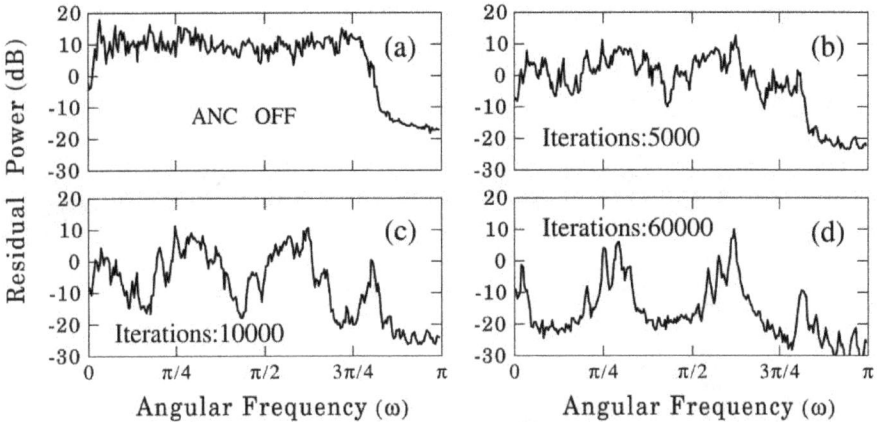

Fig. 8. Simulated power spectra of the residual noise (average of 20 times) at the error sensor #1, in the case of CASE[2,2,2,2] ANC system, when $(B = I)$. (a) Before adaptive processing; (b), (c) during adaptive processing, the numbers of the adaptive iterations are 5000 and 10000, respectively; (d) after adaptive processing, the number of the adaptive iterations are 60000. Similar results were obtained at the error sensor #2.

5. Conclusions

This chapter has shown that the convergence characteristics of the filtered-x LMS algorithm, the Delayed-x LMS algorithm and the MEFX LMS algorithm in the time domain could be analysed in the frequency domain with much less computation and a better understanding of the physical meaning. Through their analysis in the frequency domain instead of time domain, the convergence characteristics are subject to the eigenvalues of the power spectrum matrix R, whose size is much smaller than that in the time domain. Another advantage is that the determinant of the power spectrum matrix R can be expressed by the product of the input spectra and the determinant of the matrix $C^H C$ and BB^H in each frequency bin. The effect of multiple secondary paths has been investigated in detail in the case of the time invariant. It is found that the convergence characteristics of the MEFX LMS algorithm are affected by the determinant of $C^H C$, or the smallest eigenvalues of $C^H C$. However, since the transfer matrix B generally can't be measured prior to ANC cancellation, it is necessary to consider the influence of the correlation among the output of the reference sensors, which can be measured prior to ANC application. If the correlation among the output of the reference sensors is small over the whole frequency range, the convergence speed becomes fast. Simulation on the time-domain MEFX algorithm has been carried out and its convergence characteristics are evaluated in the frequency domain.

6. References

[1] Nelson, P. A., & Elliott, S. J. (1992). *Active Control of Sound,* Academic Press Inc. San Diego, Canada.

[2] Elliott, S. J., & Nelson, P. A. (1993). Active Noise Control, *IEEE signal processing magazine,* October 12-35.

[3] Elliott, S, (2001). Signal processing for active control, Academic Press, London.

[4] Widrow, B., & Stearns, S. D. (1985). *Adaptive Signal Processing,* Prentice-Hall Inc.

[5] Burgess, J.C. (1981). Active adaptive sound control in a duct: A computer simulation, *J. Acoust. Soc. Am.,* 70(3), pp. 715-726.

[6] Elliott, S. J., Stothers, I. M., & Nelson, P. A. (1987). A multiple error LMS algorithm and its application to active control of sound and vibration, proc. *IEEE Trans. Speech Audio Processing,* ASSP-35(10), pp.1423-1434.

[7] Elliott, S. J., & Boucher, C. C. (1994). Interaction between multiple feedforward active control system. *IEEE Trans. Acoustic. Speech Signal Process on speech and audio processing,* SAP-2 (4), pp.521-530.

[8] Chen, G., Abe, M., & Sone, T. (1995). Evaluation of the convergence characteristic of the filtered-x LMS algorithm in frequency domain, *J. Acoust. Soc. Jpn (E),* 16(6), pp.331-340.

[9] Chen, G., Sone, T., Saito, N., Abe, M. & Makino, S. (1998).The stability and convergence characteristics of the Delayed-X LMS algorithm in ANC systems, *Journal of Sound and Vibration,* 216(4), 637-648, pp. 637-648.

[10] Chen, G., Abe, M., & Sone, T. (1996). Effects of multiple secondary paths on convergence properties in active noise control systems with LMS algorithm, *Journal of Sound and Vibration,* 195, pp.217-228.

[11] Chen, G., Wang, H., Chen K., & Muto, K., (2008). The Influences of Path Characteristics on Multichannel Feedforword Active Noise Control System, *Journal of Sound and Vibration*, Vol.311, No.3, pp.729-736.

[12] Cowan, C. F. N., & Grant, P. M., (1985). *Adaptive filters*, Prentice-Hall Inc. Englewood Cliffs, New Jersey.

[13] Haykin, S. (1991). *Adaptive Filter Theory, 2nd ed*. Prenticehall Englewood Cliffs, Newyork.

[14] Widrow, B., McCool, J. M., Larimore, M. G. & Johnson. C. R., (1962) Stationary and nonstationary learning characteristics of the LMS adaptive filter, Proc. IEEE, 64 pp. 1151-1162.

[15] Chen, G., Abe, M., & Sone, T. (1996). Improvement of the convergence characteristics of the ANC system with the LMS algorithm by reducing the effect of secondary paths, *Journal of the Acoustical Society of Japan(E)*, Vol.17(6), pp.295-303.

[16] Kabel, P. (1983). The stability of adaptive minimum mean square error equalizers using delayed adjustment. *IEEE Transactions on Communications* COM-31, pp.430–432.

[17] Qureshi, S. K. H. & Newhall, E. E. (1973). An adaptive receiver for data transmission of time dispersive channels. *IEEE Transactions on Information Theory*, IT-19, pp.448–459.

[18] Elliott, .S. J., Baucher, C. C. & Nelson, P. A. (1991). The effects of modeling errors on the performance and stability of active noise control system. *Proceedings of the Conference on Recent Advances in Active Control of Sound and Vibration*, pp. 290–301.

[19] Saito, N. & Sone, T. (1996). Influence of modeling error on noise reduction performance of active noise control system using Filtered-x LMS algorithm, Vol. 17, pp.195–202. *Journal of the Acoustical Society of Japan (E)*.

[20] Kim, C. Abe, M. & Kido, K. (1983) Cancellation of signal picked up in a room by estimated signal. *Proceedings of Inter-noise 83*, pp.423–426.

[21] Saito, Y. Abe, M., Sone, T. & Kido, K. (1991). 3-D space active noise of sounds due to vibration sources. *Proceedings of the International Symposium on Active Control of Sound and Vibration*, pp. 315–320.

[22] Park, Y. & Kim, H. (1993). Delayed-x algorithm for a long duct system. *Proceedings of Inter-Noise 93*, pp. 767–770.

[23] Feintuch, P. L., Bershad, N. J., & Lo, A. K. (1993). A frequency domain model for "filtered" LMS algorithms-stability analysis, design, and elimination of the training mode, *IEEE Trans. Signal Process*, Vol. 41(3), pp. 1518-1531.

[24] Kurisu, K. (1996). Effect of the difference between error path C and its estimate C for active noise control using Filtered-x LMS algorithm. *Proceedings of Inter-Noise 96*, pp. 1041–1044.

[25] Chen, G., Wang, H., Chen K., & Muto, K.,(2005). A preprocessing method for multichannel feedforward active noise control, *Journal of Acoustical Science and Technology*, Vol. 26(3), pp.292-295.

[26] Wang, H., Chen, G. , Chen K., & Muto, K., (2006). Blind preprocessing method for multichannel feedforward active noise control. *Journal of Acoustical Science and Technology*, Vol. 27(5), pp.278-284.

6

Advanced Vibro-Acoustic Techniques for Noise Control in Helicopters

Emiliano Mucchi[1], Elena Pierro[2] and Antonio Vecchio[3]
[1]*Engineering Department, University of Ferrara*
[2]*DIMeG, Politecnico di Bari*
[3]*ARTEMIS Joint Undertaking, Brussels*
[1,2]*Italy*
[3]*Belgium*

1. Introduction

Originally, aircraft noise was not much of an issue because of the overarching requirement of improving vehicle performance in critical operational conditions. For many years, the design of aircraft and helicopter has been challenged by the need to improve aeromechanical performance, leaving little rooms for considering additional design parameters such as noise, vibrations and passenger comfort. More recently, a paradigm shift is being recorded that includes vibro-acoustics comfort in the set of critical design parameters. In the particular case of helicopters, this is mostly a consequence of a shift in the definition of the typical helicopter mission: in the past helicopters were mainly used in military context where flight mechanics, maneuverability and avionics are the key performance indicators; todays a trend is visible in the related industry that targets more and more civil applications, where the payload and environmental impact assume a much more significant role. In this respect, vibroacoustic comfort and more in general noise performance are becoming one of the key market differentiators: passenger comfort is included in the key performance indicators (ref. the US101 - Marine One case). In order to cope with this new and more stringent set of performance requirements, specific tools and techniques are required that allow better identifying the noise sources, the mechanics of transfer from source to target (pilot and/or passenger area) and defining an effective noise control strategy. In this scenario, the authors want to present a modus operandi to tackle these issues by several advanced experimental methodologies, since the requirement of a quieter helicopter needs a systematic study of its Noise Vibration and Harshness (NVH) behaviour. This chapter collects the results of several investigations carried out on two different helicopters: EUROCOPTER EC-135 and Agusta Westland AW-109. The proposed methodology faces these issues considering that the noise is a result of the contribution of source(s) reaching the receiver via various transmission paths. Therefore, in a noise control strategy , the designer should consider both the excitation phenomena (i.e. sources) as well as the transfer functions between the source(s) and receiver(s) (i.e. transmission paths). The sections of this chapter discuss both sources and transmission paths. In particular, Sections 2 addresses the deep analysis of transmission paths, taking into account the vibro-acoustic interaction that in cumbersome systems as helicopters is of pivotal importance; Sections 2 and 3 deal with noise and vibration sources, while Section 5 presents a technique for gaining information about sources (operational forces) and transmission paths in operational conditions. Eventually, Section 6 presents an original approach based on sound synthesis allowing designers to listen and judge the sound quality impact of structural design modifications in a process of noise control and mitigation. It is interesting to note that although the presented methodologies concern particular helicopters, they have a general meaning, since they can be applied in a large variety of fields, as in aircraft fuselage, car or track interiors , etc. This work has to be intended as an analytical review of the authors' research (Mucchi & Vecchio, 2009; 2010; Pierro et al., 2009). Due to confidentiality reasons, several vibro-acoustical quantities reported in this chapter are presented without numerical values.

2. Noise control in coupled vibro-acoustical systems

Real-life enclosures, such as aircraft fuselage and car interiors, are very often characterized by a strong fluid-structure coupling (Desmet et al., 2003; Kronast & Hildebrandt, 2000). Therefore, the acoustical response in the cavity is related to the structural excitation and a relation exists between the acoustical loading and the structural responses as well (Everstine, 1997; 1981). Two methods are generally used to characterize the vibro-acoustical behaviour of such systems.   In the first, the classical Experimental Modal Analysis (EMA), the modal parameters (natural frequencies, damping factors and mode shapes) are evaluated in laboratory, and consists in to put in vibration the system with a known excitation, out of its normal service environment. The common practice, indeed, is to make the peaks of the input force spectrum coincident with the valleys of the structural frequency response functions (FRFs). However, it is not so straightforward to get these tasks for vibro-acoustic systems because a modal analysis has a high number of theoretical and practical problems: due to the coupling, indeed, the system matrix is nonsymmetrical, yielding different right and left eigenvalues (but closely related). For this reason, care must be taken during the FRFs synthesis, but it has been proven that no matter involves in the modal model derivation itself (Wyckaert et al., 1996).   The latter and more recent technique, is the Operational Modal Analysis (OMA), which does not require the knowledge of the operating forces on the system, but considers the assumption of white noise as input. Some investigations have been already made on vibro-acoustical operational modal analysis (Abdelghani et al., 1999; Hermans & Van der Auweraer, 1999), hereafter the modal decomposition in case of vibro-acoustical operational modal analysis is reviewed, in order to point out the influence on the modal parameter estimation of the references choice, i.e. acoustical or structural, used for the Cross power spectra calculation. The complete knowledge of the vibro-acoustical response of an helicopter cabin is a crucial tool for noise control purpose, and in the OMA context, it is strongly related to the reference choice rather than the input forces. For simplicity, only an acoustical input will be considered and the cross power between acoustical and structural responses will be derived and compared with structural cross power responses. The formulation in case of structural excitation can be derived in a similar way. The theoretical results have been verified through an experimental testing on the helicopter EUROCOPTER EC-135.

2.1 Vibro-acoustical systems

The dynamical properties of vibro-acoustical systems are presented hereafter. In particular the transfer function considering as excitation both the acoustical loading and the structural force and considering as response both the structural displacement and the acoustical pressure is derived. It is well known that due to the fluid-structure interaction such systems are nonsymmetrical. Moreover, the FRF has a different form with respect to the decoupled systems. It is worth to stress that the non-symmetry is due to the choice of the variables $\mathbf{x}$, $\mathbf{p}$, $\mathbf{f}$ and $\dot{\mathbf{q}}$, being $\mathbf{x}$ the structural displacement, $\mathbf{p}$ the sound pressure, $\mathbf{f}$ the structural force and $\mathbf{q}$ the volume velocity. As stated above, due to the non-symmetry of the system, the right and the left eigenvectors, Ψ_r and Ψ_l, are different. Moreover the following relations can be proven (Ma & Hagiwara, 1991):

$$\left\{ \begin{array}{c} \Psi_{sl} \\ \Psi_{fl} \end{array} \right\}_{\lambda_h} = \left\{ \begin{array}{c} \Psi_{sr} \\ \frac{1}{\lambda_h^2} \Psi_{fr} \end{array} \right\}_{\lambda_h} \tag{1}$$

where subscript s is for the structural response location, f for the fluid (acoustical) response location, and λ_h are the roots of the system characteristic equation. In particular the following transfer functions obtained considering a structural force excitation f_j at location j, have been derived in (Wyckaert et al., 1996):

$$\frac{x_m}{f_j} = \sum_{h=1}^{n} \frac{(P_h)\,(\psi_{shm})\,\left(\psi_{shj}\right)}{(z - \lambda_h)} + \frac{(P_h)^*\,(\psi_{shm})^*\,\left(\psi_{shj}\right)^*}{\left(z - \lambda_h^*\right)} \tag{2}$$

$$\frac{p_p}{f_j} = \sum_{h=1}^{n} \frac{(P_h)\,\left(\psi_{fhp}\right)\,\left(\psi_{shj}\right)}{(z - \lambda_h)} + \frac{(P_h)^*\,\left(\psi_{fhp}\right)^*\,\left(\psi_{shj}\right)^*}{\left(z - \lambda_h^*\right)} \tag{3}$$

where x_m is the structural displacement at location m, p_p the acoustical pressure response inside the cavity at location p, P_h the modal scaling factors and n the number of modes in the frequency band of interest. Similar transfer functions can be obtained considering an acoustical excitation $\dot{q}$ at location k:

$$\frac{x_m}{\dot{q}_k} = \sum_{h=1}^{n} \frac{(P_h)\,(\psi_{shm})\,\left(\psi_{fhk}\right)}{\lambda_h^2\,(z - \lambda_h)} + \frac{(P_h)^*\,(\psi_{shm})^*\,\left(\psi_{fhk}\right)^*}{\lambda_h^2\,(z - \lambda_h^*)} \tag{4}$$

$$\frac{p_p}{\dot{q}_k} = \sum_{h=1}^{n} \frac{(P_h)\,\left(\psi_{fhp}\right)\,\left(\psi_{fhk}\right)}{\lambda_h^2\,(z - \lambda_h)} + \frac{(P_h)^*\,\left(\psi_{fhp}\right)^*\,\left(\psi_{fhk}\right)^*}{\lambda_h^2\,(z - \lambda_h^*)} \tag{5}$$

Considering $p = k$ and $j = m$ the vibroacoustical reciprocity, i.e. $(p_k/f_m) = -\left(\ddot{x}_m/\dot{q}_k\right)$, can be derived from (3) and (4). This modal description (Eqs.2-5) can be used for the parameter estimation (mode shapes, modal frequencies and damping factors) by means of standard algorithms.

Operational Modal Analysis (Peeters et al., 2007) is also called output-only modal analysis, and consists in a modal analysis without the knowledge and/or control of the input excitation. Assuming white noise as input, it enables to estimate the modal parameters from the cross power spectra (CPs) between the outputs and some responses chosen as references. In practice, the main difference with respect to the classical modal analysis is that the cross power spectra between the outputs are used to estimate the modal parameters, instead of the genuine FRFs (ratio between the system responses and the input forces). In vibro-acoustical systems the fluid-structure interaction yields both the acoustical and the structural outputs.

The power spectrum of the outputs, hence, can be obtained in three different forms, i.e. with acoustical outputs only, with structural output only and with both acoustical and structural outputs. Moreover, the input signal, supposed to be white noise, can be either structural or acoustical (or both of them simultaneously). In this section the results of the formulation of the operational modal analysis for vibro-acoustical systems are reported and discussed. For simplicity, only an acoustical input will be considered and the cross power between structural and the acoustical responses will be shown.

Let define the cross-correlation function $R_{mpk}(T)$ as the expected value $\mathbf{E}$ of the product of two stationary responses $x_m(t)$ and $p_p(t)$ due to the single $\dot{q}_k$ acoustical input at the kth degree of freedom (d.o.f.) evaluated at a time separation of T, i.e. $R_{mpk}(T) = \mathbf{E}\left[x_m(t+T)\,p_p(t)\right]$.

Assuming white noise (i.e. $\mathbf{E}\left[\dot{q}_k(\tau)\,\dot{q}_k(\sigma)\right] = \alpha_k\delta(\tau-\sigma)$, with α_k a constant value related to the input and $\delta(t)$ the Dirac delta function) and performing some calculations, one obtains the cross power spectrum between *acoustical f* and *structural s responses*:

$$\left[\mathbf{G}_{fs}(i\omega)\right] = \sum_{h=1}^{n}\left[\frac{\{\Psi_s\}_h\{\mathbf{Q}_{fs}\}_h^T}{(i\omega-\lambda_h)} + \frac{\{\Psi_s\}_h^*\{\mathbf{Q}_{fs}\}_h^H}{(i\omega-\lambda_h^*)} + \frac{\{\mathbf{Q}_{fs}\}_h\{\Psi_f\}_h^T}{(-i\omega-\lambda_h)} + \frac{\{\mathbf{Q}_{fs}\}_h^*\{\Psi_f\}_h^H}{(-i\omega-\lambda_h^*)}\right]$$

(6)

with:

$$\left(Q_{fs}\right)_{ph} = \sum_{j=1}^{n}-\alpha_k\left[\frac{(P_h)\left(\psi_{fhk}\right)(P_j)\left(\psi_{fjp}\right)\left(\psi_{fjk}\right)}{\lambda_h^2\lambda_j^2\left(\lambda_h+\lambda_j\right)} + \frac{(P_h)\left(\psi_{fhk}\right)(P_j)^*\left(\psi_{fjp}\right)^*\left(\psi_{fjk}\right)^*}{\lambda_h^2\lambda_j^2\left(\lambda_h+\lambda_j^*\right)}\right]$$

(7)

$$\left(Q_{fs}\right)_{mh} = \sum_{j=1}^{n}-\alpha_k\left[\frac{(P_j)\left(\psi_{sjm}\right)\left(\psi_{fjk}\right)(P_h)\left(\psi_{fhk}\right)}{\lambda_h^2\lambda_j^2\left(\lambda_h+\lambda_j\right)} + \frac{(P_j)^*\left(\psi_{sjm}\right)^*\left(\psi_{fjk}\right)^*(P_h)\left(\psi_{fhk}\right)}{\lambda_h^2\lambda_j^2\left(\lambda_h+\lambda_j^*\right)}\right]$$

(8)

The unstable poles $-\lambda_h$ and λ_h^* derive from the negative time lags of the cross-correlation function $R_{mpk}(T)$. Consequently they can be deleted by means of the weighted correlogram approach determining the following half spectrum:

$$\left[\mathbf{G}_{fs}(i\omega)^{++}\right] = \sum_{h=1}^{n}\left[\frac{\{\Psi_s\}_h\{\mathbf{Q}_{fs}\}_h^T}{(i\omega-\lambda_h)} + \frac{\{\Psi_s\}_h^*\{\mathbf{Q}_{fs}\}_h^H}{(i\omega-\lambda_h^*)}\right]$$

(9)

The cross power spectrum between two *structural responses s* is:

$$[\mathbf{G}_{ss}(i\omega)] = \sum_{h=1}^{n}\left[\frac{\{\Psi_s\}_h\{\mathbf{Q}_{ss}\}_h^T}{(i\omega-\lambda_h)} + \frac{\{\Psi_s\}_h^*\{\mathbf{Q}_{ss}\}_h^H}{(i\omega-\lambda_h^*)} + \frac{\{\mathbf{Q}_{ss}\}_h\{\Psi_s\}_h^T}{(-i\omega-\lambda_h)} + \frac{\{\mathbf{Q}_{ss}\}_h^*\{\Psi_s\}_h^H}{(-i\omega-\lambda_h^*)}\right]$$ (10)

with:

$$(Q_{ss})_{ph} = \sum_{j=1}^{n}-\alpha_k\left[\frac{(P_h)\left(\psi_{fhk}\right)(P_j)\left(\psi_{sjp}\right)\left(\psi_{fjk}\right)}{\lambda_h^2\lambda_j^2\left(\lambda_h+\lambda_j\right)} + \frac{(P_h)\left(\psi_{fhk}\right)(P_j)^*\left(\psi_{sjp}\right)^*\left(\psi_{fjk}\right)^*}{\lambda_h^2\lambda_j^2\left(\lambda_h+\lambda_j^*\right)}\right]$$

(11)

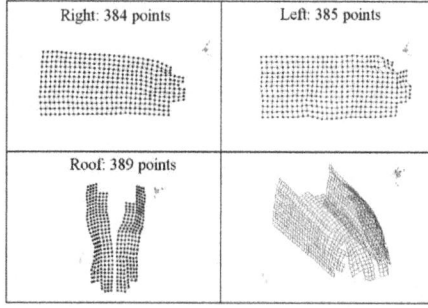

Fig. 1. Measured points of the cabin external surface.

$$(Q_{ss})_{mh} = \sum_{j=1}^{n} -\alpha_k \left[\frac{\left(P_j\right)\left(\psi_{sjm}\right)\left(\psi_{fjk}\right)(P_h)\left(\psi_{fhk}\right)}{\lambda_h^2 \lambda_j^2 \left(\lambda_h + \lambda_j\right)} + \frac{\left(P_j\right)^*\left(\psi_{sjm}\right)^*\left(\psi_{fjk}\right)^*(P_h)\left(\psi_{fhk}\right)}{\lambda_j^2 \lambda_h^2 \left(\lambda_j^* + \lambda_h\right)} \right]$$

(12)

By looking at Eqs. (6),(10) it is clear that the stable pole terms of the sum (i.e. the positive λ_h and λ_h^*) refer to the same mode shapes $\{\Psi_s\}_h$. The only differences between the two formulations are present in the output-only reference vectors $\{Q\}_h$. Moreover, rewriting Eqs. (4) in a matrix form:

$$\left[H_{fs}\left(i\omega\right)\right] = \sum_{h=1}^{n} \left(\frac{\{\Psi_s\}_h \{L_f\}_h^T}{(i\omega - \lambda_h)} + \frac{\{\Psi_s\}_h^* \{L_f\}_h^H}{(i\omega - \lambda_h^*)} \right)$$

(13)

it is evident that Eqs. (6),(10),(13), i.e. respectively the modal decompositions of OMA with an acoustical reference, OMA with a structural reference and vibro-acoustical EMA, depend on the same modal vector $\{\Psi_s\}_h$ and system poles λ_h. This means that in order to estimate the modal parameters (i.e. natural frequency and modal damping) it is possible to consider acoustical references instead of the structural ones as used in the common practice of OMA. This is useful for experimental campaigns aimed to a modal analysis of high scale vibro-acoustical structures for which it is common to take a microphone as reference inside the cavity, near the acoustical excitation. No additional reference sensors, hence, have to be located on the surface. In order to verify the practical effectiveness of the modal model obtained considering an acoustical reference (Eq.(6)), an operational modal analysis on a helicopter has been carried out, comparing the results with a classical EMA (Eq.(13)) where the input is the measured excitation (volume acceleration source).

2.2 Experimental application
A case study has been performed employing the data acquired from a test conducted on the helicopter EUROCOPTER EC-135. The EC-135 is a light twin-engine, multi-purpose helicopter with up to eight seats for pilot/s and passengers. Helicopters are typical examples of coupled vibro-acoustical systems related to a cavity environment. The experimental tasks about these systems mainly regard noise sources identification and modal parameter extraction. In this context, experimental and operational modal analyses are typically used to make the spaced peaks of the input force spectrum coincident with the valleys of the structural frequency response functions, for noise reduction purpose. These two analyses have been

EMA (45.2 [Hz])	OMA (45.7 [Hz])

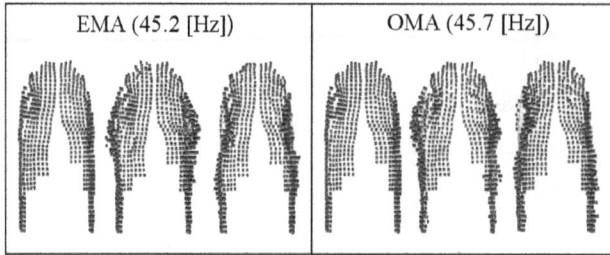

Fig. 2. First mode shape; bottom view of the helicopter cabin.

performed on the cabin external surface of the helicopter while the cavity was excited by means of an acoustical loading. The setup and the results will be shown in the following.

2.3 Test setup

For these applications, acoustical excitation is often preferred since it is usually more practicable than placing a shaker. Moreover, the acoustics of the cavity is excited in a direct way and the measurements are more efficient. In this testing campaign, indeed, a low frequency volume acceleration source (frequency range 20 − 400 Hz) has been located inside the cabin near the pilot's seat, exciting the structure with a random signal. A microphone near the loudspeaker has been used as reference. The structural responses of 1158 points on the right, left and roof part of the cabin external surface (Figure 1) have been measured through a 4x4 PU sensors array, with a 10 cm spacing between the sensors both horizontally and vertically. The PU probes, integrate in a unique casing a hot-wire particle velocity sensor and a very small pressure transducer. They are, hence, specially conceived sensors for measuring acoustic intensity. The distance between the PU probes and the surface, indeed, can be so short that the very-near-field assumption is verified, i.e. the particle velocities simply represent the velocities of the cabin surface (in the normal direction) (De Bree et al., 2004). This is due to the particle velocity level, which is just slightly dependent on the distance to the object and not on the frequency. The very-near-field is a sound field very close to a source, where the particle velocity level is elevated compared to the sound pressure level and a phase shift between sound pressure and particle velocity is observed. The two conditions of the very near field are i) the measurement distance should be closer than the structural size of the object divided by 2π and ii) the wavelength of the sound should be larger than the structural size of the object. At last, the possibility of getting with just one measurement the needed data for both a modal analysis and an acoustic intensity calculation highlights the usefulness of these PU probes in this field, instead of classical methods such as accelerometers and scanning laser vibrometers. This point will be shown more in deep hereafter.

2.4 Test results

The results obtained by the two analyses, i.e. EMA and OMA, are shown in the following. In particular the aim is to experimentally verify the effectiveness of the parameter estimation obtained considering an acoustical reference in the OMA analysis, as theoretically predicted through the comparison between the formulation derived in Section 2.1 (Eq.(9)) and the vibro-acoustical EMA (Wyckaert et al., 1996) (Eq.(13)). Tab.1 lists the results in terms of natural frequency F_n for the modes lying in the 20 − 200 Hz frequency range. In addition, the modal assurance criterion values (MAC) between the two analyses are given in Tab.1.

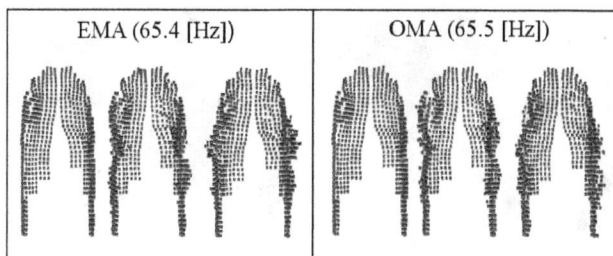

Fig. 3. Second mode shape; bottom view of the helicopter cabin.

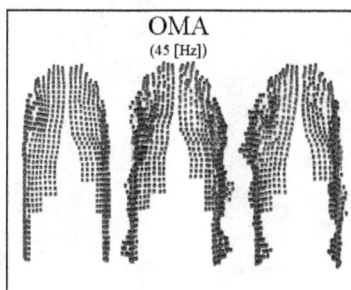

Fig. 4. OMA – first mode shape, full spectrum; bottom view of the helicopter cabin.

Mode	EMA F_n [Hz]	OMA F_n [Hz]	MAC [%]	ΔF_n [%]
1	45.2	45.7	36.5	1.22
2	65.4	65.5	63.7	0.21
3	85.3	84	68.6	1.5
4	98.3	98.3	74.5	0.02
5	108.6	108.8	82.5	0.21
6	123.2	124.4	76.4	0.88
7	139.2	139.1	70.1	0.11
8	167.6	168.5	69.2	0.52
9	188.0	189.5	47.4	0.78

Table 1. Comparison of modal parameters from EMA and OMA.

The agreement between the two analyses for both the frequencies and the damping ratios is very good. In particular, a percentage difference between the frequencies, defined as $\Delta F_n = \left(|(F_n)_{EMA} - (F_n)_{OMA}| / (F_n)_{EMA} \right) * 100$, is always less than 2 %. The correlation values between EMA-OMA listed in Tab.1 are good, except for the first mode and the last one. For what regards this first mode, the reason could be the low excitation of the structure, in fact around that value the loudspeaker has its low frequency limit. On the other hands, the higher is the frequency the more it is difficult to get high correlation values, due to the higher number of local modes. This could be the reason of the low MAC value at the last frequency listed in Tab.1. The first 2 mode shapes obtained from the two analyses are visualized in the helicopter model (bottom view) in Figures 2 and 3. The comparison is a further confirm of the agreement

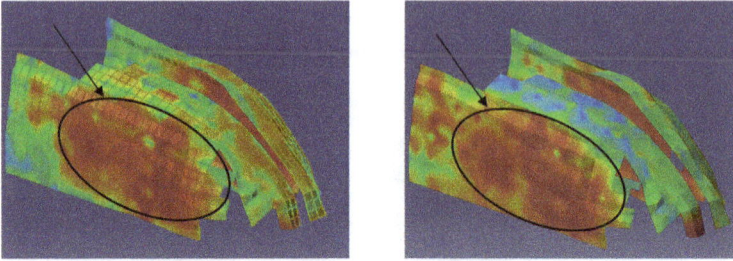

Fig. 5. 45.2 [Hz]: First mode shape on the left, intensity map on the right

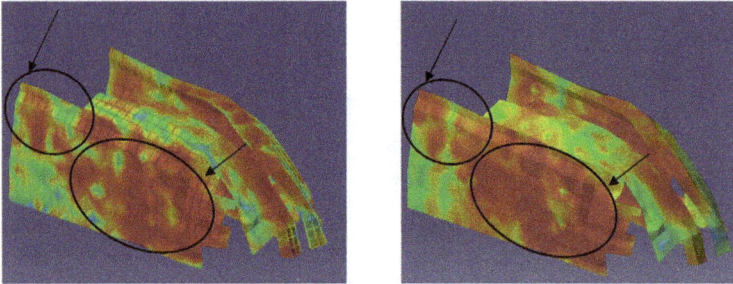

Fig. 6. 65.4 [Hz]: Second mode shape on the left, intensity map on the right

between the two analyses. In order to check the influence of the pre-processing of the CPs on the modal parameter estimation, an OMA analysis considering the CPs-full spectra, i.e. Eq.(6) as input, has been made. Figure 4 shows the first mode shape in such case. A comparison between Figure 4 (OMA full spectrum) and Figure 2 (EMA and OMA half spectrum) clearly highlights the bad numerical conditioning influence as a result of the unstable poles which are present in the full spectrum case. Figure 5 shows the comparison for the first mode (i.e. 45.2 Hz). The red zones indicate the maximum displacements for the mode shapes and the hot spots calculated with the intensity maps. Figure 5 and Figure 6 confirm the good choice of the modes. It is evident, indeed, the correspondence between the zones with high intensity values and the zones with the highest modal displacements (circled areas).

2.5 Further discussions on vibro-acoustical coupling

It is of crucial importance, for the noise control in the cabin interior, the vibro-acoustical coupling assessment between the cabin internal surface and the enclosure.

By means of different EMA on the helicopter, it has been shown that the cabin external surface is coupled to the interior as well as the internal part. Moreover, the mathematical formulation of the CPs modal decomposition in case of vibro-acoustical operational modal analysis has been presented in the previous subsections, clarifying the influence on the modal parameter estimation of the references choice, i.e. acoustical or structural, used for the CPs calculation.

The relevance of such analyses is pointed out through the results of four different experimental analyses on the helicopter EUROCOPTER EC-135, i.e. an acoustical modal analysis (AMA) of the cabin enclosure, an EMA of the cabin internal surface, an EMA and an OMA of the cabin external surface, the latter considering a microphone inside the cabin as reference for the CPs calculation, as explained before.

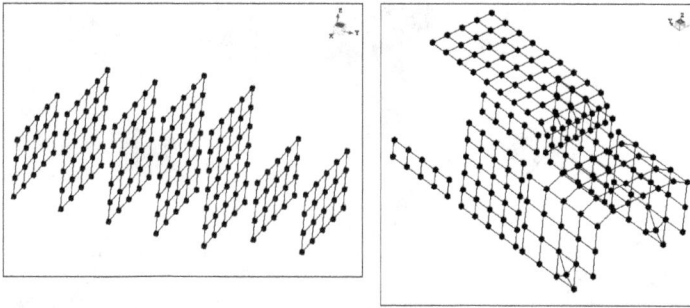

Fig. 7. Measured points in the helicopter cavity (left) by means of microphones and on the cabin internal surface (right) by means of accelerometers.

Mode	AMA-internal part		EMA-internal part		EMA-external part		OMA-external part	
	F_n [Hz]	$\zeta\%$	F_n [Hz]	$\zeta\%$	F_n [Hz]	$\zeta\%$	F_n [Hz]	$\zeta\%$
1	46.5	2.8	46.3	2.9	45.2	1.6	45.7	2.0
2	65.1	2.1	66.5	1.9	65.4	2	65.5	1.0
3	86.1	2.9	86.4	1.9	85.3	1.6	84	0.6
4	98.1	0.4	97.5	1.2	98.3	1.4	98.3	0.9
5	109.7	1.5	110.2	1.2	108.6	1.3	108.8	1.2
6	124.4	2.6	124.3	1.1	123.3	1.3	124.4	1.6
7	139.4	2.5	136.4	1.1	139.2	1.5	139.1	1.5
8	160.8	1.1	168.9	1.4	167.6	1.1	168.5	1.1
9	175.5	6.1	188.4	0.7	188	0.6	189.5	0.8
10	189	2.7	-	-	-	-	-	-

Table 2. Natural frequencies (F_n) and modal damping (ζ) for the four analyses performed on the helicopter cabin.

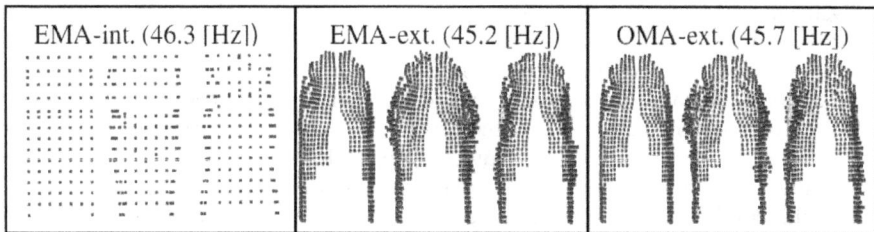

Fig. 8. First mode shape; bottom view of the helicopter cabin, internal (left) and external surface (second and third on the right).

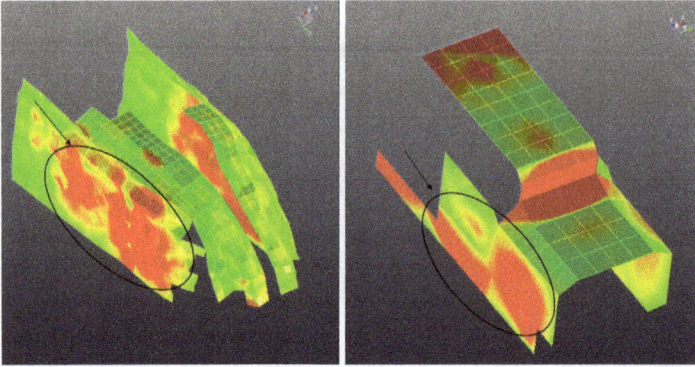

Fig. 9. First mode shape coloured map of the cabin external surface (left) (45.2 Hz) and the cabin internal surface (right) (46.3 Hz)

In the helicopter internal part two measurements were carried out aiming at performing an acoustical modal analysis of the cabin enclosure and a classical modal analysis of the cabin internal surface. Figure 7 (left) shows the 190 points measured by means of condenser microphones (1/2" prepolarized) inside the cavity and the acceleration of the cabin internal surface measured by means of accelerometers (1-10kHz of frequency range) on 195 points, in the left, right and roof part. A horizontal bar, holding five microphones spaced by 25 cm, was used to measure the acoustical responses of the enclosure. Seven measurement planes were considered, spaced by 50 cm, while the vertical spacing between the lines was 20 cm. The accelerometers were mounted on the inside walls of the cabin, measuring only the surface normal acceleration (Figure 7, right). The measurements on the interior of the cabin lead to collect two set of FRFs: pure-acoustical FRFs $[\mathrm{Pa}/(\mathrm{m}^3/\mathrm{s}^2)]$ for the cavity and acoustical-structural FRFs for the cabin internal surface $[(\mathrm{m/s}^2)/(\mathrm{m}^3/\mathrm{s}^2)]$. In Tab.2 the results for the modes lying in the 20-200 Hz frequency range are listed for each analysis, in terms of natural frequency F_n and modal damping ζ. A few remarks can be done about the modal properties (F_n and ζ) obtained from the analyses. At first, the cabin external surface is evidently acoustically coupled to the enclosure. In fact, as the first and the third columns of Tab.2 clearly indicate, the walls of the helicopter cabin (structural parts) vibrate at the same natural frequencies as the cabin internal cavity (fluid part). A further validation can be obtained by a visual inspection of the maps shown in Figure 8 and Figure 9. This implies important consequences concerning the identification of the entire coupled behaviour.

Secondly, the modal properties obtained from the OMA analysis performed by means of CPs, i.e. the cross power spectra between the external walls response and the internal acoustical reference, clearly confirm that an acoustical reference is enough efficient. However, once the coupling between the cabin external surface and the cavity has been assessed, and the most appropriate excitation method has been chosen, it is important to point out that a structural response could be taken as reference in any case; so, in this scenario a structural point fully coupled to the enclosure can be provided as reference, without any risk to loss coupled modes. Obviously, this condition is not straightforward to be a-priori satisfied, and therefore it is more clear now the convenience related to an acoustical reference, since the coupled modes will be surely identified.

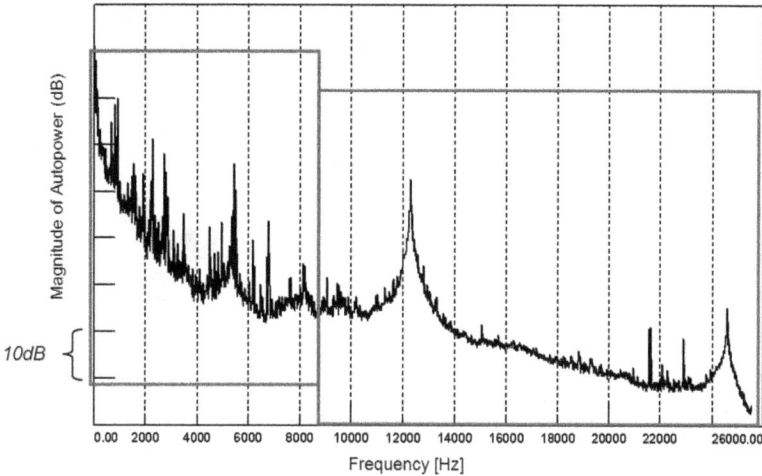

Fig. 10. Autopower of the acoustical pressure measured by the pressure probe of a PU probe in the internal surface of the helicopter cabin (roof) in the steady-state operational condition.

3. Noise control with vibro-acoustical signature analysis

Since the main vibration excitations on the helicopter units are mechanical and aerodynamical (Gelman et al., 2000), the generated noise in operational conditions can be distinguished in aerodynamic-borne and mechanic-borne noise; aerodynamical noise is not considered here, but an exhaustive analysis can be found in (Conlisk, 2001) both from an experimental as well as a numerical viewpoint. Mechanical noise is due to the contribution of the gearbox and to the contribution of the jet engines. In the case of helicopter gearboxes, there is a very wide range of shaft frequencies, and the associated meshing frequencies and bearing frequencies cover the low-medium audio frequency range. Since the power engine of a helicopter is characterized by high rotational speed (about 30000 RPM) the generated noise due to the turbine and compressor blades mainly lies in the high frequency range (10-20kHz), but the harmonics of the turbine and compressor shaft still remain in the low-medium frequency range where the human ear is strongly sensitive. Noise control in helicopter first of all requires a systematic study of its NVH (Noise Vibration and Harshness) behaviour, with special attention to helicopter operational conditions. Furthermore, sound quality is of great importance in achieving sound which is agreeable to the human ear, in fact noise annoyance not only depends on sound exposure levels. For these reasons the authors have performed a wide experimental vibro-acoustic campaign in a helicopter interior cabin. Substantially, the study has two main goals: the first goal is to give a systematic methodology for defining the NVH behaviour of an helicopter in steady-state and run up operational conditions by using efficient and fast experimental techniques; the second goal is to evaluate the importance of the noise contribution produced by the jet engines with respect to the gearbox.

3.1 Case history and experimental setup

The signature analyses presented in this work have been performed employing the data acquired from experimental tests conducted on the helicopter EUROCOPTER EC-135. The data have been acquired in operational conditions in which the helicopter was clamped on

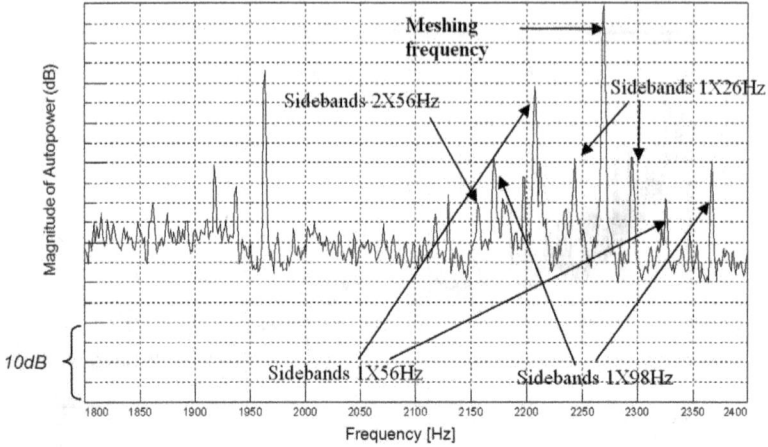

Fig. 11. Autopower of acoustical pressure. Zoom of Figure 10 around a meshing frequency (from 1800Hz to 2400Hz). Meshing frequency and relative sidebands spaced at 26Hz 56Hz, and 98Hz.

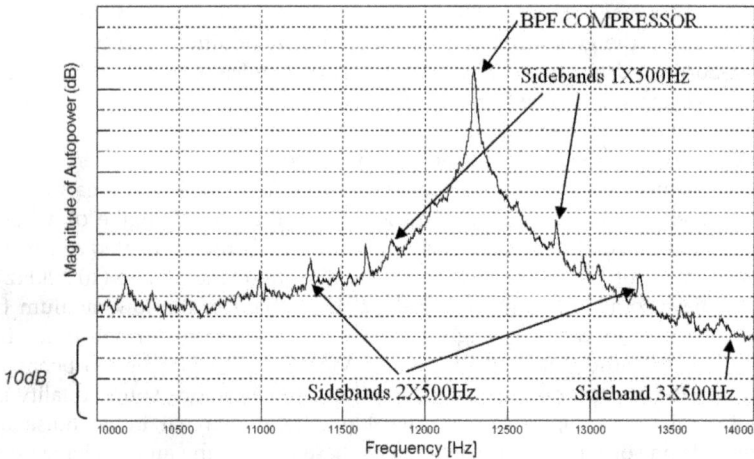

Fig. 12. Autopower of acoustical pressure. Zoom of Figure 10 around the Blade Passing Frequency of the first stage of the compressor. BPF and relative sidebands spaced of 500Hz (corresponding to the frequency of rotation of the compressor shaft).

the ground while the engines were limited to 60% of the maximum power and the two rotors giving power to the main blades and to the tail blades were activated. This operational condition can be considered representative of a possible helicopter operational condition since we are interested in the mechanical-borne noise only and not in the aerodynamical noise. Such tests were performed in steady-state operational conditions (constant main blade speed) and for a run up.

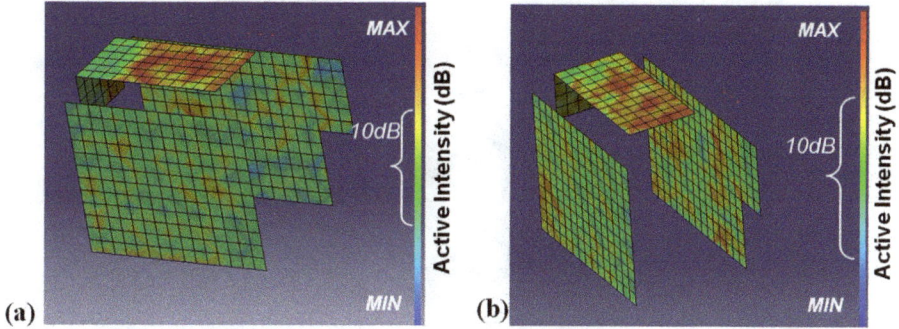

Fig. 13. Active intensity maps on the cabin internal surface at the first (a) and second (b) harmonic of BPF of the first stage of the compressor in dB scale.

The acoustical experimental tests concern measurements in the internal surface (see Figure 7, right) of the helicopter cabin by using the 4×4 PU sensor array, described in Section 2.3. The PU sensor array was positioned close to the internal walls so the very near field assumption is verified (De Bree et al., 2004). The active acoustic intensity is then evaluated as the real part of the crosspower between pressure and particle velocity. The information containing the location of the measured points (see Figure 7, right) and the active acoustic intensity data allows a 3D representation of the intensity map as shown hereafter. Simultaneously with the signals coming from the PU probes, the horizontal bar holding eight condenser microphones (described in Section 2.5) was used to acquire the noise in proximity of the right and left side of the helicopter internal cabin. Some acceleration measurements were taken too in order to obtain the operational deflection shapes (ODS) of some panels of the cabin; in particular, forty piezoelectric accelerometers (frequency range 1 to 10000 Hz) were positioned inside the cabin on the roof (30 accelerometers) and on the door panels (5 accelerometers each side), measuring only along the perpendicular direction of the panels. The ODS analysis has been performed in order to assess the results obtained by the PU probes.

3.2 Signature analysis in steady-state operational conditions
Hereafter the discussion concerning the steady-state operational conditions is presented. Figure 10 shows a typical example of the autopower of the pressure probe of a PU on the roof; two different zones can be easily distinguished: the low frequency range (0 till about 8kHz) is dominated by the rotational frequencies, meshing frequencies and relative harmonics concerning the gearbox (main transmission) whilst in the high frequency range (8kHz-25kHz) the pivotal role is played by the jet engines, and the compressor in particular (due to confidentiality reasons, the curve amplitude is represented without numerical values). In fact, the power transmission of a helicopter is divided into two main parts: one concerning gearing (gearbox) and one concerning the jet engines (composed of compressors, combustion chambers and turbines). Due to the different rotational speed of the turbines and compressor with respect to the shafts in the gearbox, the relative harmonics lie in different frequency ranges. It is a matter of fact that the turbines and compressor rotate at about 30000 RPM while the shaft of the main blades rotates 100 times slower (at about 350 RPM), In particular, for the gearbox, the meshing frequencies related to the input gear, the bevel/spur gears in some intermediate shafts, the collector gear in the pinion shaft and the spur/bevel gear

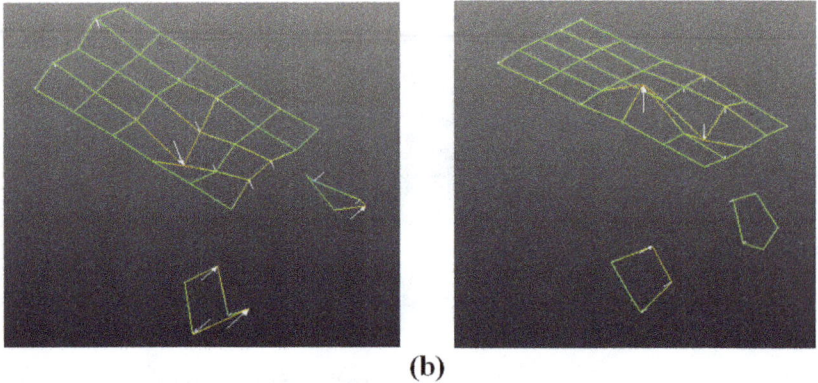

(a)   **(b)**

Fig. 14. Operational deflection shapes at the first (a) and second (b) harmonic of BPF of the first stage of the compressor obtained by acceleration measurements on the internal surface of the helicopter cabin.

in the hydraulic pump shaft lie in the low frequency range (till 3000Hz) and their main harmonics excite the medium frequency range (till about 8kHz). On the other hand, the harmonics of the blade pass frequencies and vane pass frequencies of the compressor and turbine lie in the high frequency range (10kHz-20kHz) due to the high rotational speed of the shafts and the high number of blades. Figure 11 shows a zoom of Figure 10 in the low frequency range; three sideband families spaced at a frequency corresponding to the rotational speed of the three shafts involved in such a meshing appear around each meshing frequency. Similar characteristics exhibit the harmonics related to the blade pass tones of the compressor stages; in particular Figure 12, representing a zoom of Figure 10 around 12kHz, shows the peak related to the first stage of the compressor: it is very common in helicopter jet engines for the first stage of the compressor to have the BPF component of high amplitude. Furthermore, usually it is also higher than the amplitude of the harmonics related to the turbine BPF, as confirmed in (Gelman et al., 2000). Several sidebands appear around the blade pass frequency (BPF) as occurs around the meshing frequencies; the sidebands around a certain main frequency specify the rotational frequency of the relative shaft and in this case such sidebands are spaced at 500Hz corresponding to the compressor shaft rotational speed (30000 RPM). The peak at about 25 kHz depicted in Figure 10, represents the second harmonic of the BPF of the first compressor stage.

Figure 13 shows the intensity map obtained by means of the PU probe measurements on the internal cabin in operational conditions at the frequency corresponding to the first and second harmonic of the BPF of the first stage of the compressor (the dB values are omitted for confidential reasons). The amplitude is clearly high in correspondence to the roof surface (red color) highlighting the high level of excitation coming from the jet engine location. An example of operational deflection shape (ODS) analysis performed by means of the accelerometer measurements on the cabin roof and lateral doors is represented in Figure 14; in particular the figure highlights the same behaviour as the intensity map at the frequency corresponding to the first and second harmonic of the BPF of the compressor. In fact the ODSs show large deflections in the roof panel at such frequencies. So, the intensity map and the ODS analysis can be considered very useful tools for source identification and relative quantification.

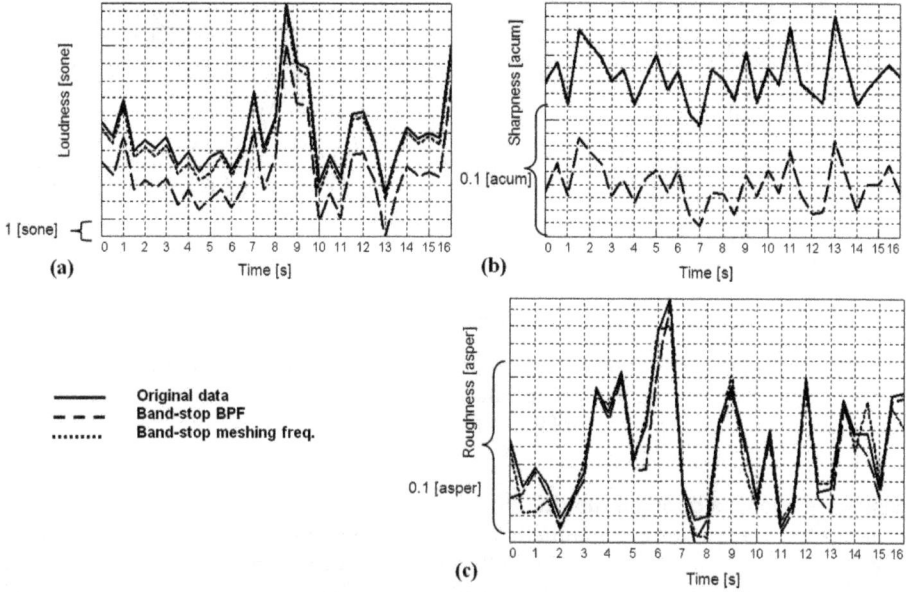

Fig. 15. (a) Loudness Stevens 6, (b) Sharpness and (c) Roughness of the acoustic pressure in the interior of the helicopter cabin (roof) in linear scale: original row data, data filtered with a band-stop filter around the BPF of the compressor of Figure 12 and data filtered with a band-stop filter around the meshing frequency of Figure 11.

Finally, by using the acoustic pressure data coming from the eight microphones, some metrics parameters (Loudness, Roughness and Sharpness) (Zwicker, 1999) are calculated. It is well known that *Loudness* is a perceptual measure of the effect of the energy content of a sound on the human ear, *Sharpness* is a measure of the high frequency content of a sound - it allows classification of sounds as shrill/sharp or dull - while *Roughness* (or harshness) is a quantity associated with amplitude modulations of tones. The curves in Figure 15 show such metrics parameters regarding the row data (solid line) and the band-stop filtered data (dotted lines); in particular the filtered data are obtained neglecting the contribution of the first harmonic of the BPF of the compressor and the first harmonic of a meshing frequency of the gearbox. The curves are obtained filtering the data by a band-stop digital filter around the meshing and blade pass frequency in Figure 11 and Figure 12, respectively. Comparing the filtered data with respect to the row data a notable difference occurs, in particular it can be seen that the Loudness and the Sharpness of the signal without the compressor harmonic is always the lowest, confirming that jet engine noise plays an important role in global perceived noise. The curves related to the Sharpness of the row data and the band-stop filtered data around the meshing frequency are overlapped: this behaviour was expected since such a meshing frequency lies in the low frequency range. The Roughness is obviously the same in the three cases because the degree of modulation is the same: the band-stop filter involves only the main frequency and not the sidebands (i.e. the modulation effect). In such a metric parameter comparison, the first meshing harmonic of the input gear in the gearbox and the BPF of the first compressor stage have been involved because they are representative of important gearbox and jet engine noise phenomena.

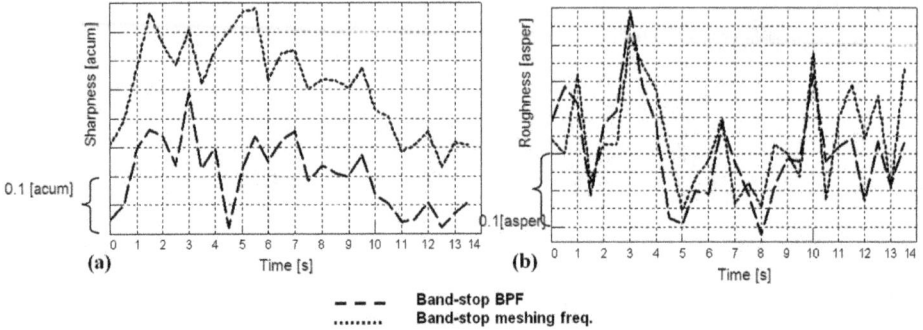

Fig. 16. Sharpness (a) and Roughness (b) of the acoustic pressure in the interior of the helicopter cabin (roof) during the run up (linear scale): data filtered with an order-stop filter around the BPF of the compressor of Figure 13 and data filtered with a order-stop filter around the meshing frequency of Figure 12.

3.3 Signature analysis during a run-up

The run up test was performed gradually increasing the main blade speed, reaching 60% of the maximum power after 14 seconds with the aim at performing a sound metric comparison as performed in the steady-state condition by filtering the contribution of the compressor BPF and the contribution of the meshing frequency of the input shaft in the gearbox, an accurate order tracking technique is hereto essential.

The idea is to apply an order filter to the run up spectrum for deleting the contribution of the meshing frequency of the input shaft and the contribution of the BPF of the compressor as done in the steady-state condition. It has to be underlined that the rotational speed of the input gear in the gearbox and the rotational speed of the compressor are different and not proportional, due to the variable transmission ratio in between. Therefore, two different order tracking analyses have to be performed, one tracking the signal with the RPM curve of the input shaft and one tracking the signal with the RPM curve of the compressor shaft. Since a tacho reference was not available, a tacholess RPM extraction method has been applied (Mucchi & Vecchio, 2010) .Then, a few sound metrics parameters (Roughness and Sharpness) are calculated from the filtered data and depicted in Figure 16; the Sharpness without the contribution of the compressor order shows lower values with respect to the curve without the contribution of the gear meshing, meaning that the compressor is responsible for the shrill sound produced by the helicopter (see Figure 16(a)). Concerning the Roughness comparison, it can be noted that, as in the steady-state condition, the values are similar (Figure 16(b)) because the order filter used does not reduce modulation effects in the signal.

4. Ultrasonic and acoustical intensity measurements for fast leakage spot detection

Leakage points are possible noise sources or locations through which noise can propagate, they are due to non-homogeneity, fractures or cavities in the material, defects during assembly (e.g. improper alignment of the seals along the cabin doors or incorrect coupling between cabin surfaces due to split rivets), etc. The detection of leakage points on the cabin surface is of high interest for NVH designers and can be a useful tool for quality and noise control. In fact, such points can cause high noise levels in the cabin since they can excite the acoustical

Fig. 17. Leakage points (in red) on the helicopter cabin found by means of the ultrasound technique (cabin right panels)

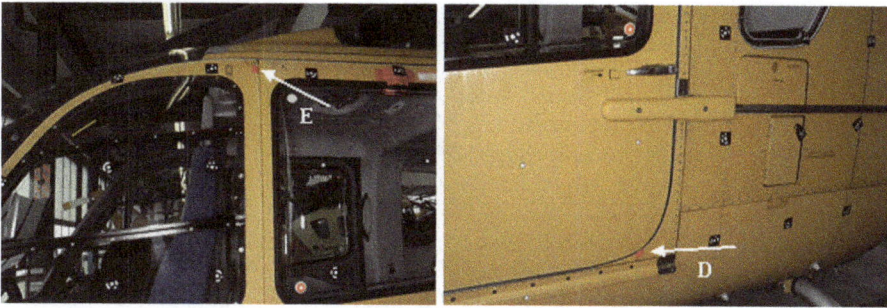

Fig. 18. Leakage points (in red) on the helicopter cabin found by means of the ultrasound technique (cabin left panels).

resonances of the enclosure and determine human discomfort during flight. Commonly, leakage point detection is performed by using an ultrasonic transmitter/receiver; in this section, the authors present how to find leakage points by means of the acoustical intensity analysis as an useful and precise tool for their localization and quantification.

The ultrasonic technique makes use of ultrasound transmitter/receiver and the acoustical intensity analysis is performed by using PU probes. This second technique is performed in operational conditions and in controlled conditions (i.e. where the excitation is measured and applied by using a volume acceleration source). Both techniques have their advantages and disadvantages that will be explained later on.

Ultrasonic sensors (Wallace, 1972) utilize transducers, which transform an electrical signal into an ultrasonic wave and vice versa. Ultrasound covers a frequency range from 20kHz to about 1 GHz, however for technical applications the range 20kHz to 10MHz is the most important one. Ultrasound enables instruments to be non invasive and also non intrusive because the acoustic wave can penetrate walls. Furthermore, high frequency sounds are more directional than lower frequency ones: this makes it easier to pinpoint the source even in the presence of other background noise. Obviously, the propagation of acoustic waves through multi-layered structures depends on the acoustics impedance mismatch at each of the boundaries the ultrasound has to pass, furthermore it depends on the attenuation of the ultrasound in the different materials and finally in some cases on the relationship between the wall thickness and the wavelength of the ultrasound waves. In particular, firstly an ultrasonic transmitter (SDT 8MS) producing an omni-directional tonal noise at the frequency of about

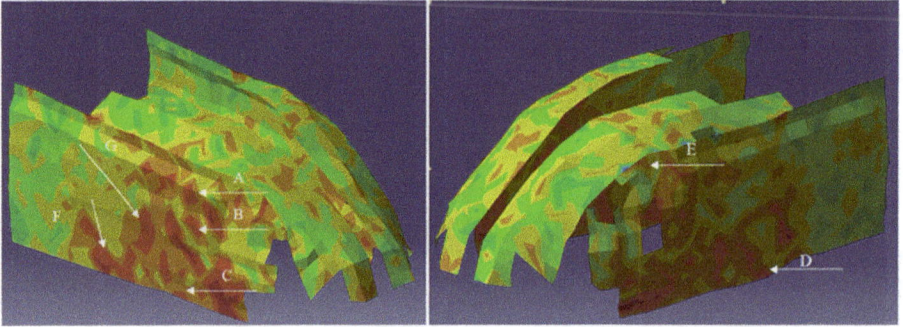

Fig. 19. Intensity map in dB of the cabin external surface at the meshing frequency of the input gear in the gearbox.

40kHz has been used inside the helicopter cabin while the ultrasonic receiver (SDT 170) was on the external surface of the cabin scanning the entire external cabin surface. Thus, the location of the leakage points can be evaluated by this first simple measurement, in fact when the receiver detects a point through which the level of ultrasonic noise exceed a threshold level, an audio sound signal occurs highlighting a leakage point. This measurement deals with the transmissibility of the tested surface to the ultrasonic waves. Obviously, the surface transmissibility is increased in correspondence of fracture or in-homogeneity on the material, malfunctioning of the seals, etc. It is important to underline that in helicopter operational conditions, the noise comes from the exterior to the interior of the helicopter cabin, on the contrary during this first measurement, as well as on the further measurements described hereafter, the noise comes from the interior (by means of the ultrasonic transmitter) and the detection is performed on the exterior. This can be considered acceptable under the assumption that the helicopter cabin satisfying the reciprocity principle, as demonstrated in (Pintelon & Schoukens, 2001). Furthermore, in order to localize the leakage points with more spatial precision, a second ultrasonic measurement is carried out exciting on the internal cabin side in correspondence of the above-detected leakage points and measuring the value of the wave crossing the cabin panel (from the interior to the exterior) by means of the same ultrasonic receiver. In this second test, an ultrasonic transmitter (SDT 200mW) producing a one-directional tonal noise at the frequency of about 40kHz has been used as excitation inside the cabin giving more accuracy in the measurement.

Figure 17 and Figure 18 show the leakage points (red dots in the figures) found by means of the above-described ultrasound technique. These points represent locations where the ultrasound wave can propagate from the interior to the exterior, in particular they are located on the boundary of the doors where the seal between the door and the cabin is not so efficient from an insulation viewpoint. The knowledge of the location of these defects is of primary importance for the designer that has to find solutions in order to improve the cabin insulation. In fact, the test is carried out exciting the interior of the cabin ad measuring in the exterior, but as stated before, due to reciprocity, the same results can be obtained exciting the exterior and measuring the ultrasound wave on the interior of the cabin. This means that such a test gives information about the locations from which the acoustic radiation can propagate on the cabin impoverishing the acoustic comfort of the passengers and pilots.

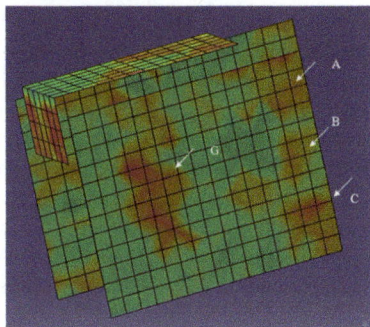

Fig. 20. Intensity map in dB of the cabin internal surface at the meshing frequency of the input gear in the gearbox

Similar results can be obtained from the intensity maps of the external surface (Figure 19, the dB values are omitted for confidential reasons). As described in Section 1.3 such a test is carried out exciting the cabin from the interior side with a flat broadband noise (white noise), then the active acoustic intensity is evaluated as the real part of the crosspower between pressure and particle velocity, being the measurements carried out by PU probes. In Section 2.4 the same measurements were used for EMA purpose obtaining the mode shapes at the natural frequencies, here the sound intensity maps are evaluate at every frequency in the measured frequency range. Such maps depicted at the natural frequency of the cabin are obviously similar to the mode shapes of the cabin external surface, due to the contribution of the particle velocity acquiring in the very near field and being representative of the structural vibration of the panels. On the other hand, the maps depicted at frequency far from the natural frequencies and for example corresponding to the tonal excitations of the gearbox (e.g. meshing frequency of the collector gear in the pinion shaft, or meshing frequency related to the input gear) highlight high intensity values in the same locations as the pictures of Figure 17 and Figure 18 shown (compare same characters on the figures). In fact the active intensity represents the energy flow that a surface radiates and therefore is very sensible to leakage point localization.

Figure 20 shows an example of intensity map on the internal surface obtained in operational conditions and depicted at the frequency corresponding to the meshing frequency of the input shaft in the gearbox. The map is able to identify the same leakage points found by the intensity map on the external surface and by the ultrasound technique (consider the corresponding characters in Figures 17, 18, 19). It is interesting to note that both the ultrasound techniques and the intensity measurements are effective tools for the identification of leakage points. The first technique is simpler to apply, the cost of the instrumentation needed is cheaper than the second techniques, but the intensity maps gives more information than ultrasonic techniques, in fact since they represent the active sound intensity they are effective for panel efficiency evaluation too (Wallace, 1972). Moreover the intensity maps can be calculated on the entire measured frequency range, allowing calculation of the energy flow dissipated by the leakage points as a function of frequency.

5. Vibro-acoustic transfer path analysis

Noise levels recorded in helicopters' cabin are severely affected by the strength and vicinity of noise sources. The jet engines, the gearbox and the rotors can be considered as separated sources - whose spectral content is strongly tonal and RPM dependant - exciting simultaneously the cabin's acoustic cavity. Under the hypothesis of linear behaviour, the total sound pressure level measurable in the cabin can be considered as the summation of a number of partial pressure contributions, each generated by one source acting separately. The mechanism responsible for transferring the mechanical energy from each sources to the target location can be either structure borne - via the mechanical joints connecting the gearbox to the helicopter's frame - or airborne - via the sound propagation in the air. Some transfer paths may cause interference at certain frequencies such that the observer does not notice anything significant- until he moves position. Transfer Path Analysis (TPA) represents a techniques used to assess the structure and air-borne energy paths among excitation source(s) and receiver location(s) in a process of noise control. TPA has been largely applied in the automotive industry that allows identifying the main transmission paths and their relative contribution to the total sound pressure level at target location. From a theoretical standpoint there is not reason why TPA should be limited to cars. An helicopter is a more complex system then a car, but this actually implies that there may be more noise sources, hence more transfer paths. An experimental TPA approach is hereby applied for the first time to the helicopter Agusta A109 (Vecchio et al., 2006), to assess noise source contribution to the cabin noise and to simulate a number of realistic noise reduction scenarios. Usually vibro-acoustic TPA involves measurements of accelerations in operational conditions and of vibration and/or vibro-acoustic transfer functions (TFs). The transfer functions can be measured using the most practical approaches – either using hammer or shaker or volume acceleration source excitation techniques. Briefly, the vibro-acoustic TPA is an analysis fully described by means of two sets of equations 14-15:

$$
\left\{ \begin{array}{c} F_1 \\ \vdots \\ \vdots \\ F_n \end{array} \right\} = \left[\begin{array}{cccc} \frac{\ddot{X}_1}{F_1} & \frac{\ddot{X}_1}{F_2} & \cdots & \frac{\ddot{X}_1}{F_n} \\ \vdots & & & \\ \frac{\ddot{X}_m}{F_1} & & & \end{array} \right]^{-1} \left\{ \begin{array}{c} \ddot{X}_1 \\ \vdots \\ \vdots \\ \ddot{X}_m \end{array} \right\} \tag{14}
$$

$$
\left\{ \begin{array}{c} p_1 \\ \vdots \\ p_q \end{array} \right\} = \left[\begin{array}{cccc} \frac{p_1}{F_1} & \frac{p_1}{F_2} & \cdots & \frac{p_1}{F_n} \\ \vdots & & & \\ \frac{p_q}{F_1} & & & \end{array} \right]^{-1} \left\{ \begin{array}{c} F_1 \\ \vdots \\ F_n \end{array} \right\} \tag{15}
$$

where $\{\ddot{X}_1 ... \ddot{X}_m\}^t$, $\{F_1 ... F_n\}^t$, and $\{p_1 ... p_q\}^t$ are the acceleration, the operating force and acoustic response vectors, respectively. Equation 14 brings into play the relation between the operating forces transmitted along the paths and the structural accelerations caused by these forces, while the second set of equations relates the acoustic responses, e.g. the noise inside a cabin, and the operating forces. Hence, it is pretty clear that by exploiting the information that the first set of equations carries it is possible to compute the acoustic responses from the second one. Indeed, it is much easier to measure the accelerations of a structure rather than the forces; these accelerations can then be employed in order to compute the operational forces which substituted in the second set of equations will lead to the final result. On the left hand side of equation 14, there are the operating forces. As it can be seen, it is necessary to invert the matrix linking the accelerations and the forces. This constitutes the biggest computational effort of the TPA.

In the helicopter under study, the vibro-acoustic TPA is performed considering the gearbox as the source and the cabin cavity as the receiver location. The gearbox is connected to the cabin roof by means of two front struts, two rear struts and the anti-torque plate through four bolts leading to have eight structural paths. The transfer function matrix (equation 14) has been obtained by exciting the structure with an impact hammer and measuring the acceleration responses. The acceleration vector of equation 14 (i.e. $\{\ddot{X}_1 ... \ddot{X}_m\}^t$) has been calculated in flight operational conditions, so the operational forces can be calculated. Concerning equation 15, the vibro-acoustic transfer functions have been obtained taking advantage of the vibro-acoustic reciprocity (i.e. $p_i / F_j = -\ddot{X}_j / \dot{q}_i$, where $\dot{q}_i$ is the volume velocity at location j). Therefore the TFs have been calculated exciting by means of two volume acceleration sources working at different frequencies, 0 - 400 Hz and 400 - 4000 Hz, and measuring the responses using the accelerometers located in the eight structural paths. The sources were placed in the helicopter cabin, close to the pilot seat. Experimental data are used to implement a numerical TPA model. The model points out the most critical transfer paths and paves the way to a number of simulations that allows predicting the noise reduction achievable in the helicopter cabin for a given reduction of the source strength and as result of structural modifications. The analysis focused on the connecting points between the gearbox and the helicopters frame.

The spectral analysis of the computed operational loads combined with a spectral analysis of the sound pressure levels measured in the helicopter mock-up cabin during simulated operating conditions leads to the identification of a number of critical frequency tones that show a very efficient noise transmission mechanism in the helicopter cabin. Out of a list of critical frequencies, three main tones are hereby identified that are responsible for generating the largest contribution to the cabin noise spectrum. These tones correspond to the rotor shaft rotation and two of the gearbox meshing frequencies. The noise path analysis will then focus on those frequencies only. The analysis of the operational load spectra shows that the transmission path corresponding to the rear left and rear right struts and anti-torque plate exhibits the highest levels and provides the major contribution to the cabin noise at the identified critical frequencies. This means that any action aiming at reducing cabin noise recorded at those frequencies should focus on the anti-torque plate.

Once the main noise transmission paths contributing to the cabin noise are identified for the most critical noise frequency, the TPA model can be used to simulate the noise reduction that can be achieved if structural modifications would be implemented on the helicopter structure or on the gearbox in a process of noise control.

Two simulation scenarios are thus presented. One consists of introducing a modification in the TPA models; this can be easily done by editing the FRF of a selected transmission path and, e.g., zeroing the FRF amplitude in correspondence of the selected frequency. This is equivalent to simulating an active noise control system that induces a modification in the FRF of a specific transmission path. In order to select the best target frequency to be treated by the noise reduction simulation, a simple spectral analysis is carried out that shows the effect of zeroing the cumulatively the previously identified critical tones and compute the noise reduction that results from that action. In Tab. 3 it is shown that suppressing the two main gearbox meshing frequency tones for the most relevant transmission paths results into 5 dB(A) Sound Pressure Level (SPL) reduction as the SPL decreases from 111 dB(A) to 106 dB(A); yet suppressing the 5 highest tones results into 1 dB(A) more noise reduction (Tab. 3). Therefore, noise comfort improvement can be achieved with active control systems acting on the anti-torque plate hosting the gearbox. A second scenario consists of simulating a source modification that results into a frequency shift of a selected critical frequency. The considered source is the gearbox. The target frequency is again the meshing tone. This corresponds to simulate a design modification (gear diameter, number of teeth of a number of gears, etc) in

Original configuration	Suppression of the 2 main freq. tones	Suppression of the 5 main freq. tones	Shift of a meshing frequency in the gearbox
111 dB(A)	106 dB(A)	105 dB(A)	106 dB(A)

Table 3. Noise reduction cumulative effect after modifications of the TPA model.

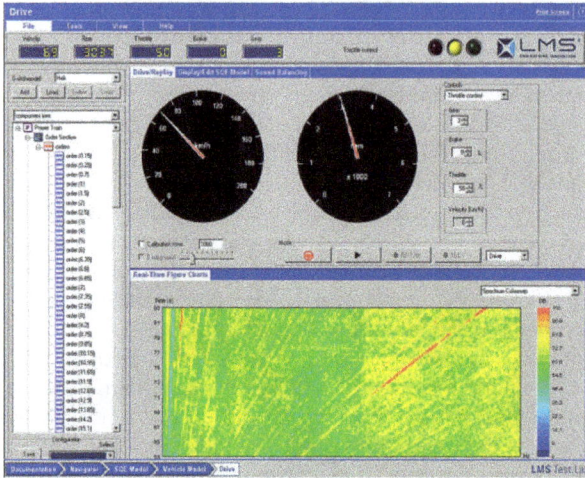

Fig. 21. Virtual Helicopter Sound Environment.

the gearbox resulting into a shift of the meshing frequency. The TPA model is then run again to compute a new set of partial pressure contributions to the interior cabin noise. 5 dB(A) reduction in the SPL can be observed.

6. Target sound design: The virtual helicopter sound environment

A sound synthesis approach was developed for helicopter noise (Vecchio et al., 2007). The noise synthesis is based on a compact and sound-quality-accurate model which is identified from measured noise data. A sound synthesis model consists of a number of tonal components and third octave bands that describe the broadband noise. The ground-reflection coefficient and the time-delay between the direct and reflected sound are also taken into account to characterize the typical interference pattern peculiar of flyover noise. For the few aircraft sounds studied so far, impressive synthesis results were achieved. Almost no differences could be heard between the synthesized and measured sounds. Acoustic engineers can easily modify model components and assess the impact on the human perception. This way they can design target sounds with improved sound quality and suggest guidelines for future design improvements. The sound synthesis helps better understand sound quality differences among various types of helicopters manoeuvres and forms an excellent basis to design target sounds with improved noise signature. A software environment (the Virtual Helicopter Sound Environment, see Figure 21) was developed where it is possible to virtually drive an helicopter and in the meantime, study the impact different modifications have on the noise perception and the sound quality. The tool was first designed for interior car sounds, but some modifications were implemented to cope with aircraft and helicopter noise.

Some of the tool's features are very relevant for the case under study: once the fundamental noise components have been suitably modelled, the sound can be synthesized and replayed according to any RPM evolution. Sound synthesis can then be recorded and stored on a data file. The advantage of the Virtual Helicopter Sound environment is here evident: one can easily modify model components and evaluate the impact on the human perception. A large variety of so-called "what if" scenarios can be played. For example, what happens if the dominant tonal components are reduced with 3 dB? Or what is the sound quality impact if the levels of some low or high frequency third octave noise bands are changed? Or what happens with our sensation of sharpness when the rapid amplitude variations of the high-frequency tonal components are smoothed? By playing some of these "what if" scenarios, one can design target sounds with improved sound quality and suggest engineering guidelines for noise control.

7. Concluding remarks

The work contributes to define a systematic experimental procedure aimed at identifying the noise sources, the mechanics of transfer from source to target (pilot and/or passenger area) and defining an effective noise control strategy in helicopters. This activity has led a number of conclusions:

1   Since the vibro-acoustical coupling between the cabin enclosure and the internal and external surfaces, is one of the cabin interior noise responsible, its assessment is necessary in a process of noise control;

2   Sound quality parameters and acoustics intensity analysis are effective tools for defining the relative importance of the noise produced by the jet engines with respect to the gearbox and for a fast and efficient localization of noise and vibration sources and leakage points in the helicopter cabin;

3   The TPA allows pointing out a subset of critical frequency tones that are transmitted into the helicopter cabin and a sub-set of transmission paths contributing the most to the cabin noise. The results show that noise comfort improvement can be achieved with active control systems acting on the anti-torque plate hosting the gearbox and with modifications of the meshing frequency in the gearbox;

4   The sound synthesis approach can be considered as an innovative technology, allowing engineers and designers to listen to their models and judge the sound quality impact of structural design modifications. This way they can design target sounds with improved sound quality and suggest guidelines for noise control.

8. References

M. Abdelghani, L. Hermans, H. Van der Auweraer (1999). A state space approach to output-only vibro-acoustical modal analysis, in: Proceedings of the 17th International Modal Analysis Conference, Vol. 3727, Kissimmee, FL, pp.1789-1793

A.T. Conlisk (2001). Modern helicopter aerodynamics, Progress in aerospace sciences, Vol. 37, pp. 419-476

H.E. De Bree, P. Leussink, T. Korthorst, H. Jansen, T. Lammerink, M. Elwenspoek (1996). The Microflown: a novel device measuring acoustical flows, Sensors and Actuators SNA054/1-3, pp. 552-557

H-E. de Bree, V.B. Svetovoy, R. Raangs, R. Visser (2004). "The very near field: Theory, simulations and measurements of sound pressure and particle velocity in the very near field", in: Proceedings of ICSV11, Eleventh International Congress on Sound and Vibration, St. Petersburg, Russia

W. Desmet, B. Pluymers, P. Sas (2003). Vibro-acoustic analysis procedures for the evaluation of the sound insulation characteristics of agricultural machinery cabins, Journal of Sound and Vibration 266, 407-441

G. C. Everstine (1997). Finite Element Formulations of structural acoustics problems, Computers and Structures 65, 307-321

G. C. Everstine (1981). Structural acoustic analogies for scalar field problems, International Journal of Numerical Methods in Engineering 17, 471-476

L.M. Gelman, D.A. Kripak, V.V. Fedorov, L.N. Udovenko (2000). Condition monitoring diagnosis methods of helicopter units, Mechanical system and signal processing, Vol. 14,No. 4, pp. 613-624

L. Hermans and H. Van der Auweraer (1999). Modal Testing and Analysis of Structures under Operational Conditions: Industrial Applications, Mechanical Systems and Signal Processing 13, 193-216

M. Kronast, M. Hildebrandt (2000). Vibro-acoustic modal analysis of automobile body cavity noise, Sound and Vibration 34. 20-23

Z.-D. Ma and I. Hagiwara (1991). Sensitivity analysis methods for coupled acoustical-structural systems. Part I: Modal sensitivities, AIAA J. 2, 1787–1795

E. Mucchi, A. Vecchio (2009). Experimental transfer path analysis on helicopters, Proceedings of the International Conference on Acoustics (NAG/DAGA2009), Rotterdam, Netherlands, 23-26 March, pp. 788-791

E. Mucchi, A. Vecchio (2010). Acoustical signature analysis of a helicopter cabin in steady-state and run up operational conditions, Measurement 43, 283-293

B. Peeters, H. Van der Auweraer, F. Vanhollebeke, P. Guillaume (2007). Operational modal analysis for estimating the dynamic properties of a stadium structure during a football game, Shock and Vibration – Special Issue: Assembly Structures under Crowd Dynamic Excitation 14, 283-303

E. Pierro, E. Mucchi, L. Soria, A. Vecchio (2009). On the vibro-acoustical operational modal analysis of a helicopter cabin - Mechanical Systems and Signal Processing, Volume 23, Issue 4, Pages 1205-1217

Pintelon, R., Schoukens, J. (2001). System Identification: a Frequency Domain Approach, New York: IEEE Press

A. Vecchio, C. Urbanet, F. Cenedese (2006). Experimental Noise Transfer Path Analysis on Helicopters, Proceedings of the International Conference on Noise and Vibration Engineering, 18-20 September, Leuven. Belgium

A. Vecchio, K. Janssens, C. Schram, J. Fromell, F. Cenedese (2007) Synthesis of Helicopters Flyover Noise for Sound Quality Analysis, Proceedings of the 13th AIAA/CEAS Aeroacoustics Conference, 21-23 May, Rome, Italy

Wallace, C.E. (1972). Radiation resistance of a rectangular panel, The Journal of the Acoustic Sociecy of America, Vol.51, N. 3, Part2

K. Wyckaert, F. Augusztinovicz, P. Sas (1996). Vibro-acoustical modal analysis: Reciprocity, model symmetry and model validity, Journal of Acoustical Society of America 100, 3172-3181

E. Zwicker, H. Fastl (1999). Psycho-acoustics, Facts and Models, Springer

7

Vibration Source Contribution Evaluation of a Thin Shell Structure Based on ICA

Wei Cheng[1,2], Zhousuo Zhang[1,*] and Zhengjia He[1]

[1]State Key Laboratory for Manufacturing Systems Engineering, Xi'an,
[2]Department of Mechanical Engineering, University of Michigan, Ann Arbor,
[1]PR China
[2]USA

1. Introduction

As submarines navigate underwater, it is difficult for satellites, anti-submarine aircrafts and warships to detect them. However, the radiated noises produced by the equipment of submarines pose a serious threat to the concealment, and directly influence the operational performance even survivability. Therefore, the reduction and control of vibration and noise is an important work to improve the survival capability and operational performance. As the influences of structural transmission, some components of vibration sources will be changed as they go through mechanical structures, which means the measured vibration signals on the shell are the mixed signals of all the sources. Therefore, it is a challenge but important work to effectively identify the sources and evaluate the source contributions.

The radiated noise of submarines is mainly produced by the diesel engines, and it transmits from the engine bases to the shell according to the hull. One basic method to reduce the radiated noise is based on improving the hull structure to reduce vibration transmission. In the past decades, much research work is devoted to vibration transmission characteristic analysis of different structures, such as beams (Lee et al., 2007), girders (Senjanovic et al., 2009), rafts (Niu et al., 2005), casings (Otrin et al., 2005), panels (Lee et al., 2009), plates (Xie et al., 2007; Bonfiglio et al., 2007) and shells (Efimtsov & Lazarev, 2009). Some studies dedicated to the responses of whole ship hull, such as free vibration analysis of thin shell (Lee, 2006), insertion loss prediction of floating floors (Cha & Chun, 2008) and structural responses of ship hull (Iijima et al., 2008). Another method based on the active control over vibration and noise is also deeply studied in recent years, such as controlling high frequencies of vibration signals by structure modification (Tian et al., 2009), active vibration control using delayed position feedback method (Jnifene, 2007), high frequency spatial vibration control for complex structures (Barrault et al., 2008), and active vibration isolation of floating raft system (Niu, et al., 2005). However, all these techniques are static analysis method, and the radiated noise can be reduced limitedly as the strength requirements of hulls and indispensability of diesel engines.

Aiming at the active control over vibration and noise, a novel approach based on independent component analysis (ICA) is proposed in this paper, which identifies the vibration sources

---

*Corresponding author

from the mixed signals, and quantitatively evaluates the source contributions. Firstly the vibration signals at the different positions are measured, and the radiated noise is evaluated according to the signal energy. Secondly the signals which radiate the noise significantly are selected as the mixed signals, and the source signals contained in the mixed signals are extracted by an improved ICA method. Thirdly, the vibration signals on the engine bases are measured as the source signals, and the vibration sources are identified according to correlation analysis. Lastly, the contributions of each source are quantitatively calculated according to the mixing mode of the independent components. Therefore, the vibration sources which have big contributions can be online identified and controlled if necessary, which provides a novel approach for active control over the vibration and noise.

Compared with other signal processing method, the independent component analysis can reveal the basic sources contained in the mixed signals. ICA is firstly proposed (Jutten & Herault, 1991) and applied in blind source separation (Comon, 1994). A well known ICA algorithm called fixed-point algorithm based on fourth order cumulant is proposed (Hyvarinen & Oja, 1997), and latter fast fixed-point algorithm based on negentropy is proposed (Hyvarinen, 1999), and then is further improved (Hyvarinen, et al., 2001). Currently, ICA is widely used in image feature extraction and recognition (Hu, 2008; Correa et al., 2007), biological signal analysis and feature extraction (Ye et al., 2008; Xie et al., 2008), fault feature extraction (Zuo et al., 2005), astronomical data analysis (Moussaoui et al., 2008), and data compression (Kwak et al., 2008). However, as a statistical signal processing method, a major problem of ICA is that the reliability of the estimated independent components is not known, and the separated components may be different in the repeatedly calculations. Therefore, the independent components are difficult to be explained. A good way to solve this problem is introduced according to clustering evaluation, and the stability of the separated components can be significantly enhanced.

This paper is organized as follows. In section 1, the motivation and research status are introduced. In section 2, the basic theory of ICA is introduced. In section 3, the quantitative calculation of source contributions method based on the enhanced ICA algorithm is proposed. In section 4, the stability of the enhanced ICA algorithm is validated by a comparative numeric study. In section 5, the proposed method is applied to quantitatively calculate the source contributions of a thin shell structure. In section 6, we give the conclusions and discussions.

2. Independent component analysis

2.1 Basic theory of BSS and ICA

Assume that n sources $S = [s_1, s_2, ..., s_n]^T$ exist at the same time, and m mixed signals $X = [x_1, x_2, ..., x_m]^T$ which are composed by these sources are obtained in different places. And thus each mixed signals can be described as:

$$x_i = \sum_{i=1}^{n} a_{ij}s_j + n_i \; i = 1, 2, ..., m, \; j = 1, ..., n \tag{1}$$

Where x_i is the ith mixed signal observed in the position i, s_j is the jth source signal, a_{ij} is the mixing coefficient, and n_i is the noise of ith mixed signal.
The mixed signal can be also described as:

$$X = AS + N \tag{2}$$

Where A is the mixing matrix, and N is the noise matrix.

Blind source separation (BSS) can be described as follows: in the condition that the mixing matrix and source signals are unknown, BSS obtains the estimates of sources $Y = [y_1, y_2, ..., y_n]^T$ according to separating matrix W and mixed signals. That is

$$Y = WX = WAS = GS \tag{3}$$

Where G is a global matrix.

2.2 Assumptions of independent component analysis

As a result of unknown source signals and mixing mode, information that can be used by ICA is only the mixed signals observed by the sensors, which will give multiple solutions of the problem. Therefore, some assumptions are necessary to get a definite solution:

1. Each source signal $s_j (j = 1, ..., n)$ is zero mean and real random variable, and all the source signals are mutually statistic independent at any time.

$$p_s(s) = \prod_{j=1}^{n} p_j(s_j) \tag{4}$$

Where $p_s(s)$ is the probability density of source signals S, and $p_j(s_j)$ is the probability density of source signal s_j.

2. The number of sources are less than the number of mixed signals($n \leq m$).
3. Only one source is allowed to have a guassian distribution.

2.3 Separating criterion based on negentropy

Assume that one random variable is composed of some independent variables according to superposition. According to the central limit theorem, the superposed variable will tend to a Gaussian distribution if the independent variables have the same limited means and variances. Therefore, Gaussian feature are always used to determine whether the separating components are independent or not.

Assume that a source signal s has a probability density $p(s)$, and its negentropy is defined as follow equation:

$$Ng(s) = H(s_{Gauss}) - H(s) \tag{5}$$

Where s_{gauss} is a signal of Gaussian distribution and has the same variance with s. $H(\bullet)$ is the information entropy of a signal.

$$H(s) = -\int p(s) \lg p(s) ds \tag{6}$$

As the probability density distribution function is unknown, the independence of each separated signal is commonly measured by an approximate equation

$$Ng(s) \propto [E\{G(s)\} - E\{G(s_{gauss})\}]^2 \tag{7}$$

Where $E(\bullet)$ is a mean function, and $G(\bullet)$ is a nonlinear function. The commonly used $G(\bullet)$ are logarithmic function and exponential function

$$G(u) = \frac{1}{a} \lg \cosh(au) \tag{8}$$

$$G(u) = -\exp(-u^2 / 2) \tag{9}$$

Where $1 \le a \le 2$.

The independent component analysis method based on negentropy has obvious advantages, such as simple concept, fast computing speed and good stability. It can be also used to determine whether the separating process should be stopped or not according to non-gaussian feature of the separated components.

2.4 Framework of fast fixed-point algorithm

Fast fixed-point algorithm based on the negentropy criterion is a typical independent component analysis algorithm. Its calculation process includes two steps: signal preprocessing and extracting components one by one. The projection pursuit method is applied to extracting the independent components, and the framework of fast fixed-point algorithm can be described as eight steps:

1. Mean and whiten the mixed signals X, remove the redundant information, and thus obtain the preprocessed signals Z.
2. Set the number p of independent component esxtracted for each time.
3. Set the initial iteration value $u_p(0)$, and let $\left\| u_p(0) \right\|_2 = 1$.
4. Iterative calculation

$$u_p(k+1) = E[zg(u_p^T(k)z)] - E[g'(u_p^T(k)z)]u_p(k) \tag{10}$$

where g is the differential of G

5. Orthogonal calculation

$$u_p(k+1) = u_p(k+1) - \sum_{j=1}^{p-1} <u_p(k+1), u_j > u_j \tag{11}$$

6. Normalizing calculation

$$u_p(k+1) \leftarrow \frac{u_p(k+1)}{\left\| u_p(k+1) \right\|_2} \tag{12}$$

7. Determine whether u_p is convergent. If not, return to step (4).
8. Repeat step (3), and else stop calculation.

3. Quantitatively calculate source contributions

To quantitatively calculate source contributions, a novel method based on an enhanced ICA algorithm and priori information is proposed in this paper. The enhanced ICA algorithm extracts independent components by running a single ICA algorithm for many times, and selects the optimal components as the optimal independent components according to clustering analysis (Hamberg & Hyvarinen, 2003; Hyvarinen et al., 2004). The proposed method separates mixed signals into independent components by the enhanced ICA algorithm, and calculates source contributions according to the mixing matrix. Priori information was employed to further enhance the separating performance, because priori information is proved to be able to weaken the uncertainty of problems (Ma, et al., 2006).

3.1 Framework of fast fixed-point algorithm

For linear superposition model, fast fixed-point algorithm has good separating performance. However, most algorithms based on ICA are stochastic, and their results may be different in repeatedly calculations, so the outputs of a single run can not be trusted   (Hamberg & Hyvarinen, 2003; Hyvarinen et al., 2004). One reason for this problem is that engineering data is not strictly complied with the blind source separation model, and the actual calculation algorithm may converge to local minima rather than the overall minimum value. The other reason is the statistical error of estimated components which is caused by the finite sample size.

To enhance the stability of ICA, an enhanced ICA algorithm is constructed based on clustering evaluation. At first the independent components (ICs) are extracted according to a single ICA algorithm for many times, and then the reliability of each IC is estimated by clustering analysis and the optimal Ics are selected as the best solutions. Basic framework of the enhanced ICA algorithm is as follows:

1. Parameters of a single algorithm are set, such as orthogonalization approach and the nonlinearity function.
2. The single ICA algorithm is executed for certain times with different initial values to produce more results in the different condition.
3. All the ICs are clustered according to their mutual similarities, and then these ICs are divided into several clusters. According to the average-linkage clustering criterion, the clustering center which has the largest relevances with other components is selected as the optimal results.

Assume that the mutual correlation coefficient between estimated components y_i and y_j is ρ_{ij}, and the cluster validity index I_q can be defined as follows:

$$I_q(k) = \frac{1}{|C_k|^2} \sum_{i,j \in C_k} \rho_{ij} - \frac{1}{|C_k||C_{-k}|} \sum_{i \in C_k} \sum_{j \in C_{-k}} \rho_{ij} \tag{13}$$

where C is all the IC set, C_k is the IC set in the kth cluster, C_{-k} is its complementary set, and $|C_k|$ is the size of the kth cluster.

By the clustering evaluation method, the ICs produced by the enhanced ICA algorithm are clustered into different clusters, and the separating performance can be indicated by the tightness of ICs. The more tightness they are, the better the separating performance will be. After clustering, the cluster center is selected as the optimal ICs.

3.2 Priori information

In the applications, some mixed signals do not follow the linear superposition model. Therefore, priori information should be used to weaken the uncertain problems. For sources $S = [s_1, s_2]^T$, it can be inferred that the joint distribution $p_{s_1,s_2}(s_1, s_2)$ of the two source signals must be included in a rectangle as

$$p_{s_1,s_2}(s_1,s_2) = p_{s_1}(s_1)p_{s_2}(s_2) \tag{14}$$

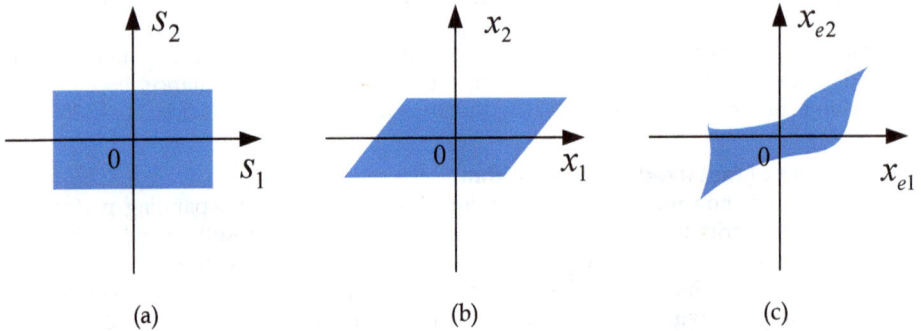

(a)   (b)   (c)

Fig. 1. Joint distributions of different signals

The joint distribution of S is shown in Fig. 1(a). Mixed signals are obtained according to S and mixing matrix A, which can be shown as

$$X = AS = [x_1,x_2]^T \tag{15}$$

The joint distribution of X is shown in Fig. 1(b). Nonlinear mixed signals $X_e = [x_{e1},x_{e2}]^T$ are obtained by nonlinear mixing mode, and their joint distribution is shown in Fig. 1(c).
Theorem 1: for transformation

$$\begin{cases} p_1 = h_1(e_1) \\ p_2 = h_2(e_2) \end{cases} \tag{16}$$

Where h_1 and h_2 is an analytic function.
If the parallelogram border on the plane (e_1,e_2) is transformed into the plane (p_1,p_2), and these parallelogram borders are not parallel to the related coordinate axis, there have real constants a_1 、 a_2 、 b_1 and b_2

$$\begin{cases} h_1(u) = a_1u + b_1 \\ h_2(u) = a_2u + b_2 \end{cases} \tag{17}$$

The theorem 1 provides a transmission from nonlinear mixed signals to linear mixed signals based on priori information, and this method does not need the assumption of independence. The limited border of source signals provides additional information, which can optimize the nonlinear parts of the mixed signals. Therefore, the separating performance of the proposed method is further enhanced.

3.3 Quantitative evaluation of source contributions
From the definition of whitening, the covariance matrix R_{xx} can be described as

$$R_{xx} = E[xx^T] = E[Qxx^TQ^T] = QR_{xx}Q^T = I_n \tag{18}$$

Where Q is a whitening matrix, and I_n is a unit covariance matrix. From the equation (18), it can be seen that the mixed signals $X = [x_1, x_2, ..., x_n]^T$ are transformed into unrelated signals $Z = [z_1, z_2, ..., z_n]^T$ by whitening.

According to the definition of orthogonal transformation, there exists follow equation

$$UU^T = U^T U = I_n \tag{19}$$

$$Y = UZ \tag{20}$$

$Y = [y_1, y_2, ..., y_n]^T$ are independent components with unit variances. From the mathematical model of ICA, mixed signals X can be obtained by Y.

$$X = AS = AG^{-1}Y = \hat{A}Y \tag{21}$$

Each components of X are composed by all the components of Y according to the mixing matrix $\hat{A}$. Each combination coefficients of the mixing matrix $\hat{A}$ reveal the contribution of the related independent components. Therefore, the quantitative calculation of source contribution problem in essential is how to obtain the mixing matrix $\hat{A}$ effectively.

4. Simulation experiment analysis

4.1 Validation of separating performance
When the rotational parts, such as gears and rolling bearings occur faults (teeth broken, peeling or rubbing), the vibration signals measured on the shell will be frequency and amplitude modulated (He et al., 2001), and thus the source signals(teeth broken, peeling or rubbing) can not be well revealed just by the measured signals. As the composite faults occur, the vibration signal will be coupled together, and thus it is difficult to reveal the conditions of each parts. Therefore, some vibration signals of typical fault features are selected to test the separating performance of different methods.

Four source signals are selected: $s_1(t)$ is a white noise signal, $s_2(t)$ is a frequency modulated signal, $s_3(t)$ is an amplitude modulated signal, and $s_4(t)$ is a both amplitude and frequency modulated signal. The data length of t is 1000 and the step is 1. The generating functions of sources and mixing matrix are listed as follows.

$$S(t) = \begin{bmatrix} s_1(t) \\ s_2(t) \\ s_3(t) \\ s_4(t) \end{bmatrix} = \begin{bmatrix} n(t) \\ \sin(0.2 \times t) \times \cos(15 \times t) + \sin(2 \times t) \\ \sin(0.3 \times t) \times \sin(5 \times t + \sin(t)) \\ \sin(0.3 \times t \times \sin(0.5 \times t)) \end{bmatrix}$$

$$A = \begin{bmatrix} 0.63 & 0.77 & 0.54 & 0.65 \\ 0.94 & 0.72 & 0.78 & 0.83 \\ 0.88 & 0.93 & 0.84 & 0.32 \\ 0.98 & 0.62 & 0.54 & 0.95 \end{bmatrix} \tag{22}$$

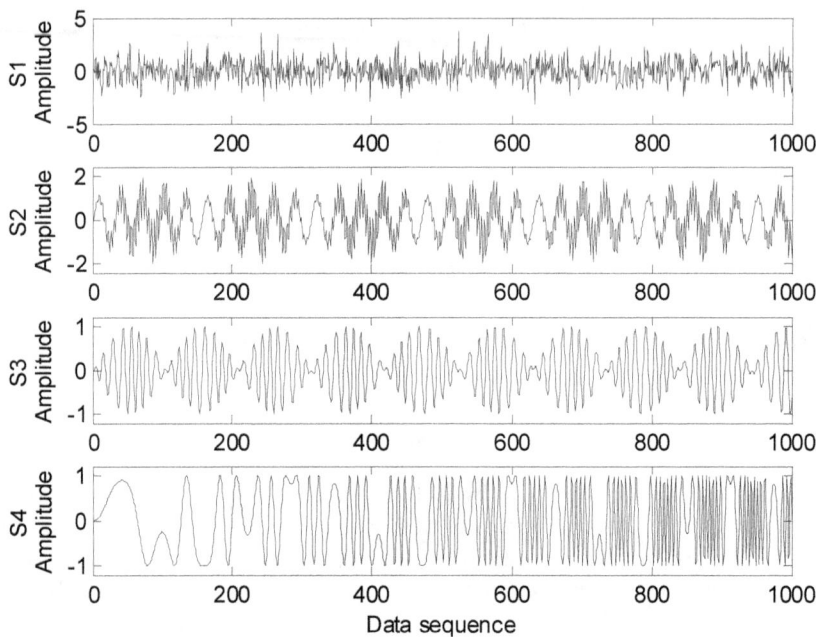

Fig. 2. Waveforms of the source signals

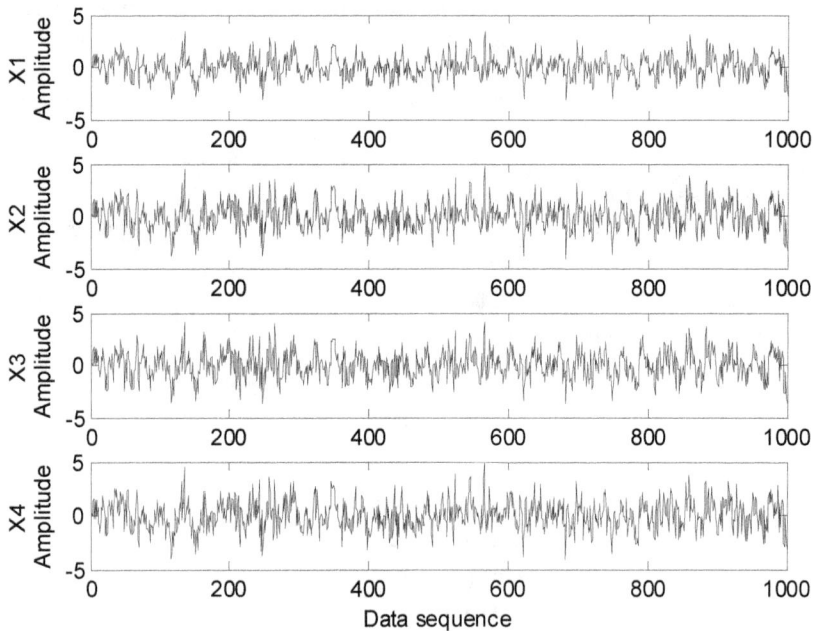

Fig. 3. Waveforms of the mixed signals

Waveforms of the sources are shown in Fig.2, which indicates that each source is of obviously different waveform features. The mixed signals are composed of four sources by mixing matrix with linear superposition, and the mixed signals are shown in Fig. 3.

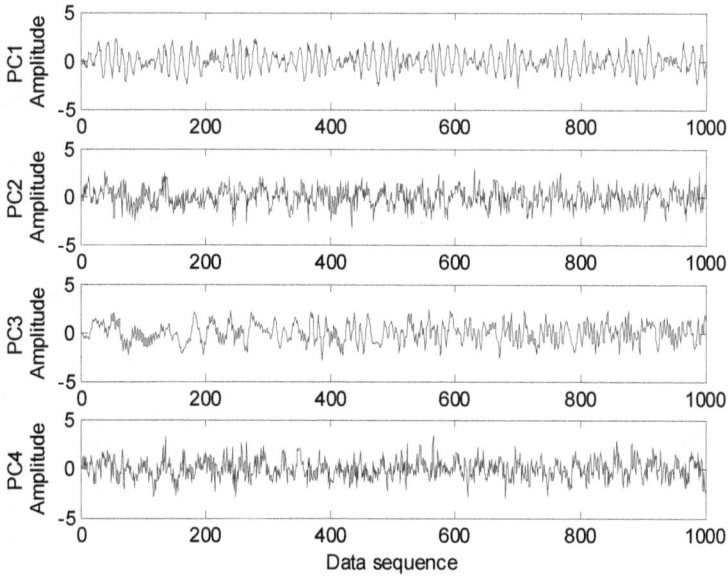

Fig. 4. Waveforms of principal components

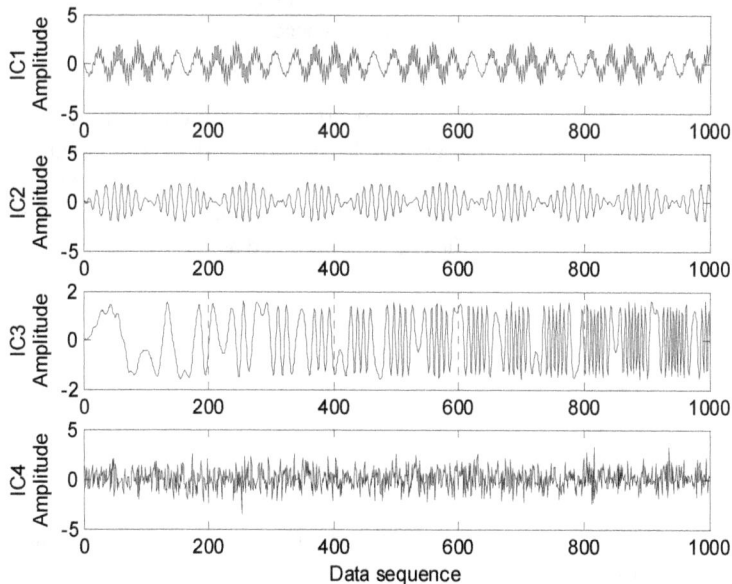

Fig. 5. Waveforms of independent components by fast fixed-point algorithm

Three different BSS methods are applied to separate the mixed signals, including principal component analysis based on second-order cumulant, fast fixed-point algorithm based on negentropy, and the enhanced ICA algorithm based on clustering optimization. The waveforms of the separated components are shown in Fig. 4 - Fig. 6 respectively.

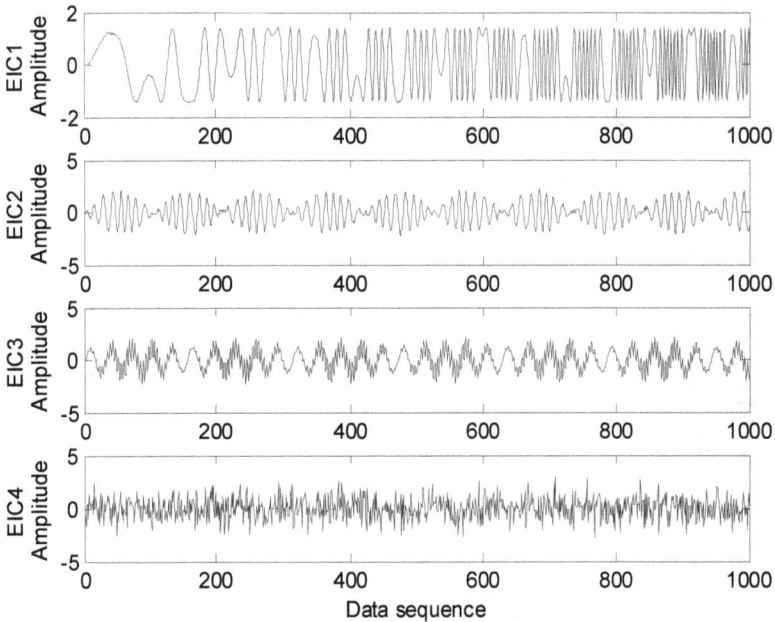

Fig. 6. Waveforms of the separated components by the enhanced ICA

From Fig. 4 to Fig. 6, it can be clearly seen that principal component analysis only separate s_3, and the other sources are not well separated. The independent components separated by the other two algorithms are obvious, and the waveform information of the sources is well separated. Correlation analysis is employed to quantitatively validate the separating performance, and the correlation coefficient ρ_{sy} is defined as follows.

$$\rho_{sy} = \frac{\sum_{k=1}^{n} s(k)y(k)}{\sqrt{\sum_{k=1}^{n} s^2(k) \sum_{k=1}^{n} y^2(k)}} \tag{23}$$

Where $s(k)$ is the source signal, $y(k)$ is the independent component, and k is the data sequence.

In the numeric studies, the correlation matrices Ω_{sy} between the separated components and the sources are listed as follows:

1. Ω_{sy} of principal component analysis

$$\Omega_{sy} = \begin{bmatrix} 0.2684 & 0.2317 & \underline{0.8932} & 0.1903 \\ 0.6154 & \underline{0.5829} & 0.1837 & 0.4735 \\ 0.1280 & 0.5709 & 0.3356 & \underline{0.7673} \\ \underline{0.7299} & 0.5298 & 0.2363 & 0.3942 \end{bmatrix}$$

2. Ω_{sy} of the fast fixed-point algorithm

$$\Omega_{sy} = \begin{bmatrix} 0.0687 & \underline{0.9996} & 0.0092 & 0.0429 \\ 0.0835 & 0.0120 & \underline{0.9988} & 0.1561 \\ 0.0984 & 0.0404 & 0.1527 & \underline{0.9952} \\ \underline{0.9986} & 0.0689 & 0.0793 & 0.0988 \end{bmatrix}$$

3. Ω_{sy} of the enhanced ICA algorithm

$$\rho_{sy} = \begin{bmatrix} 0.0873 & 0.0402 & 0.1532 & \underline{0.9997} \\ 0.0833 & 0.0185 & \underline{0.9913} & 0.1535 \\ 0.0647 & \underline{0.9980} & 0.0407 & 0.0418 \\ \underline{0.9859} & 0.1245 & 0.1267 & 0.0809 \end{bmatrix}$$

Comparing the three Ω_{sy} by different methods, the correlation coefficients between each components and related source signals of fast fixed-point algorithm and the enhanced ICA algorithm are more than 0.98, which means waveform information implied in the mixed signals is well extracted. Therefore, these two algorithms have good separating performance. However, the correlation coefficient between the second component and s_2 is only 0.58, but the other correlation coefficients are even up to 0.61, which means the source information is not well separated. Therefore, principal component analysis fails to separate the sources effectively in this case.

4.2 Stability validation of the enhanced ICA algorithm

As a statistical signal processing method, fast fixed-point algorithm may produce different results in repeated executions, and thus the separated components will be unreliable. To illustrate this problem clearly, another case is given in this section.

The source signals are: s_5 is a white noise signal, s_6 is a sinusoidal signal, s_7 is a triangle wave signal, and s_8 is a square wave signal. The generating function of sources and the mixing matrix A are as follows:

$$S(t) = \begin{bmatrix} s_5(t) \\ s_6(t) \\ s_7(t) \\ s_8(t) \end{bmatrix} = \begin{bmatrix} n(t) \\ sin(0.4 \times \pi \times t) \\ sawtooth(0.5 \times \pi \times t, 0.5) \\ square(0.6 \times \pi \times t, 50) \end{bmatrix} \quad A = \begin{bmatrix} 0.65 & 0.75 & 0.65 & 0.60 \\ 0.95 & 0.70 & 0.75 & 0.85 \\ 0.88 & 0.90 & 0.80 & 0.32 \\ 0.90 & 0.40 & 0.42 & 0.95 \end{bmatrix} \tag{24}$$

The waveforms of source signals and the mixed signals are shown in Fig. 7 and Fig. 8 respectively. The fast fixed-point algorithm is repeatedly executed for more than 30 times, and more than 60% of the results show that the ICs are accurate. However, the other 40% of the results are different from the source signals obviously. One of the inaccurate results is shown in Fig. 9, which indicates that the source information is not well separated.

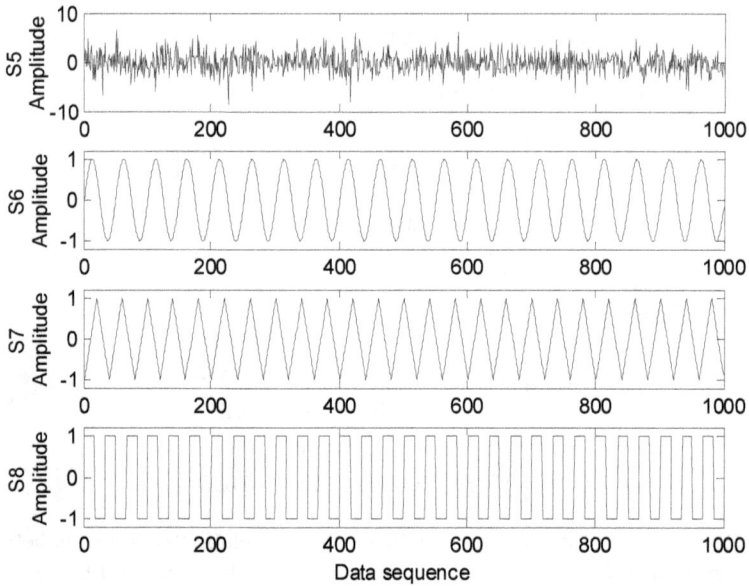

Fig. 7. Waveforms of the source signals

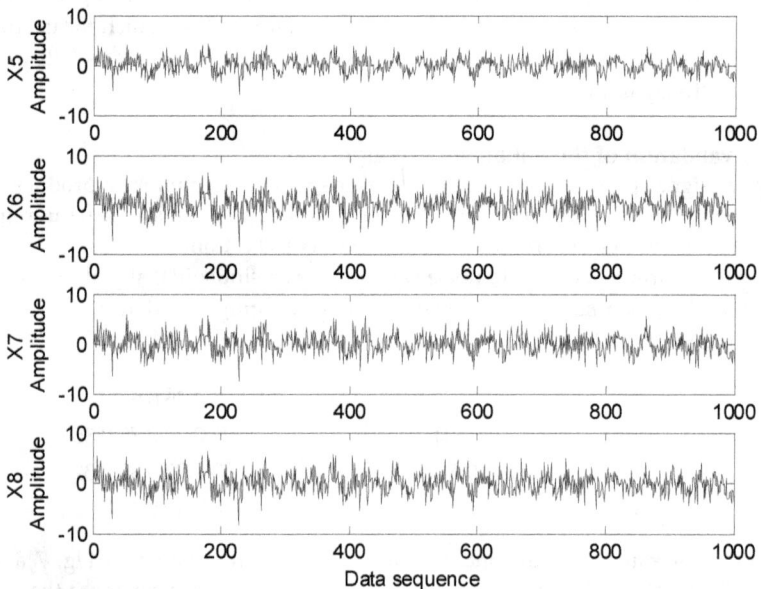

Fig. 8. Waveforms of the mixed signals

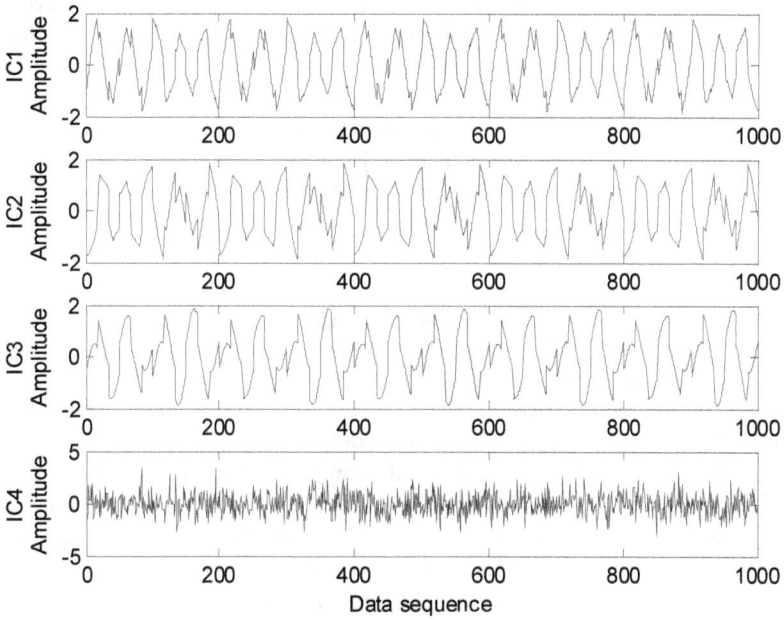

Fig. 9. Waveforms of one inaccurate separation by fast fixed-point algorithm

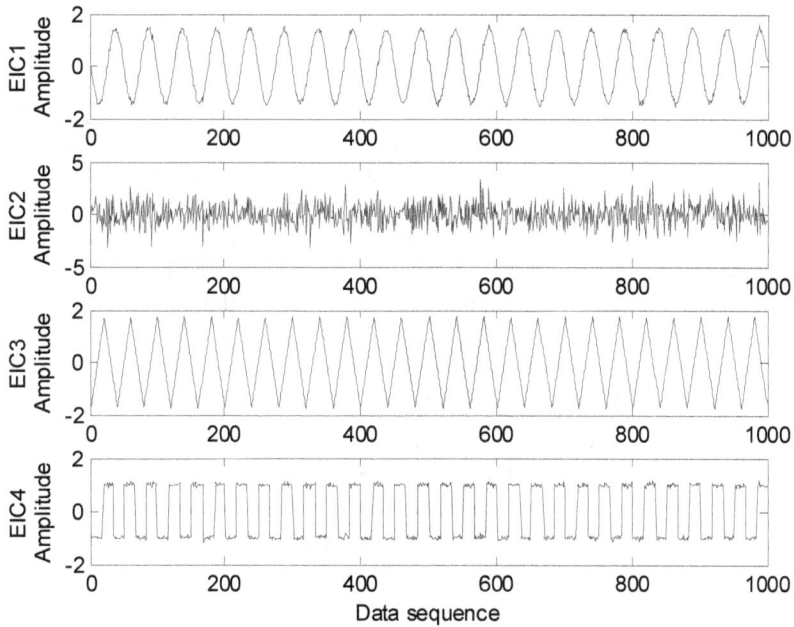

Fig. 10. Waveforms of separated components by the enhanced ICA algorithm

To enhance the stability of traditional ICA, the enhanced ICA algorithm is constructed. The enhanced ICA algorithm is repeatedly executed for 20 times, and all the result is stable. The waveforms of the separated components by the enhanced ICA algorithm are shown in Fig. 10, which indicates that the source information is well extracted.

It can be concluded that fast fixed-point algorithm may produce inaccurate components because it is a statistic signal processing method. To enhance the stability, an effective way is to execute the single ICA algorithm for certain times, and evaluate the separated components by clustering analysis. Therefore, the enhanced algorithm based on clustering optimization is of high precision and stability.

4.3 Quantitative evaluation of source contributions

From the basic theory of blind source separation model, it can be explained that some complicate signals are composed of several independent components, and the combination relationship of the independent components reveals their contributions. Therefore, source contributions can be obtained according to the mixing matrix.

We take the case in section 4.1 to show the numeric studies of source contribution evaluation. After the separation, a mixing matrix A' of the independent components for the mixed signals is obtained. As the source signals and the independent components have different scales. We add a factor to let the independent components have the same energy with the related sources, and thus get a modified mixing matrix $\hat{A}$.

$$A = \begin{bmatrix} 0.63\ 0.77\ 0.54\ 0.65 \\ 0.94\ 0.72\ 0.78\ 0.83 \\ 0.88\ 0.93\ 0.84\ 0.32 \\ 0.98\ 0.62\ 0.54\ 0.95 \end{bmatrix} \leftrightarrow \hat{A} = \begin{bmatrix} 0.65\ 0.75\ 0.55\ 0.62 \\ 0.95\ 0.70\ 0.75\ 0.85 \\ 0.88\ 0.90\ 0.80\ 0.32 \\ 0.94\ 0.60\ 0.52\ 0.95 \end{bmatrix}$$

Compared the mixing matrix A with the mixing matrix $\hat{A}$ calculated by the proposed method, the real proportion of each source to x_1 are 0.63, 0.77, 0.54 and 0.65, while the contributions calculated by the proposed method are 0.65, 0.75, 0.55 and 0.62, which means the relative error between the calculated values and the real values is less than 4.6%. The calculated results of source contributions to x_2 , x_3 and x_4 are also close to the real values, and the relative errors are all less than 4.8%. Therefore, the comparative results show that it is effective to quantitatively evaluate the source contributions just from the mixed signals.

5. Source contribution evaluation of a thin shell structure

5.1 Introductions of the thin shell structure

In this application, a thin shell structure is set up to validate the effectiveness of the proposed method. Three motors are installed in the thin shell structure to simulate the vibration sources, and blocks are added in each output shaft of the motors to generate eccentric vibration. Magnetic brakes are installed in the output shaft of these motors for a variable load. Rubber springs are applied to support the entire structure, and they can also eliminate the environmental influences. The structural diagram of the thin shell structure is shown in Fig.11, and the photo of the entire system is shown in Fig.12.

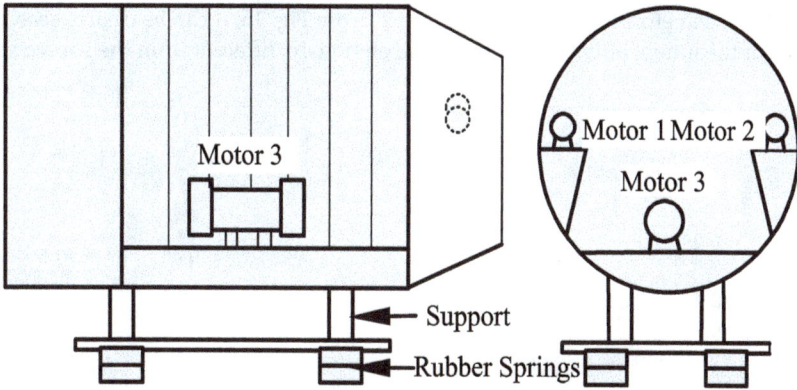

Fig. 11. The structural diagram of the thin shell structure

Fig. 12. The photo of the entire system

Acceleration sensors are used to sample the vibration signals. Three sensors are installed in the bases of each motor to measure the sources, and nine sensors are installed on the left inside shell to measure the mixed signals. In the testing, the sampling frequency is 16384 Hz, the data length is 16384, and the unit of measured signals is g ($1.0g = 9.8m/s^2$). The rotational speeds of motor 1, motor 2, and motor 3 are respectively 1350 rpm, 1470 rpm, and 1230 rpm.

5.2 Blind source separation of the mixed signals

To calculate the source contributions quantitatively, a direct method is to separate the mixed signals and calculate the contributions by the mixing matrix. Three vibration signals on the inside shell are selected as the mixed signals, and their waveforms are shown in Fig. 13. The enhanced ICA algorithm is applied to separate the mixed signals, and three independent components extracted are shown in Fig. 14. Three source signals are measured on the motor bases, and their waveforms are shown in Fig. 15. From Fig. 15, it can be clearly seen that the waveforms of the independent components are obviously different from the source signals.

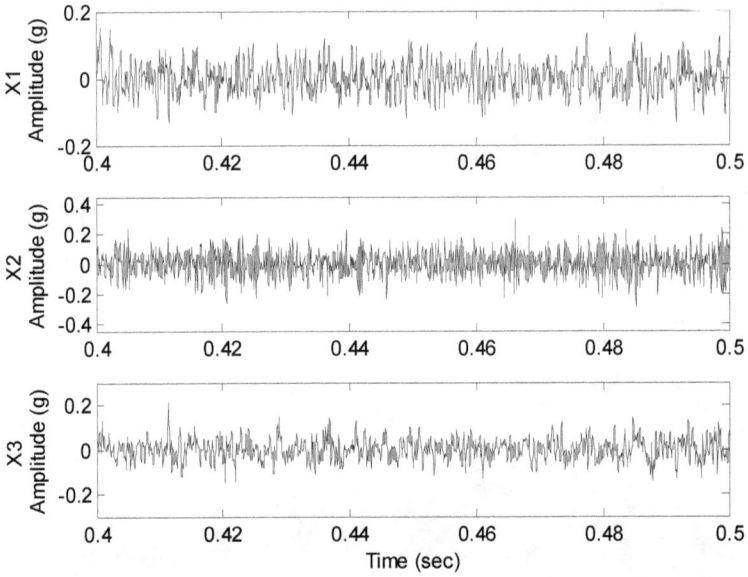

Fig. 13. Waveforms of the mixed signals on the inside shell

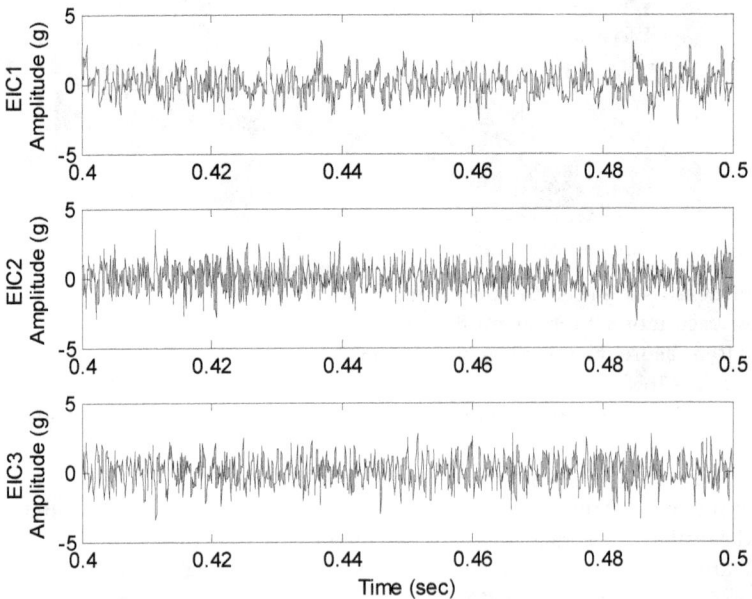

Fig. 14. Waveforms of the ICs by the enhanced ICA algorithm

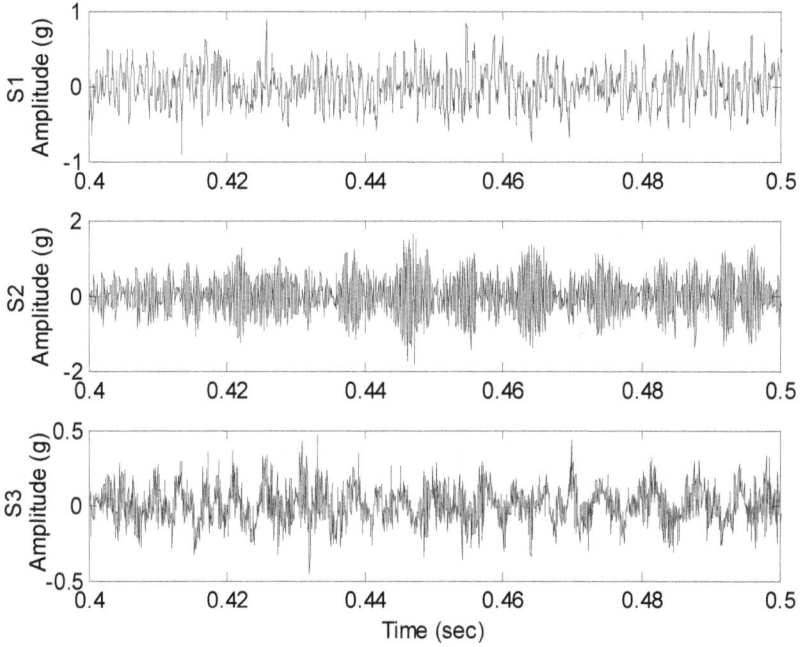

Fig. 15. Waveforms of the source signals from three motor bases

The correlation matrix Ω_{sy} is as follows.

$$\Omega_{sy} = \begin{bmatrix} 0.2955 & 0.2471 & 0.4269 \\ 0.2614 & 0.4235 & 0.1621 \\ 0.3376 & 0.3898 & 0.2217 \end{bmatrix}$$

Ω_{sy} shows that the correlation coefficients between independent components and source signals are relatively small, and the maximum coefficient is only 0.4269, which indicates that the source information is not well separated. Therefore, the separated components extracted cannot be regarded as the sources for calculating source contributions.

5.3 Source contribution evaluation based on priori information
In this application, source signals can be obtained from the motor bases. Source contribution based on priori information is proposed and it can be described as follows: four signals are selected as the mixed signals, one of them is measured inside shell, and the other three signals are measured from the motor bases. The waveforms of the mixed signals are shown in Fig.16. The independent components are extracted by the enhanced ICA algorithm. The optimal independent components are selected by clustering evaluation, and waveforms of the optimal independent components are shown in Fig.17. Obviously three independent components have the similar waveforms with three source signals.

Correlation matrix Ω_{sy} between independent components and source signals are:

$$\Omega_{sy} = \begin{bmatrix} 0.2907 & \underline{0.8903} & 0.0795 \\ \underline{0.8397} & 0.1041 & 0.3083 \\ 0.2717 & 0.1094 & \underline{0.9539} \end{bmatrix}$$

The correlation matrix Ω_{sy} shows that the correlation coefficient between IC 1 and source signal 2 is 0.8397, IC 2 and source signal 1 is 0.8903, and IC 3 and source signal 3 is up to 0.9539, which shows that three source signals are well extracted from the mixed signals. Therefore, with the priori information, the source information can be well separated in the real applications, which means the priori information can further improve the separating performance. After the separation, the mixing matrix is calculated by the enhanced ICA algorithm. The mixing matrix $\hat{A}$ is as follow:

$$\hat{A} = \begin{bmatrix} -0.1299 & -0.0097 & 0.0440 \\ -0.1971 & 0.0358 & 0.1656 \\ 0.1424 & 0.0318 & 0.0708 \\ -0.0313 & -0.1375 & 0.0275 \end{bmatrix}$$

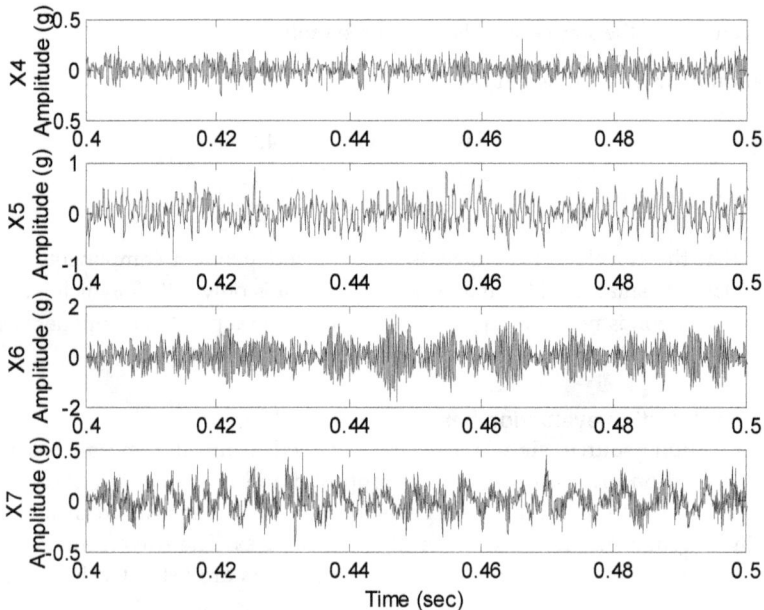

Fig. 16. Waveforms of the mixed signals

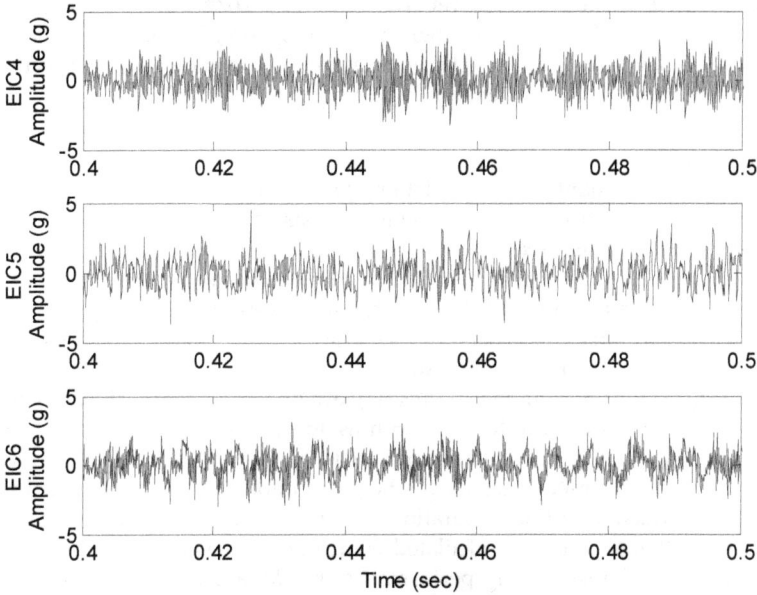

Fig. 17. Waveforms of ICs by the enhanced ICA algorithm with priori information

The mixing matrix $\hat{A}$ shows that the mixed signals are composed of three ICs, and their proportional contributions are 0.1299, 0.0097 and 0.0440. The three ICs represent the source signals from three motors, and the percentage contributions of motor 1, motor 2 and motor 3 are 23.97%, 70.75% and 5.280% respectively. The percentage contribution indicates that IC 2 (motor 3) has the largest vibration contribution, which means that mixed signal 1 mainly comes from the motor 3. Therefore, motor 3 should be controlled or the vibration reduction equipments should be adapted in the transmission path to reduce vibration.

Contributions	Motor 1	Motor 2	Motor 3
Measurement	16.79%	5.66%	77.55%
Without priori information	47.19%	11.93%	40.88%
Proposed Method	23.97%	5.28%	70.75%

Table 1. Percentage contributions

In the measuring point on the shell, the real vibration contributions are measured in the condition that only one motor is running at the given speed (motor1 - 1350 rpm, motor2 - 1470 rpm and motor3 - 1230 rpm respectively), and thus the contribution of the related motor to the same location on the shell can be measured one by one. The energy contributions of motor 1, motor 2 and motor 3 are 5.41, 1.82 and 24.98 respectively. The real contributions also show that motor 3 has the largest contribution. This paper also gives the source contributions calculated by the enhanced ICA without priori information. The percentage contributions are listed in Tab .1. Tab. 1 shows that the real contribution of motor 3 is up to 77.55%, which means that motor 3 gives a large vibration contribution to the shell. The contribution of motor 3 calculated by the enhanced ICA method without priori information is 40.88%, which has a relative error of 36.67%. The relative error of the proposed method is 6.8%, and the other two motor contributions also show that the proposed method has high accuracy. Therefore, it can be concluded that the accuracy of the proposed method has been greatly enhanced by priori information.

5.4 Discussions

1. The key process of quantitative calculation of source contributions is that mixed signals are well separated. Principal component analysis in the simulation does not give a satisfied result, and fast fixed point algorithm has good separating performance for linear superposition but it is not stable for repeatedly calculation. The enhanced ICA algorithm executes the single ICA algorithm for several times, and selects the optimal ICs by clustering analysis. Therefore, separating performance and reliability of the enhanced ICA algorithm are enhanced.
2. Most of the mixed signals in the engineering are nonlinear mixed signals, and thus it is a challenge work for most ICA algorithms to separate the sources accurately. To overcome this challege, priori information is employed to weaken the uncertainty problem of nonlinear mixed signals. By the priori information, the separated signals can be better separated, and thus separating performance can be further enhanced, which can improve the accuracy of calculated contributions. Therefore, it provides another way to enhance the separating performance of ICA according to making full use of priori information.

6. Conclusions

As the influences of transmission paths, the vibration signals on the shell are influenced significantly, so it is a very challenging task to identify the sources and quantitatively evaluate the source contributions, which is important for vibration reduction and control.

In this paper, a novel method to quantitatively evaluate the source contributions based on the enhanced ICA algorithm and priori information is proposed. The enhanced ICA method provides a powerful way to effectively and reliably separate sources from the mixed signals. In the simulations, the comparative results show that principal component analysis cannot deal with some typical mechanical signals. Fast fixed point algorithm has strong separating performance for linear superstition signals, but the reliability is not very good because it is a statistical signal processing method. The enhanced ICA algorithm evaluates the separating components by clustering analysis and selects the optimal ICs as the best results. Therefore, the separating performance and reliability are significantly enhanced. The error of contributions calculated by the enhanced ICA algorithm is less than 4.8%, which indicates that the proposed method is of high accuracy.

The proposed method is applied to evaluate the source contributions of a thin shell structure. The priori information is discussed in two different separation process. The case that does not use priori information has an obvious error of 36.67%. Compared with the real values by measurement, the comparative results show that the proposed method obtains a high accuracy with priori information, and the relative error is less than 7.18%. Therefore, the proposed method can effectively evaluate source contributions, which can provide reliable referances for vibration reduction and monitoring.

7. Acknowledgement

This work was supported by the key project of National Nature Science Foundation of China (No. 51035007), the project of National Nature Science Foundation of China (No. 50875197) and National S&T Major Project (No. 2009ZX04014-101).

8. References

Lee SK. Lee, Mace B.R., Brennan MJ. (2007). Wave propagation, reflection and transmission in curved beams. *Journal of Sound and Vibration*, vol.306, no.3-5: pp.636–656, ISSN: 0022-460X

Senjanovic. I, Tomasevic. S, Vladimir. N. (2009). An advanced theory of thin-walled girders with application to ship vibrations. *Marine Structures*, vol.22, no.3: pp.387–437, ISSN: 0951-8339

Niu. JC, Song. KJ, Lim .CW. (2005). On active vibration isolation of floating raft system. *Journal of Sound and Vibration*, vol.285, no.1-2: pp.391–406, ISSN: 0022-460X

Otrin. M, Boltezar. M. (2009). On the modeling of vibration transmission over a spatially curved cable with casing. *Journal of Sound and Vibration*, vol.325, no.4-5: pp.798–815, ISSN: 0022-460X

Lee. YY, Su. RKL, Ng. CF. (2009). The effect of modal energy transfer on the sound radiation and vibration of a curved panel theory and experiment. *Journal of sound and vibration*, vol.324, no.3-5: pp.1003-1015, ISSN: 0022-460X

Xie. SL, Or. SW, Chan. HLW. (2007). Analysis of vibration power flow from a vibrating machinery to a floating elastic panel. *Mechanical Systems and Signal Processing*, vol.21, no.1: pp.389–404, ISSN: 0888-3270

Bonfiglio. P, Pompoli. F, Peplow. AT. (2007). Aspects of computational vibration transmission for sandwich panels. *Journal of Sound and Vibration*, vol.303, no.3-5: pp.780–797, ISSN: 0022-460X

Efimtsov. BM, Lazarev. LA. (2009). Forced vibrations of plates and cylindrical shells with regular orthogonal system of stiffeners. *Journal of Sound and Vibration*, vol.327, no.1-2: pp.41–54, ISSN: 0022-460X

Li. XB. (2006). A new approach for free vibration analysis of thin circular cylindrical shell. *Journal of Sound and Vibration*, vol.296, no.1-2: pp.91-98, ISSN: 0022-460X

Cha. SL, Chun. HH. (2008). Insertion loss prediction of floating floors used in ship cabins. *Applied Acoustics*, vol.69, no.10: pp.913–917, ISSN: 0003-682X

Iijima. K, Yao. T, Moan. T. (2008). Structural response of a ship in severe seas considering global hydroelastic vibrations. *Marine Structures*, vol.21, no.4: pp.420–445, ISSN: 0951-8339

Tian RL, Pan. J, Peter. JO., et al. (2009). A study of vibration and vibration control of ship structures. *Marine Structures*, vol.22, no.4: pp.730-743, ISSN: 0951-8339

Jnifene. A. (2007). Active vibration control of flexible structures using delayed position feedback. *Systems & Control Letters*, vol.56, no.3: pp.215-222, ISSN: 0167-6911

Barrault. G, Halim. D, Hansen. C, et al. (2008). High frequency spatial vibration control for complex structures, *Applied Acoustic*, vol.69, no.11: pp.933-944, ISSN: 0003-682X

Jutten. C, Herault. J. (1991). Blind separation of sources. Part I: An adaptive algorithm based on neuromimatic architecture. *Signal Processing*, vol.24, no.1: pp.1-10, ISSN: 0165-1684

Comon. P. (1994). Independent component analysis, a new concept? *Signal Processing* , vol.36, no.3: pp.287-314, ISSN: 0165-1684

Hyvarinen. A, Oja. E. (1997). A fast fixed-point algorithm for independent component analysis. *Neural Computation*, vol.9, no.7 : pp.1486-1492, ISBN: 0-262-58168-X

Hyvarinen. A. (1999). Fast and robust fixed-point algorithm for independent component analysis. *IEEE Transactions On Neural Network*, vol.10, no.3 : pp.626-634, ISSN: 1045-9227

Hyvarinen. A, Karhunen. J, Oja. E. (2001). In : *Independent component analysis*. John Wiley and Sons. ISBN : 0-471-40540-X, New York, USA

Hu. HF. (2008). ICA-based neighborhood preserving analysis for face recognition. *Computer Vision and Image Understanding*, vol.112, no.3: pp.286-295, ISSN: 1077-3142

Correa. N, Adali. T, Vince. Calhoun. VD. (2007). Performance of blind source separation algorithms for fMRI analysis using a group ICA method. *Magnetic Resonance Imaging*, vol.25, no.5: 684–694, ISSN : 0730-725X

Ye. YL, Zhang. ZL, Zeng. JZ. (2008). A fast and adaptive ICA algorithm with its application to fetal electrocardiogram extraction. *Applied Mathematics and Computation*, vol.205, no.2: pp.799-806, ISSN : 0096-3003

Xie. L, Wu. J. (2006). Global optimal ICA and its application in MEG data analysis. *Neurocomputing*, vol.69, no.16-18: pp.2438-2442, ISSN: 0925-2312

Zuo. MJ, Lin. J, Fan. XF. (2005). Feature separation using ICA for a one-dimensional time series and its application in fault detection. *Journal of Sound and Vibration*, vol.287, no.3: pp.614-624, ISSN: 0022-460X

Moussaoui. S, Hauksdottir. H, Schmidt. F. (2008). On the decomposition of Mars hyper-spectral data by ICA and Bayesian positive source separation. *Neurocomputing*, vol.71, no10-12: pp.2194–2208, ISSN : 0925-2312

Kwak. N, Kim. C, Kim. H. (2008). Dimensionality reduction based on ICA for regression problems. *Neurocomputing*, vol.71, no.13-15: pp.2596– 2603, ISSN: 0925-2312

Himberg. J, Hyvarinen. A. (2003). Icasso: Software for investigating the reliability of ICA estimates by clustering and visualization. *IEEE XIII Workshops On Neural Networks for Signal Processing*, pp. 259-268, ISBN: 0-7803-8177-7, Toulouse, France

Himberg. J, Hyvarinen. A, Esposito. F. (2004). Validating the independent components of neuroimaging time-series via clustering and visualization. *Neuroimage*, vol.22, no.3: pp.1214-1222, ISSN: 1053-8119

Ma. JC, Niu. YL, Chen. HY. (2006). In : *Blind Signal Processing*. National Defense Industry Press, pp.150-160, ISBN: 9787118045079, BeiJing, China

He. ZJ, Zi. YY, Zhang. XN. (2006). In : *Modern signal processing technology and its application*, Xi'an Jiaotong University Press, pp.4-9, ISBN: 9787560525464, Xi'an, China

8

Psychophysiological Experiments on Extent of Disturbance of Noises Under Conditions of Different Types of Brain Works

Sohei Tsujimura[1] and Takeshi Akita[2]
[1]Institute of Industrial Science, The University of Tokyo
[2]Tokyo Denki University
Japan

1. Introduction

For quantitative evaluation of human conditions, subjective evaluation obtained through psychological measurements is generally used for assessment of an acoustic environment. It is also important that human reactions to surrounding environments be objectively evaluated from physiological aspect.

In the field of architectural environment, Authors reported that auditory information processing was affected by the priority of processing in the brain (Akita et al., 1995). From this finding, it was demonstrated that the Electroencephalogram (EEG) was effective for objective evaluation as a physiological index. In the field of psychophysiology, the distinct EEG theta rhythm from the frontal midline area observed during performance of mental tasks such as continuous addition was defined as the frontal midline theta rhythm (Fmθ) (Brazier et al., 1952; Ishihara et al., 1972). In the studies on Fmθ, it was found that Fmθ was associated with concentration on the task, with higher Fmθ values indicating higher concentration (Ishihara et al., 1975; Ishihara et al., 1976; Ishihara et al., 1991; Suetsugi et al., 2002). As the previous studies on the influence of various noises on task performance, for example, Hashimoto et al. investigated the effects of familiar noises with a moderate sound level of 60 dB L_{Aeq} on physiological responses, task performance and psychological responses (Hashimoto et al., 1999). From the results, it was shown that the effect of noise on task performance depended on the kind of noises, it strongly influences the task performance as the task became increasingly more complex.

Author has studied relationships between extent of disturbance caused by sounds and EEG during simple calculation task up to now (Tsujimura et al., 2006; Tsujimura et al., 2008). As the finding in the paper, information processing mechanism of cerebral cortex relevant to disturbance was reported (Figure 1) (Tsujimura et al., 2006). Moreover, to clear an effect of meaning of noise on task performance or extent of disturbance, we have investigated the effect of meaningless and meaningful noises on extent of disturbance caused by noises and task performance of two kinds of brain works (Tsujimura et al., 2007).

In this paper, in order to find influence of noise including verbal information on the extent of disturbance or EEG during different types of brain works, the psychophysiological experiment was conducted. The relationships between extent of disturbance and EEG were examined during each brain work.

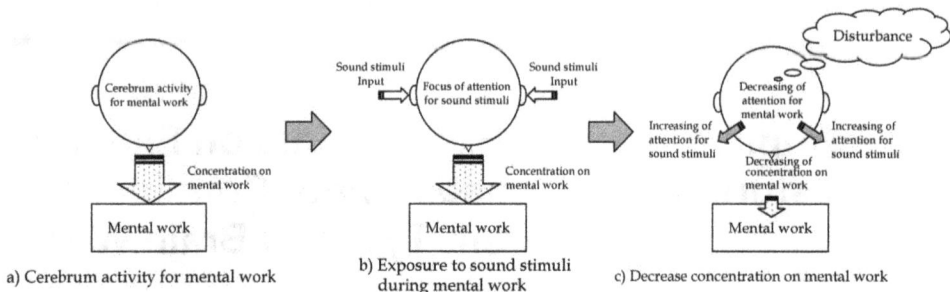

a) Cerebrum activity for mental work

b) Exposure to sound stimuli during mental work

c) Decrease concentration on mental work

Fig. 1. Information processing mechanism of cerebral cortex relevant to disturbance (Tsujimura et al., 2006)

2. Subjective experiment

In a hemi-anechoic room, the subject performed two kinds of brain works during exposure to three kinds of noises. The EEG was measured, and extent of disturbance was judged by him/her under each experimental condition.

2.1 Experimental system

The experimental system was constructed in a hemi-anechoic room. The block diagram of the experimental system is shown in Figure 2. The distance between a loudspeaker and subject was 2.0 meters. In this experiment, the 8-channel bioelectrical amplifier (DIGITEX Lab., BA1008) was set up under following conditions: The sensitivity was 50 microvolt. The time constant was 0.1 second. The 30 Hz treble cutoff filter was used.

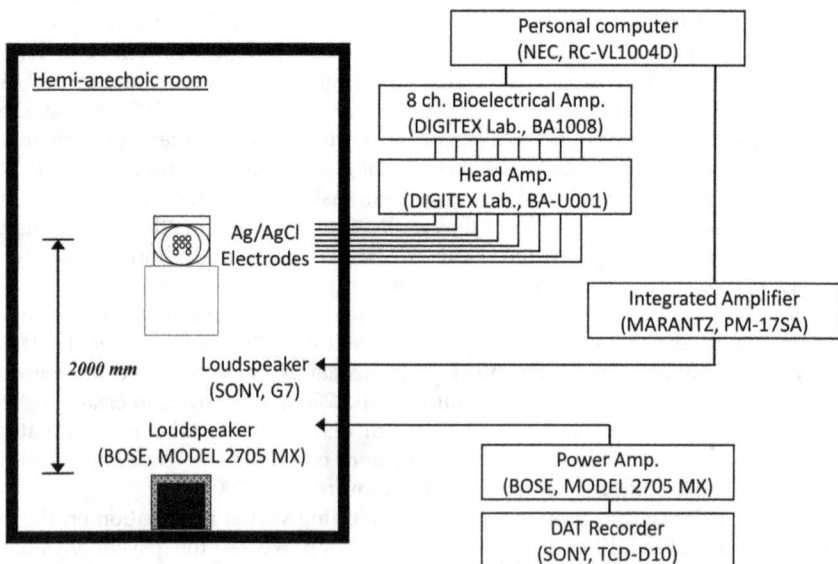

Fig. 2. Block diagram of experimental system

2.2 Mental works

Two kinds of brain works used in this experiment were described as follows. Mental arithmetic manipulation of double digits and memorization of Chinese characters were used in this experiment as brain works (Yoshida et al., 2007). In the mental arithmetic manipulation of double digits, subject was instructed to continue mental arithmetic of double digits (arithmetic addition and subtraction of double digits) for 5 minutes. In the memorization of Chinese characters, subject was instructed to memorize as many Chinese characters as he/she could in 2 minutes, and then write the answers on answer sheets in 3 minutes. The examples of brain works were shown in Figure 3-a) and b). The subject was given an instruction that he/she does not move his/her heads during the experiment.

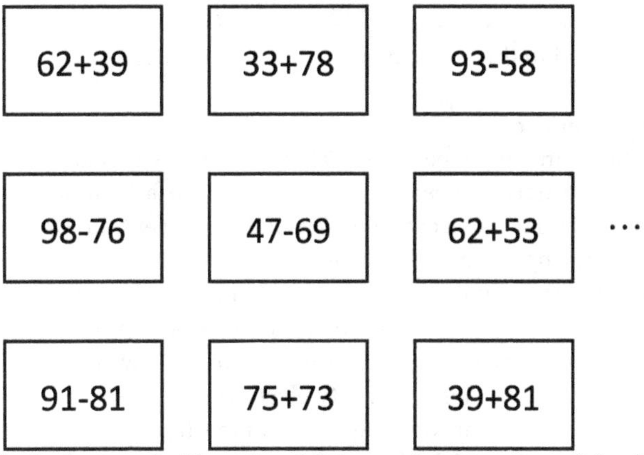

62+39	33+78	93-58
98-76	47-69	62+53	⋯
91-81	75+73	39+81

a) Mental calculations (arithmetic addition and subtraction of double digits)

合	業	覚	憲	間
乗	解	首	殺	戸
宗	修	兄	告	囲
右	暑	急	青	共
机	衣	回	京	結

b) Memorization of Chinese characters

Fig. 3. Examples of brain works in this experiment

2.3 Noise

In this experiment, news broadcast was used as meaningful noise, and white noise was used as meaningless noise. These noises were presented to the subject at 50 dB $L_{Aeq, 5min}$. This experiment consisted of three experimental conditions – two kinds of noises and one with pink noise as a background noise (The pink noise was presented to the subject at 30 dB $L_{Aeq, 5min}$). The experimental conditions in this experiment are shown in Table 1.

Meaning of noises	The kinds of noises	$L_{Aeq, 5min}$ [dB]
Meaningful noise	News broadcast	50
Meaningless noise	White noise	
	Pink noise (BGN)	30

Table 1. Noises were used in this experiment

2.4 Experimental procedure

19 subjects participated in this experiment; 11 males and 8 females. They were Japanese university students with normal hearing ability. Firstly, in a hemi-anechoic room, subject was asked to sit on chair. The subject was instructed to concentrate on the brain work, and performed the work during exposure to noise described above Table 1. The EEG of each subject was measured with or without noise during brain work. In order to keep a constant background noise in hemi-anechoic room, the pink noise was presented to subjects at $L_{Aeq, 5min}$ 30 dB during this experiment. The measurement of EEG was based on the ten-twenty electrode system (Jasper, 1958). The ten-twenty electrode system is shown in Figure 4. Inoue et al. reported that there was a part of the generation mechanism of Fmθ in the paramedian line area of cerebral cortex of frontal lobe in each cerebral hemisphere, firstly, Fmθ occurred at the paramedian line of frontal lobe (F_{p1}, F_{p2}, F_3, F_4, C_3, C_4, P_3, and P_4 positions) (Inoue et al., 1984). From these findings, in our study, EEG was measured by electrodes at F_z, C_z, F_{p1}, F_{p2}, F_3, F_4, C_3 and C_4 positions using A_1 and A_2 as references. A measured EEG passed through a 30 Hz treble cutoff filter, and then it was amplified by an 8ch bioelectrical amplifier.

After the measurement of EEG, the subject was asked to evaluate the extent of disturbance in performing the brain work caused by each of noise shown in Table 1 except for background noise condition by the rating scale method using ten-step category. Thus, they evaluated the extent of disturbance four times in all. This category scale used in the evaluation of extent of disturbance is shown in Figure 5. The extent of disturbance was evaluated throughout the brain work.

2.5 Analysis of EEG

In the field of psychophysiology, Ishihara, et al. have defined the distinct theta rhythm (6-7 Hz) of EEG observed at frontal midline area (C_z position described above Figure 4) during performance of mental tasks such as continuous addition and taking an intelligence test as frontal midline theta rhythm (Ishihara et al., 1976; Inoue et al., 1984). It was found that frontal midline theta rhythm was associated with concentration on a mental task (Ishihara et al., 1996). From these findings, in this study, potentials of EEG of 6-7 Hz (in what follows "Fmθ" was described) was analyzed by use of FFT .

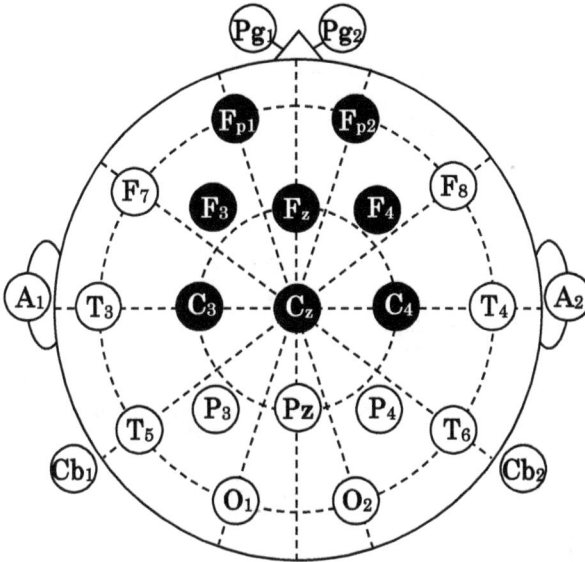

Fig. 4. Points of measurement based on the ten-twenty electrode system (Jasper, 1958)

Not especially disturbed   Extremely disturbed

1   10

Fig. 5. Category scale of extent of subjective disturbance used in this experiment

3. Result and discussion

3.1 Effect of noise on extent of disturbance

The relationships between noise conditions and extent of disturbance for each brain work
are shown in Figure 6. The averaged evaluation of extent of disturbance for 19 subjects is
used in Figure 5. Two-way ANOVA in which the kind of brain work and noise were used as
factors was calculated. From the results, a main effect of noise was significantly different in
both works (Mental arithmetic manipulation of double digits: $F (2, 54) = 35.52$, $p<0.01$,
Memorization of Chinese characters: $F (2, 54) = 37.94$, $p<0.01$). In both works, multiple
comparison (Tukey's HSD test) was calculated. The extent of disturbance showed significant
differences between each noise conditions in both works ($p<0.01$).

In both works, the extent of disturbance obtained on exposure to news broadcast was higher
than that of white noise. In regard to news broadcast, the extent of disturbance in
memorization of Chinese characters was higher than that of mental arithmetic manipulation
of double digits. From these results, it was suggested that noise including verbal
information have effects on extent of disturbance more than that of nonverbal information
during brain works. The effects of noise including verbal information such as news
broadcast varied according to the kinds of brain works.

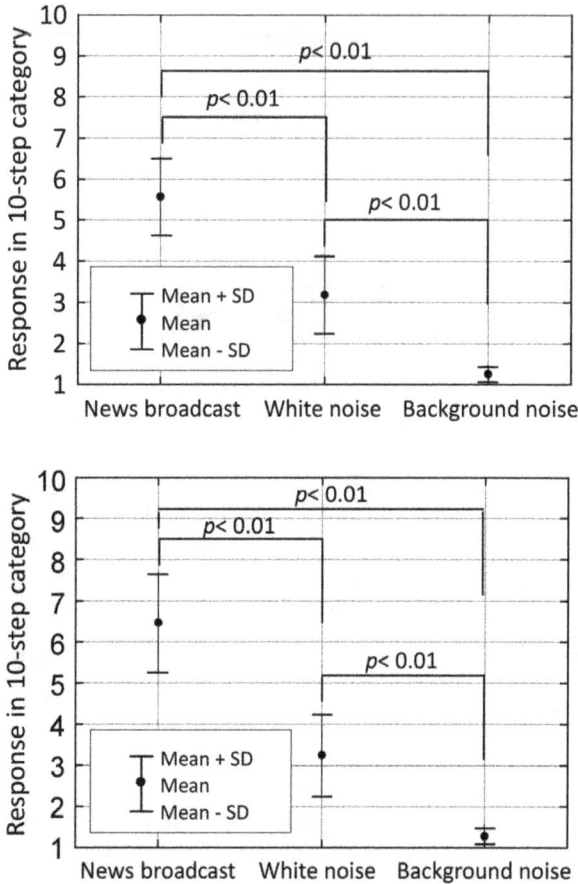

Fig. 6. Relationship between noise conditions and extent of disturbance for each brain work. a) Mental arithmetic manipulation of double digits, b) Memorization of Chinese characters. The y-axis ranges from 1 = 'not especially disturbed' to 10 'extremely disturbed'. The extent of disturbance of noise was not evaluated in the condition of no noise (background noise condition) because there was no noise targeted for evaluation.

3.2 Effect of noise on task performance

The relationships between noise conditions and task performance for each brain work are shown in Figure 7. From the results shown in Figure 6, in memorization of Chinese characters, task performance of exposure to news broadcast seemed to be lower than those of white noise and background noise. Two-way ANOVA in which the kind of brain work and noise were used as factors was calculated. From the results, in regard to task performance, a main effect of the kind of works and noise conditions were not significantly different.

These results show that noises used in this experiment have no influence on task performance of mental arithmetic manipulation of double digits, and they have some effect

on that of memorization of Chinese characters. The news broadcast influence on task
performance of memorization of Chinese characters to a greater degree than white noise. It
was suggested that noise including verbal information adversely affected task performance
of work containing verbal information such as memorization of Chinese characters.

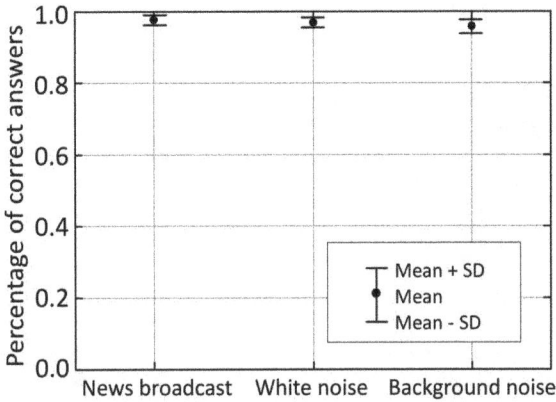

(a) Mental arithmetic manipulation of double digits

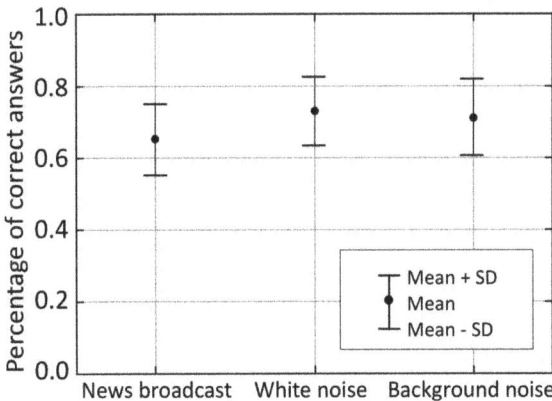

(b) Memorization of Chinese characters

Fig. 7. Relationship between noise conditions and task performance for each brain work. (a)
Mental arithmetic manipulation of double digits, (b) Memorization of Chinese characters.

3.3 Effect of noise on EEG

The relationships between noise conditions and Fmθ power for each brain work are shown
in Figure 8 and Figure 9. The mean of Fmθ power for 19 subjects was used in Figure 8 and 9.
Two-way ANOVA in which the kind of brain work and noise were used as factors was
calculated. From the results, in mental arithmetic manipulation of double digits, a main
effect of noise conditions showed significant differences ($F_{(1, 36)} = 5.22$, $p < 0.05$). Also, in the
case of writing in the answers for memorization of Chinese characters, a main effect of noise

conditions showed significant differences (F (1, 36) = 6.64, $p < 0.05$) however a main effect of the kind of works and the interaction of the kind of brain work and noise were not significantly different. In these works, Tukey's HSD test was calculated. Fmθ power showed significant differences between the conditions of exposure to news broadcast and white noise ($p < 0.01$).

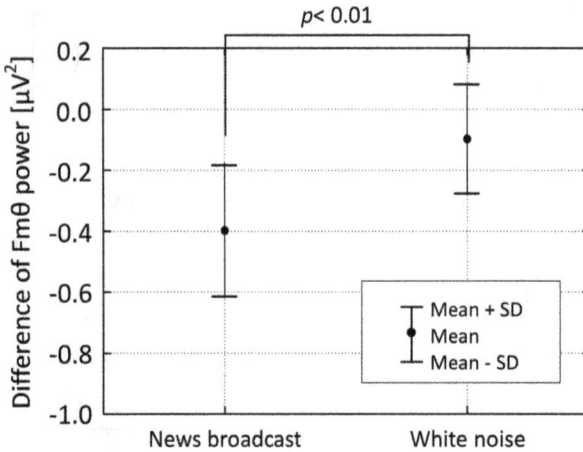

Fig. 8. Difference of Fmθ power between the condition of background noise and each noise condition (Mental arithmetic manipulation of double digits).

These results showed that there was a tendency to decrease Fmθ power when the subjects were exposed to noise. It was found that Fmθ power more decrease especially when the subjects were exposed to news broadcast. In the results of Two-way ANOVA and Tukey's HSD test, for memorization of Chinese characters, a main effect of noise conditions showed significant differences under the experimental condition which subjects wrote in the answers. This means that noise generated during the writing in the answers have an effect on Fmθ power more than that of during memorization of Chinese characters, from the result, it was suggested that memorization take more concentration than remembering.

The relationships between Fmθ power and extent of disturbance in performing each brain work are shown in Figure 10. As the result of Figure 9, it was indicated that Fmθ potentials decrease as the extent of disturbance becomes large. Ishihara, et al. reported that Fmθ was easily detected when the subjects were able to concentrate on a task (Ishihara et al., 1996). This means that Fmθ power was strongly associated with intensity of concentration on the task. The extent of disturbance increases as Fmθ power determined concentration on a task decreases. Therefore, it was found that concentration was one of factors that triggered the extent of disturbance for task. Additionally, it was suggested that there is a possibility that extent of disturbance was estimated from Fmθ power.

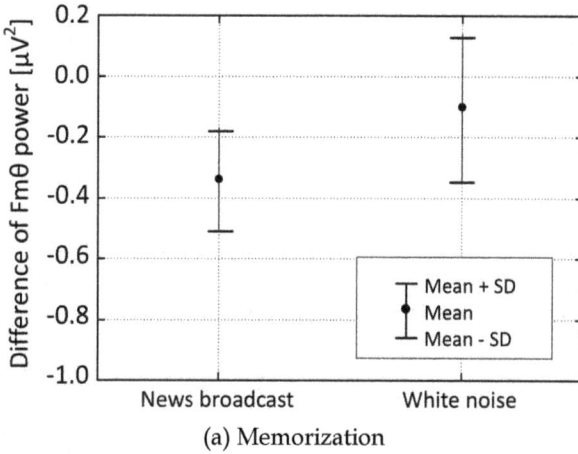

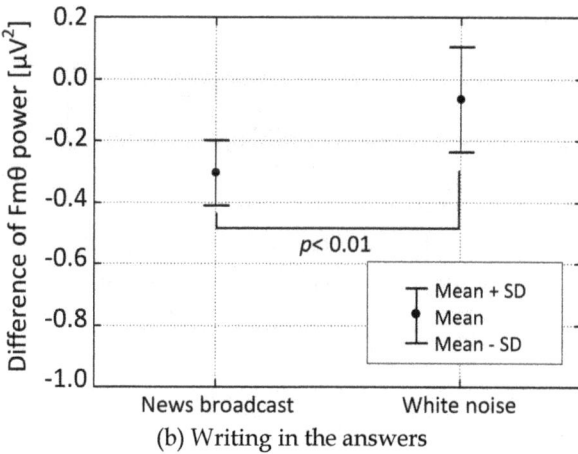

Fig. 9. Difference of Fmθ power between the condition of background noise and each noise condition (Memorization of Chinese characters). (a) Memorization of Chinese characters, (b) Writing in the answers. The y-axis shows the difference (variation) between the Fmθ potentials measured on the condition of no noise (background noise condition) and each noise condition.

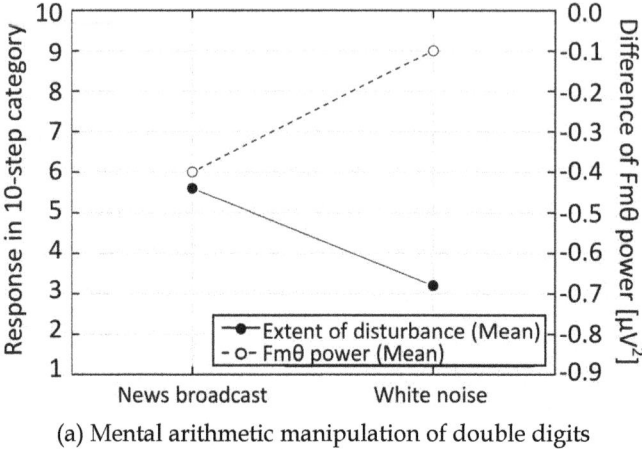

(a) Mental arithmetic manipulation of double digits

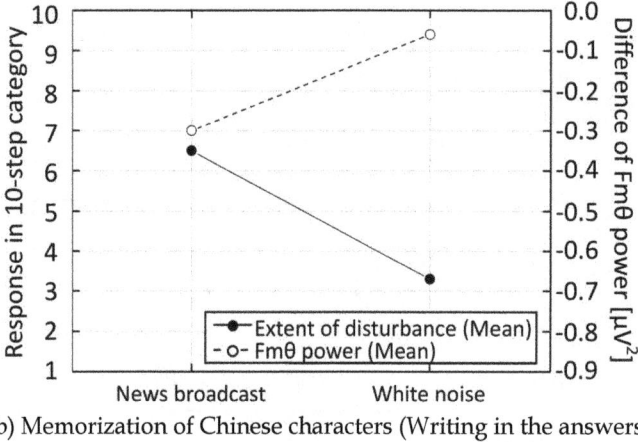

(b) Memorization of Chinese characters (Writing in the answers)

Fig. 10. Relationship between difference of Fmθ power and extent of disturbance.

4. Conclusion

In this paper, in order to find an influence of noise including verbal information on extent of disturbance or EEG under different types of brain works, we investigated: (1) effects of news broadcast and white noise on the extent of disturbance or EEG; (2) differences of them due to kinds of brain works; (3) relationship between difference of Fmθ power and extent of disturbance. From their results, it was suggested that: (1) noise including verbal information have effects on extent of disturbance more than that of nonverbal information; (2) noise including verbal information adversely affected task performance of work containing verbal information such as memorization of Chinese characters; (3) noise generated during the writing in the answers have an effect on Fmθ power more than that of during memorization of Chinese characters, memorization take more concentration than remembering; (4) there is a possibility that the extent of disturbance was estimated from Fmθ power.

These findings in our study are useful in the improvement of sound environment for office and classroom of school in the field of noise control. Furthermore, they could help to propose the development of environmental design method to upgrade creativity in office environment and the effective improvement method of sound environment for open-plan classroom. As the study applied to learning environment these findings in this report, authors have investigated the effect of indoor sound environment in a classroom on learning efficiency. In the study, the percentage of correct answers of each task, subjective judgment on degree of disturbance and the power level of frontal midline theta rhythm (Fmθ) were measured under the three types of sound environmental conditions (no-noise, air-conditioning noise, talking noise) in anechoic room (Tsujimura et al., 2010). From the results, it was confirmed that there is a negative correlation between Fmθ power and extent of disturbance, it was shown that Fmθ power was applicable to measurement of evaluation of learning efficiency as physiological index. Furthermore, it was found that noise generated during the writing in the answers have an effect on Fmθ power more than that of during memorization. It was very important findings that noise generated during unspread of idea interfere with that work in the field of noise control.

5. Acknowledgment

This research was partially supported by the Ministry of Education, Culture, Sports, Science and Technology, Grant for Young Scientists (B), 19760407, 2007.4-2010.3.

6. References

Akita, T.; Fujii, T.; Hirate, K.; Yasuoka, M. (1995). A research on auditory information processing in the brain when a person is at his task by means of measurement and analysis of auditory evoked potential, *Journal of Architecture,Planning and Environmental Engineering*, Vol.507, pp.61-69.

Brazier, M., A., B. & Casby, J., U. (1952). Crosscorrelation and autocorrelation studies of electroencephalographic potentials, *Electroencephalography and Clinical Neurophysiology*, Vol.4, pp.201-211.

Ishihara, T.; Yoshii, N. (1972). Multivariate analysis study of EEG and mental activity in juvenile delinquents, *Electroencephalography and Clinical Neurophysiology*, Vol.33, pp.71-80.

Ishihara, T.; Izumi, M. (1975). Fmθ and imaginative mental activities, *Clinical Electroencephalogram*, Vol.17, No.6, pp.381-384.

Ishihara, T.; Izumi, M. (1976). Distribution of the Fmθ on the scalp in mental calculation, resting and drowsy states, *Clinical Electroencephalogram*, Vol.18, No.10, pp.638-644.

Ishihara, T. (1991). One the appearance factors of the Fmθ induced by the 12 rarious task, *Clinical Electroencephalogram*, Vol.33, No.2, pp.96-100.

Suetsugi, M.; Mizuki, Y.; Ushijima, I.; Watanabe, Y. (2002). The relationship between rhythmic activities during a mental task and sleep spindles: a correlative analysis, *Progress in Neuro-Psychopharmacology and Biological Psychiatry*, Vol.26, pp.631-637.

Hashimoto, Y.; Nii, Y.; Naruse, T. (1999). The effects of traffic noise on physiological responses, task performance and psychological responses, *Journal of Architecture,Planning and Environmental Engineering*, Vol.515, pp.25-31.

Tsujimura, S.; Yamada, Y. (2006). A fundamental study into degree of disturbance by the sounds under the mental task using the electroencephalogram, *Journal of Architecture,Planning and Environmental Engineering,* Vol.608, pp.67-74.

Tsujimura, S.; Yamada, Y. (2008). A study on the degree of disturbance by sound under mental tasks using electroencephalogram, *Noise Control Engineering Journal,* Vol.56, pp.63-70.

Tsujimura, S.; Yamada, Y. (2007). A study on the degree of disturbance by meaningful and meaningless noise under the brain task, *Proceedings of the 19th International Congress on Acoustics 2007.*

Yoshida, H.; Ise, S. (2007). An attempt of quantitative evaluation of meaningful noise during brain work, *Proceedings of Spring Meeting Acoustical Society of Japan,* pp.389-390.

Jasper, H. H. (1958). The ten-twenty electrode system of the International Federation, *Electroencephalography and Clinical Neurophysiology,* Vol.10, pp.371-375.

Inoue, T.; Ishihara, T.; Shinosaki, K. (1984). Generating mechanism for frontal midline theta activity, *Clinical Electroencephalogram,* Vol.26, No.10, pp.795-798.

Ishihara, T.; Hirohashi, H.; Niki, Y.; Tsuda, K. (1996). Fmθ and its related theta rhythm during mental activity, *Clinical Electroencephalogram,* Vol.38, No.6, pp.369-374.

Tsujimura, S.; Ueno, K. (2010). effect of sound environment on learning efficiency in classrooms, *Journal of Architecture,Planning and Environmental Engineering,* Vol.75, No.653, pp.561-568.

9

Synergistic Noise-By-Flow Control Design of Blade-Tip in Axial Fans: Experimental Investigation

Stefano Bianchi[1], Alessandro Corsini[1] and Anthony G. Sheard[2]
[1]Sapienza University of Rome,
[2]Fläkt Woods Ltd,
[1]Italy
[2]UK

1. Introduction

The increasing concern on noise emission had recently inspired the definition of a regulatory framework providing standards on eco-design requirements for energy-using products and noise levels, i.e. the European Directive 2005/32/EC (European Parliament, 2005). The compliance to standards within the fan industry appears more stringent than for other sectors because of their use in ventilation systems entailing the direct human exposure to noise emission. This driver demands the elaboration of fan design solutions and technologies in which the exploitation of noise control strategies must not affect aerodynamic performances.

To summarise, the main generation mechanisms of aerodynamic noise in low-speed axial fans are: turbulent inflow, self noise (turbulent or laminar boundary layers, boundary layer separation), trailing edge noise, secondary flows (Fukano *et al.*, 1986), and tip leakage related noise.

Among these aerodynamic phenomena, Inoue and Kuroumaru (1989), Storer and Cumpsty (1991) and Lakshminarayana *et al.* (1995) investigated the role of tip leakage flows in compressor rotors and demonstrated the three-dimensional and unsteady nature, and the influence on aerodynamic losses and noise generation. In the context of low-speed turbomachines, Akaike *et al.* (1991) pointed out that the vortical structure near the rotor tip in industrial fans is one of the major noise generating mechanisms. The tip leakage noise can be one of the most significant sources correlated to the broadband spectral signature (Longhouse, 1978; Fukano *et al.*, 1986; Kameier & Neise, 1997). Kameier and Neise (1997) highlighted that, in addition to the broadband influence, tip leakage flows could be responsible for narrowband tones at frequencies below the blade passing frequency in coincidence with a tip vortex separation.

During the last decades, noise control has emerged as a new field of research as the literature demonstrates (Gad-el-Hak, 2000; Joslin *et al.*, 2005), and scholars have proposed noise reduction strategy classifications, which distinguish between passive, active and reactive devices. A number of research programmes envisioned designs tailored to a synergistic noise and flow control by incorporating structural (passive) or flow (active) technologies in synergy with those mechanisms leading to a reduction of the noise sources' effectiveness. Two textbook examples of this include: i) the chevron mixer for jet noise reduction among the structural-passive technology (Saiyed *et al.*, 2000), and ii) the trailing edge blowing in turbomachinery noise suppression in the active-flow technology family (Brookfield & Waitz, 2000).

According to Thomas *et al.* (2002) and Joslin *et al.* (2005), we can base a possible classification of flow and noise control technologies on the nature of the linkage between the underlying flow physics and the noise generation mechanisms. According to the exploitation of the flow control, the overall effect on performance or noise can be productive, due to the direct or indirect linkages, or counterproductive (Joslin *et al.*, 2005).

In the framework of industrial turbomachinery, the noise control techniques documented in the open literature are mostly passive systems. In terms of the flow to noise control relationship, we can regard them as exploiting a direct linkage targeting the flow features responsible for a significant noise generation. Researchers mostly base the passive or preventive control concepts on the rotor's (blade's) geometrical characteristics or its environment (casing) to control the generation mechanisms and the fluctuating forces without sacrificing aerodynamic performance. Researchers usually accomplished this goal by either reducing the leakage flow rate or by enhancing the primary-secondary flow momentum transfer. Researchers first reported casing treatments in the shroud over compressor blades in the early 1970s and with grooves (Takata & Tsukuda, 1977; Smith & Cumpsty, 1984), and stepped tip gaps (Thompson *et al.*, 1998). They found them to improve the stable operating range by weakening the tip leakage flow. In his pioneering work, Longhouse (1978) introduced rotating shrouds attached to the rotor tips to reduce tip leakage vortex noise, implying that the vortical flow near the tip is closely related to the fans' noise characteristics as well as the aerodynamics. Specifically associated with fan technology, researchers have proposed recirculating vanes and annular rings as anti-stall devices (Karlsson & Holmkvist, 1986) which commercial applications routinely utilise today. When examining the rotor blade tip, a number of contributions demonstrated the viability of anti-vortex appendages such as Quinlan and Bent's (1998) investigations, or by industrial patents for ventilating fans (Karlsson & Holmkvist, 1986; Longet, 2003; Mimura, 2003; Uselton *et al.*, 2005). Recent systematic numerical and experimental research efforts have renewed the interest on the incorporation of flow and structural control methodologies in blade design to reduce noise emission. Akturk & Camci (2010) reported a programme of work on novel tip platform extensions for energy efficiency and acoustic gains, in view of the implementation of the control capability on tip leakage swirl to ducted lift fans for vertical take-off and landing applications (Akturk and Camci, 2011a, b).

Corsini and co-workers (Corsini & Sheard, 2007; Corsini *et al.*, 2007; Bianchi *et al.*, 2009b, c) investigated the application of profiled end plates to the blade tips on compact cooling fans. They based the novel design concept on the shaping of end-plates according to the control of chord-wise evolution of leakage vortex rotation number (Corsini & Sheard, 2007; Corsini *et al.*, 2007).

This chapter outlines new concepts for the design of blade tip end-plate in subsonic axial fan rotors based on the tip leakage vortex control. Because of the role that organised structures in turbulent flow could play in the noise generation process, controlling these eddies may be one of the keys to noise suppression (Ffowcs Williams, 1977). Experimental and numerical studies proved the pay-off resulting from the adoption of tip leakage flow control technologies (Bianchi *et al.*, 2009b, c; Corsini *et al.*, 2010), and the authors speculated about the role of leakage vortex bursting phenomenon on the aerodynamic and aeroacoustic performance of such a fan class. The vortex breakdown is an intriguing and practically important phenomenon occurring in swirling flows, and depending on the application, could be a productive or counter-productive flow feature. For this reason, scholars have been interested in its control for two decades. This is an active research area that primarily the aeronautics field continues to explore, for example delaying delta wing vortices or accelerating trailing tip vortices (Spall *et al.*, 1987), and studying combustor, valve or cyclone behaviour (Escudier, 1987). The proposed design concept rationale advocates the linkage between the end-plate's chordwise shape and the augmentation-diminution of tip leakage vortex near-axis swirl (Jones *et al.*, 2001; Herrada & Shtern, 2003). Reconfiguring the end-plate at the tip implements the exploitation of this direct noise-by-flow control in order to influence the momentum transfer from the leakage flow and to force some waviness into the leakage vortex trajectory, as the delta-wing platform design (Srigrarom & Kurosaka, 2000) also shows. The authors have deliberately designed the variation of thickness to control the chordwise evolution of the leakage vortex rotation number, here chosen as the metric for the vortex swirl level. The different design criteria under scrutiny (Corsini & Sheard, 2007, 2011; Corsini *et al.*, 2009a) are intended to control the leaked flow and to induce a subtraction-addition of near-axis momentum to the tip vortex. As such, the new end-plate design concepts comprise passive control of tip vortex swirl level to enhance the mixing of coherent tip vortical structures with a favourable correlated sound field modulation (Corsini *et al.*, 2009a; Corsini and Sheard, 2011).

Section 2 describes the test fans. Section 3 illustrates the rationale of the proposed end-plate thickness distribution concepts. Section 4 presents the experimental methodology that the authors used in the investigation, while Section 5 illustrates the comparative assessment of the end-plate aerodynamics and aeroacoustics to prove the technical merits of a passive control strategy for controlling leakage flow and rotor-tip self noise. The chapter concludes, in Section 6, with a discussion of the findings and a summary of the major conclusions.

2. Axial fans

The authors conducted the present research on a family of commercially available cooling fans. The in-service experiences indicated that this family of fans gives good acoustic performance with respect to the state-of-the-art configurations. The modified ARA-D type blade sections were single-parameter airfoils originally conceived for propellers. Clearance between the blade tips and the casing was constant at 1% of blade span. The impeller had a mechanism for varying the blade-pitch angle to customise the load, airflow, and stalling properties. Table 1 shows the blade profiles geometry and the fan rotor specifications. Data in Table 1 refer to the *datum* fan family, coded AC90/6.

The authors operated the fan in a custom-built casing made from rings of cast and machined steel. The studied blade configurations, for *datum* and rotors fitted with tip end-plates, feature a high tip pitch angle, i.e. 28 degrees, measured, as is customary in industrial fan practice, from the peripheral direction. The authors selected this angular setting which corresponds to the fan unit peak performance. Notably, the datum fan features a peak pressure rise of about 300 Pa at 5 m^3/s with a design pressure rise of point of 270 Pa at 7 m^3/s and 0.77 of total efficiency.

	AC90/6 datum fan	
Blade geometry	hub	tip
ℓ / t	1.32	0.31
stagger angle (deg)	54	62
camber angle (deg)	46	41
Rotor specifications		
blade count	6	
blade tip pitch angle (deg)	28	
hub diameter D_h (mm)	200.0	
casing diameter D_c (mm)	900.0	
rotor tip clearance τ (% span)	1.0	
rated rotational frequency (rpm)	900 – 950	

Table 1. AC90/6 fan family specifications. *Datum* blade geometry and rotor specifications

3. Design of tip end-plates for noise-by-flow control

3.1 Background studies on tip leakage vortex

Corsini and Sheard (2007) and Corsini *et al.* (2007, 2010) have assessed the performance gains related with the development of end-plate technology, consisting of constant thickness end-plate originally designed at Fläkt Woods Ltd for an industrial fan designated AC90/6/TF. The authors studied numerically the leakage flow evolution to isolate the mechanisms that govern the improvement in the aeroacoustic signature. The numerical investigation tool was based on an original parallel Finite Element flow solver (Corsini et al., 2005). Despite the steady-state conditions (Corsini *et al.*, 2007, 2010), RANS was adopted as an effective tool for this investigation because it is capable of capturing details of the vortex structures at the tip (Escudier & Zehnder, 1982, Inoue & Furukawa, 2002). One of the main findings was that in rotor AC90/6/TF, the leakage flow control that the aerodynamic appendages at the blade tip gave rise along the operating line, in the near-design operation, to a tip leakage vortex bursting.

A multi-criteria analysis assessed the occurrence of tip leakage vortex breakdown in the state-of-the-art fan rotor (Corsini *et al.*, 2010). First, the visualization of three-dimensional streamlines at the tip revealed the presence of a large bubble-type separation with the flow reversal being indicative of the critical bursting phenomenon (Lucca-Negro & O'Doherty, 2001). According to Leibovich (1982), the appearance of the vortex bursting in the bubble form can be interpreted as a consequence of a sufficiently large swirl level and this form of the vortex breakdown is increasingly steady with the level of swirl (Escudier & Zehnder, 1982, Inoue & Furukawa 2002). Second, the presence of a stagnation point along the vortex axis indicated the onset of the vortex collapse in accordance with the observation of Spall et al. (1987), who first identified it as a necessary condition for the breakdown to appear. As a last criterion, the most distinctive feature was the tip leakage vortex helicity inversion occurring as a result of the counter-rotation of the vortex exiting from the bursting region, accepted as evidence of vortex breakdown in compressor rotors (Inoue & Furkawa, 2002). In particular this evidence could be interpreted, according to Leibovich (1982) explanation of the vortex breakdown on the basis of the behaviour of the azimuthal vortex component, as a consequence of a change in the angle between the velocity and vortex vectors.

In order to suppress the tip leakage vortex-bursting phenomenon in the fan rotor fitted with a base-line tip feature, the authors proposed new end-plate design concepts to exploit a passive control on vortex breakdown formation (Corsini & Sheard, 2007, 2011).

3.2 Variable thickness end-plate design concept

The *Variable Thickness End-plate* design concept accomplishes the control on vortex breakdown onset advocating the use of a chordwise variation, in particular diminution, of the end-plate thickness. The authors designated the fan rotor fitted with this end-plate geometry as AC90/6/TFvte.

To suppress the tip-leakage vortex-bursting phenomenon in the operation of the AC90/6/TF fan rotor, the authors recently proposed a new end-plate design concept. The proposed configuration controls vortex breakdown by means of an end-plate of variable chord-wise thickness. The first improved concept's aim was to enhance near-axis swirl ((Jones *et al.*, 2001; Herrada & Shtern, 2003) by reconfiguring the end-plate at the tip with a view to influencing the momentum transfer from the leakage flow and to force some waviness into the leakage-vortex trajectory. According to the analysis of tip-leakage vortices in terms of a Rossby number based swirl metric (Spall *et al.*, 1987), the end-plate's shape depends on the definition of a safe rotation number chordwise gradient. The investigation's rationale was to compute the end-plate thickness distribution by combining a simplified law for the tip-gap pressure drop with a stability criterion of tip leakage vortex prescribed by the vortex rotation's safe chordwise distribution or Rossby number.

Notably, as Corsini *et al.* (2009b) also explained, the adoption of variable thickness end-plate resulted in improved pressure rise capability and efficiency owing to the control on tip vortex bursting. The counterpart of this aerodynamic pay-off was that the presence of leakage related organised vortical structure in TFvte end-plate associated with a slightly higher sound A-filtered power level SWL.

3.3 Multiple vortex-breakdown end-plate design concept

In accordance with the background experiences on tip leakage vortex swirl behaviour, the second end-plate design criterion, named *Multiple Vortex Breakdown*, advocates the linkage between the end-plate geometry and the modulation of tip leakage vortex near-axis swirl. We can interpret this direct linkage flow-to-noise control by deliberately designing the shape of the end-plate at the tip to control the chordwise evolution of the leakage vortex rotation number.

The design criterion is to control the leaked flow in order to induce a sequence of subtraction of momentum transfer to the tip vortex up to a near-critical condition, followed by an addition of near-axis momentum (Corsini *et al.*, 2009a). As such, the new end-plate design concept exploits a passive control of tip vortex swirl level based on a succession of breakdown or bursting conditions able to enhance the mixing of coherent tip vortical structures with a favourable modulation of the correlated sound field.

The rationale is to derive an explicit correlation among the tip end-plate thickness, the tip vortex kinematics, the local blade loading, and the magnitude of the leakage flow. In the vein of the flow mechanism under control, which we base on the modulation of the tip vortex rotation number, the authors built the analytical law for the end-plate geometry advocating the definition of the Rossby number as the problem's key metric.

3.4 Tip end-plate aerodynamics

To complete the illustration of the passive noise concepts, Figure 1 outlines the main background findings on improved tip design. In detail, Figure 1.a shows the chordwise distribution of the tip leakage vortex swirl level at design condition. The authors selected the Rossby number Ro swirl metric definition in acccordance with the interpretive criterion on vortex breakdown Ito and co-workers (1985) proposed. The figure gives the evolution of the Rossby number on the chord fraction for four different end-plate configurations, namely the base-line with constant thickness (AC90/6/TF), the variable thickness end-plate (AC90/6/TFvte) and the MVB end-plate design (AC90/6/TFmvb) plotted against the *datum* one. Moreover, Figure 1.b shows a sketch of the blade tip for the AC90/6 class of fan unit.

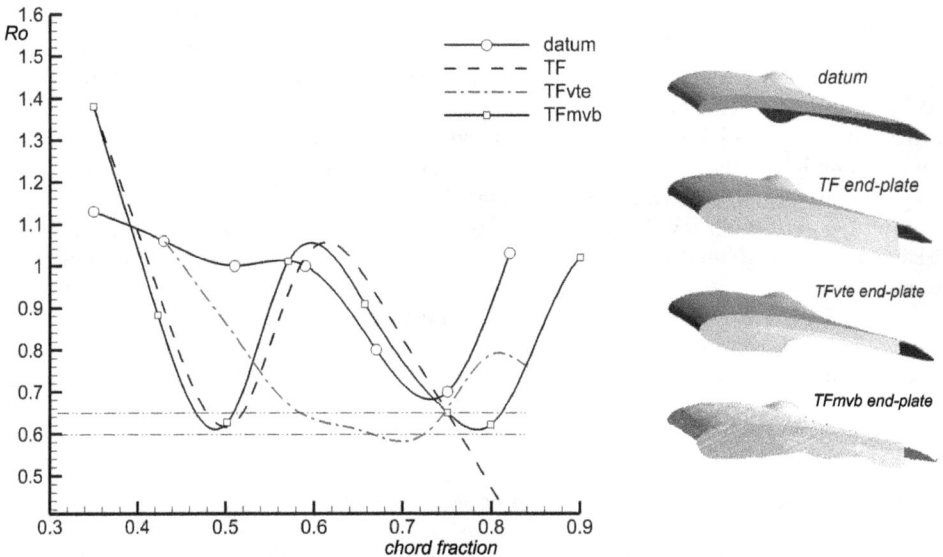

Fig. 1. Background on improved tip concepts, chordwise evolution of tip leakage vortex Rossby number (left), and end-plate geometries (right)

The Ro distributions, Figure 1 (left), indicate that the AC90/6/TF fan features a vortex breakdown in coincidence with the attainment of the Rossby number's critical value. The critical Ro number values that the open literature mentions range from 0.64 for the breakdown of a confined axi-symmetric vortex (Uchida *et al.*, 1985), to 0.6 for wing tip vortices with bubble or spiral type vortex breakdown (Garg & Leibovich, 1979). Figure 1 (right) depicts the end-plate geometries (Corsini & Sheard, 2007, 2011).

3.4.1 Fan performance

The main performance parameters were the fan total pressure and the efficiency. The authors measured the static and dynamic pressure for this study in the Fläkt Woods test rig at Colchester (UK). They based the performance measurements on pressure taps, equally spaced on the casing wall, and a standard Pitot-probe. The authors mounted the probe on a traverse mechanism fixed to the test rig's outer wall. The Furness digital multichannel micro-manometer (Model FC012, Furness Controls Ltd, UK) had 2 kPa range and a resolution of 1 Pa on pressure data. The pressure measurements accuracy was ± 0.5% of read data. The authors calculated the efficiency as the ratio between the air power (computed on either static or dynamic pressure rise) and the electric power. They measured the absorbed electric power with an AC power analyser with an accuracy of 0.24% of read data.

Figure 2 plots the curves of the total pressure and the efficiency based on total pressure for the datum fan rotor.

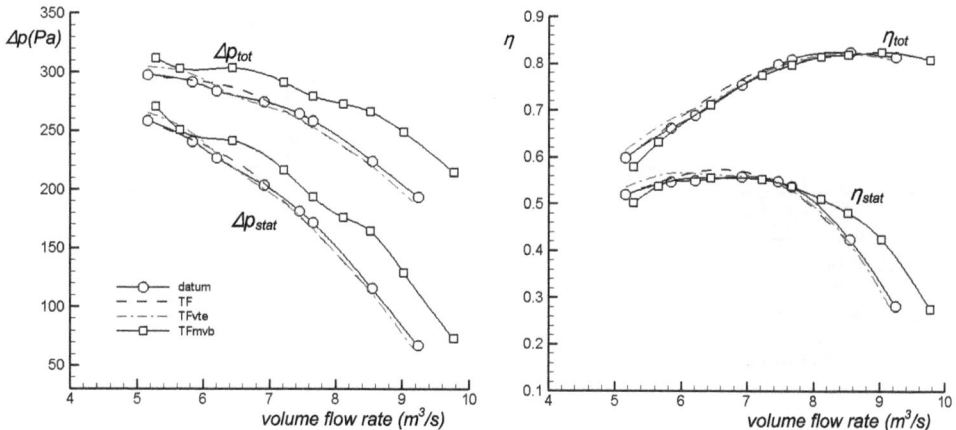

Fig. 2. Measured pressure (left) and efficiency (right) characteristic curves

4. Methodology

4.1 Noise measurements

The authors designed this study's methodology to achieve the following specific objectives: (i) to characterise the azimuthal distribution of the radiated sound pressure level (Lp) of the various fan-blade configurations; and (ii) to investigate the directivity of the Lp spectra and (iii) speculate about the aerodynamic origins of the noise. In pursuit of these aims, the study measured the azimuthal far-field fan noise radiated from the exhaust flow according to the BS848-2.6:2000 standard (ISO 2000), which is equivalent to the ISO 10302:1996 standard.

The authors conducted the experiments in the Fläkt Woods anechoic chamber at Colchester (UK). The chamber is an anechoic facility for noise emission certification according to the BS848-2.6:2000 standard (ISO 2000). The anechoic chamber cut-off frequency is 25 Hz. Bianchi et al. (2009a) report additional details on the experimental setting. The authors conducted measurements at the fan exhaust. Table 2 gives the main specifications of the near- and far-field probes.

The authors acquired microphone signals with a 01-dB Symphonie digital signal processor. In all cases, the authors measured signals for 2.85×10^4 rotor revolutions. The balancement of rotors ensures a vibration peak no greater than 4 mm/s.

Pre-amplifier	Brüel & Kjær equipment connected to a 2 channels Symphonie acquisition system
Near-field microphone	GRAS 40AG
Diameter	2.7 mm
Frequency range	0 Hz – 20 kHz
Sampling frequency	50 kHz
Dynamic range	27 dB – 160 dB
Far-field microphone	Brüel & Kjær 4954
Frequency range	0 Hz – 20 kHz
Sampling frequency	50 kHz
Dynamic range	Up to 165 dB
Sound analyser	01dB-Metravib dBFA Suite

Table 2. Specifications of the near- and far-field microphones

The authors operated the fan in a custom-built casing made from cast and machined steel rings. Clearance between the blade tips and the casing was constant at 1% of blade span. The authors used a 2.5-kW, direct coupled-induction 400-volt (AC), 3-phase CM29 motor to drive the rotor at a constant speed of 945±5 rpm. The blade tip's speed was 44.3 m/s. Under these conditions, the blade-passing frequencies (BPFs) for all the tested configurations were in the range 93.75±3.125 Hz. To isolate the aerodynamic noise, the authors conducted a preliminary motor test to establish its spectral signature so as to allow for a correction of subsequent noise measurements. In all cases, the authors measured the signals for 30 s. They repeated each measurement three times. The error was in the range of 0.1-0.2 dB at 1kHz, as given by the calibration certification on the microphones and the acquisition system according to ISO IEC60651. Figure 3 shows a photograph of the test-rig, together with the set-up of the near- and far-field probes.

The authors installed the fan downstream from the plenum chamber, using a type-A configuration, with a free inlet blowing on the motor and on the four struts, as is customary for the in-service operation. They mounted a bell-mouth on the fan inlet. The shape of this bell-mouth was aerodynamically optimised to give uniform and unseparated flow into the fan. The rotor centre line was 2 m from the floor, in an arrangement similar for compact cooling fans. The authors acoustically treated the downstream and upstream plena, which connected the fan to the outside environment, to minimise both incidental noise from the induced air stream, and external noise transmission from the outside environment. A louver in the top of the facility's inlet section enabled the authors to make variations in both the fan head and flow rate. With the far-field microphone set at the same height as the rotor centre, the authors measured far-field noise at a distance of two fan diameters from the outflow sections (Leggat & Siddon, 1978), in accordance with the prescribed BS 848 standard for far-field noise testing in a semi-reverberant environment (ISO 2000).

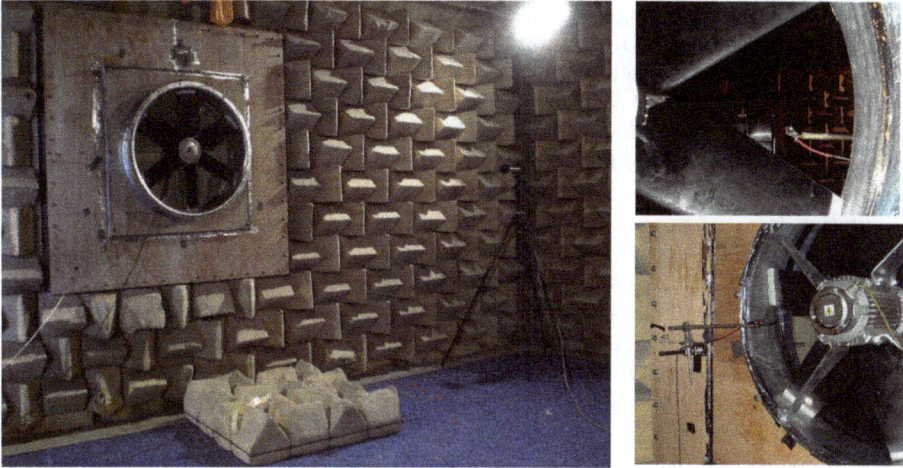

Fig. 3. Test-rig and instrumentations

The authors measured the noise at several azimuthal locations from -90° to +90° from the rotor exhaust centre line with an angular increment of 30°. Air-speed measurements at the far-field microphone recorded negligible effects for all microphones.

4.2 Noise source dissection

The causal analysis utilises simultaneous measurements of the hydrodynamic pressure in the near-field and of the acoustic pressure in the far-field to separate aerodynamic noise sources at fan outlet. The authors based the adopted experimental technique in the present study, to varying degrees, on previous fan (Leggat & Siddon, 1978; Bianchi *et al.*, 2009b), radial pumps (Mongeau *et al.*, 1995), and turbofan engines (Kameier & Neise, 1997; Miles, 2006) studies. The experimental technique consisted of a pressure-field microphone traversing the blade's span with simultaneous pressure measurements in the far-field. The authors simultaneously recorded the signals from the near-field microphone and the far-field microphone on two separate channels of a 01-dB Symphonie digital signal processor.

The use of a cross correlation technique during data analysis enabled isolation of the near-field noise sources' signature from the far-field noise signal, thus avoiding error due to 'pseudo sound' (defined as turbulence-generated noise that a recording can pick up in the near-field, but which decays so quickly that it cannot be responsible for noise emission to the external environment). Correlation revealed a causal relationship between individual noise-source phenomena and the overall radiated sound in a given direction, thereby yielding quantitative information about the distribution of acoustic sources, their local spectra, and the degree of coherence. If a strong harmonic coupling between source and far-field spectra existed, it was apparent that the resulting correlation function did not decay quickly, but was periodic in nature.

Concerning the location of the near-field probe, the limit of the causal analysis in the present work is that all the measured aerodynamic sources in the near-field larger than the correlation length resulted in poor matching with the far-field and the cross-correlation did not highlight. Notably, the rotor blades under scrutiny operate with a tip-based Reynolds number range of order 7.0×10^5, typically featuring an irregular vortex shedding. In agreement with numerous investigators who have reported on the correlation length dependence on flow conditions (Blake 1986), in a Reynolds number range 3.3×10^3 to 7.5×10^5 the correlation length is proportional to the length scale of the shed vortices d i.e. $\Lambda_3 \approx d/2$. Past numerical studies in axial flow fans (Corsini *et al.*, 2009b) showed d to be 10% to 20% of the blade span, when referring to the vortices shed by rotor blade passages. Accordingly, the value of the correlation length in the present test configuration is $\Lambda_3 \approx 7\%$ of the blade mid-span chord. In order to fulfill the correlation length threshold, the authors mounted the near-field test-rig setup microphone at 10% of the mid-span chord from the blade trailing edge, on a 10-mm wide traversing arm which was capable of moving the microphone along the whole blade span. The probe traversed the blade span in constant steps, each of which was 2% of the blade span. A rubber panel isolated the traversing arm from vibration in the chamber. During the measurements, a nose-cone windshield protected the microphone diaphragm, upon which the authors conducted a preliminary test to quantify its self-induced noise and thus enabled them to include a correction factor in data-processing calculations. The acoustic effects compensated for the measurement technique that the near-field microphone presence induced (Bianchi *et al.*, 2009b).

With the far-field microphone set at the same height as the rotor centre, the authors measured far-field noise at a distance of 2.3 fan diameters from the outflow section. Air-speed measurements at the far-field microphone recorded negligible airflow for all of these microphone positions, with the exception of the one at 0°. At such a distance, as Leggat & Siddon (1978) also found, we can ignore the direct influence of the flow over the microphone. Moreover, the researchers corrected for wind noise using a correction factor that the microphone manufacturer provided (Brüel & Kjær, 2006).

5. Experimental investigations

The aim of the family of blades experimental investigations which the authors designed according to the proposed synergistic passive noise control concepts, is to speculate on their aeroacoustic pay-off. The authors explore the nature of the noise-to-flow productive linkage by investigating the space distribution of the noise's directivity pattern and within the rotor blade span in the near-field. The presentation of the experimental findings includes: (i) integral Lp analysis of the directivity for the different configurations; (ii) spectral analysis of directivity of the rotor noise sources in the radiated field; and (iii) a description of the near-field sources responsible for the aerodynamic noise along the blade span.

The authors examined the fan's acoustic performance when fitted with each of the three blade-tip configurations (*datum*, TF, TFvte, TFmvb) under near-design operating conditions, with a volume flow rate of $V = 7 \text{ m}^3/\text{s}$.

5.1 Overall noise emission and directivity

The authors derived the Lp auto-spectra of the instantaneous pressure signals in the far-field under the tested operating condition, and then integrated it to calculate the Lp's azimuthal radiation patterns at seven angular positions on the rotor axis plane. Figure 4 shows the integrated Lp value of the *datum* fan plotted against those of the rotors fitted with the end-plates. It is evident that the peak levels aligned with the fan axis for the *datum*, TF and TFvte fans. Remarkably, the TFmvb fan features a nearly isotropic radiation pattern with a large attenuation (as compared to all the other fan blades) of the axial level. As expected (Leggat & Siddon, 1978), the maxima in the near-axis direction for the *datum*, TF and TFvte fans indicated a significant dipolar noise source in the fan outlet. In the near-axis region, TF had a maximum Lp level of 82 dB, similar to the *datum* and the TFvte maxima (81 dB); whereas, the TFmvb peak was at 71 dB. According to previous directivity studies in a different test-rig set-up (Bianchi *et al.*, 2011a), the tendency toward the attenuation of in-axis noise correlated with the presence of coherent swirling structures in the exhaust flow. TFvte noise control design is even more magnified in the TFmvb end-plates.

When moving away from the fan axis, all the directivity patterns of the modified impellers differed from that of the *datum* fan. While the *datum*, the TF and the TFvte fans featured a dipolar-like emission, the TFmvb directivity confirmed its isotropic radiation. In a jet-noise sense, the TFmvb far-field directivity is a consequence of the shifting of low frequency noise sources (radiating axially) to the high frequency range (radiating laterally) which we can consider as the acoustic counterpart of the control implemented on the tip leakage vortex (enhanced mixing) and shed vorticity (reduction of 3D blade separation). This interpretation is routed on the causal relationship between small-scale random turbulent structures and high-frequency sound with nearly-isotropic radiation, and large-scale coherent eddies with directional patterns according to their frequency scales.

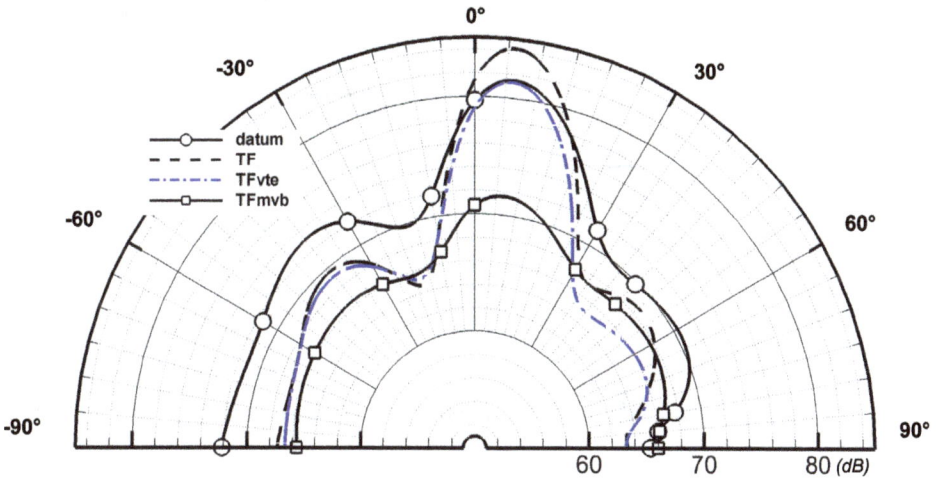

Fig. 4. Directivity of the integrated Lp for the different blade geometries.

The analyses showed that the dipolar noise characteristic of the *datum*, and original TF rotors changed in the TFmvb as a result of the superposition of reduced intensity longitudinal dipole and a lateral quadrupole. As Shah *et al.* (2007) previously stated, we can attribute this quadrupole-like noise source to the turbulent mixing of the swirling core flow which scatters the dipolar source related to the presence of a coherent vortical structure.

To give additional hints on the overall acoustic performance, Figure 4 shows polar diagrams correlating the A-filtered sound power level (LwA) and the specific noise level

(Ks) when varying the airflow power in the fan operating range. We define the specific noise level as Ks = Lw − 10 log[V (Δp_{tot})²] (dB). The polar representation clearly shows the pay-off in the noise emission of the TFmvb, when compared with the datum and the TF and TFvte rotors.

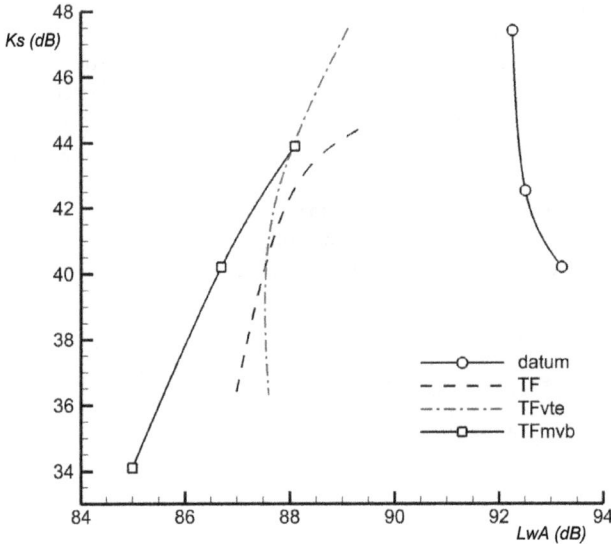

Fig. 5. Comparison of A-filtered overall sound power level (LwA) – specific noise level (Ks) polar plot.

Notably, the *datum* fan features a unique behaviour entailing, when increasing the power, a specific noise level reduction and an augmented sound power level, LwA. This suggests that the increased aerodynamic share in the sound generating mechanism, due to the actual fan operations, radiates more efficiently to the far-field and enhances the emitted noise in the audible frequency range. In contrast, both the fan blade fitted with end-plates show a difference on the sound power-specific power levels. In particular, the TF end-plate shows that the reduction of specific noise level with aerodynamic power correlates to an attenuation of the emitted noise according to a power law Ks ∝ (LwA)ⁿ with n = 0.5. Instead, the TFmvb speeds-up the reduction in sound levels featuring a nearly linear distribution of exponent n = 1.

5.2 Fourier analysis

In order to give additional hints on the directivity of noise spectra, Figure 5 compares the Lp narrow-band auto-spectra at three azimuthal positions in the far-field, namely 30° (top), 60° (mid) and 90° (bottom) angles to the fan axis. The authors mapped the Lp spectra in the frequency range 25 Hz to 10kHz.

It was confirmed that the noise radiation had a specific directional characterisation, irrespective to the particular geometry, both in the tonal noise and in the broadband components. The authors recorded the major acoustic emissions as expected, near the fan centre-line in the 30° position, in accordance with the dipolar-like noise signatures from early studies on low-speed fans (Wright, 1976) (Leggat & Siddon, 1978). Clearly, as the noise direction path moves farther from the axis line, two phenomena became evident: i) the Lp attenuation is mostly concentrated above 1 kHz, and ii) some of the fans featured an spectral tone noise enrichment as the aerodynamic broadband noise reduced its influence and allowed the tonal sources to emerge.

The directivity map of the Lp far-field autospectra at 30° shows that that all the tested fan blades, with the exception of the TFmvb one, had similar spectral behaviour with a plateau from the second fundamental harmonic to 1 kHz and in the narrowband where the only remarkable difference is the presence of emerging tones for the datum Lp over the tenth BPF. The TFmvb spectrum, on the opposite, featured an amplification of the second BPF, usually considered as a signature of the tip leakage vortex self-noise. Here the Lp of the TFmvb was 7 dB higher than the TF, and more than 10 dB higher than *datum* and TFvte. TFmvb tones were also evident on the third and the sixth BPFs which coincide with equivalent TF end-plate peaks. In the high frequency range, the TFmvb featured a steeper Lp attenuation (6 dB/kHz against 3.2 dB/kHz for the other blade geometries in the range 1kHz to 4 kHz). Moreover, the spectrum suddenly lose the higher harmonic tones in contrast to TF and TFvte spectra where tonal components emerge up to 8 kHz.

Similar noise frequency distribution appeared for the microphone at 60°. Nonetheless, there are noticeable small differences: i) a general reduction in the global noise level, ii) a further enrichment of the noise's tonal structure, and iii) the occurrence of an abrupt rise of Lp in the TF distribution above 5 kHz. Moving to the side of the fan exhaust at 90°, the Lp spectra changed remarkably, over the entire frequency range, in terms of Lp magnitude (i.e. at 1 kHz, Lp levels are comparable to those recorded at 30°) and broad-band attenuation.

In the spectra's low frequency portion, the TFmvb again featured a dominant tone on the second and third BPFs, with some degree of similarity to the TF spectrum only. Notably, after 300 Hz, the TFmvb spectrum lost any significant tonal behavior. The authors found similar spectra modifications in the broad-band at frequencies higher than 1 kHz for the *datum* and the TFvte. A different behavior was evident in the TF Lp spectrum, which again featured a significant rise in the broad-band. The authors attributed this noise increase to a particular phenomenon that they had studied in the TF fan (Corsini et *al.*, 2010). Close to the TF blade' trailing edge, the tip-leakage vortex collapsed and produced a 'bubble-type' separation that indicates vortex breakdown. The separated flow turned into a counter clockwise vortex under the influence of trailing-edge leakage flow streams. This results in a noise emission increase in the frequency range of interest as a consiquence of chaotic turbulent flow which is usually located in the mid-high frequency range.

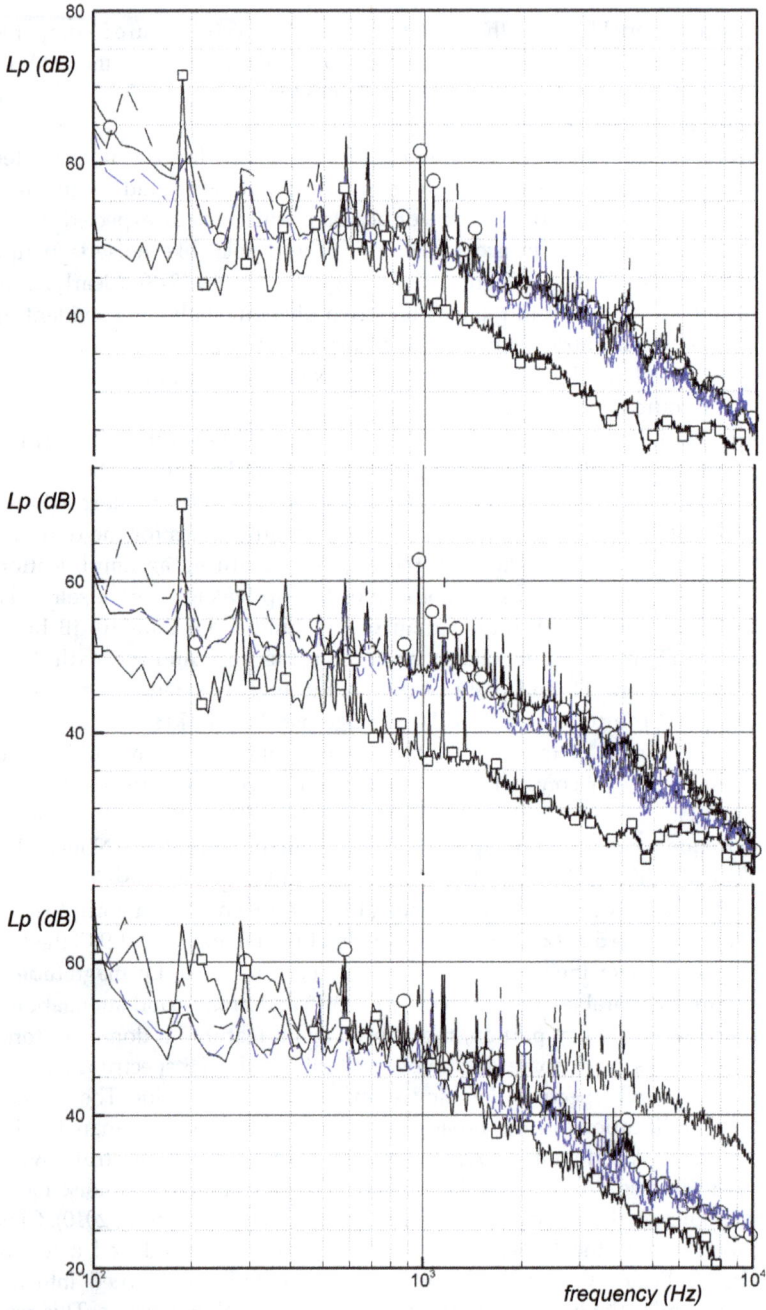

Fig. 6. Auto-spectra of the Lp for the different blade geometries at three azimuthal position: 30° (top), 60° (mid) and 90° (bottom)

The authors then calculated the Lw and Lw(A) power spectra which Figure 6 shows. The Lw spectra (Figure 6, top) demonstrated that, when integrating the pressure on the whole hemisphere, the noise that the different geometries emit for frequencies lower than 1 kHz likely had the same power level. In particular, the Lw spectra show distinctive features on the overall tonal behaviour, up to the 10th BPF harmonic. Notably, Figure 6 shows a differentiation of the *datum* from the rotors' spectra fitted with the end-plates, which gave evidence of background noise power level reduction and of the broad band noise level (about 10 dB difference to the *datum*). Concerning the A weighted noise power level (Figure 6, bottom), the differences are even more little. TF and TFvte spectra are pretty much super-positioned and they showed minimal differences in the emitted LwA. The TFmvb is similar to them for frequencies up to 1 kHz and then became quieter. Usually it was 10 dB quieter than the others, at the average, up to -14 dB in frequencies less than 5 kHz and -8 dB for the frequencies up to 10 kHz.

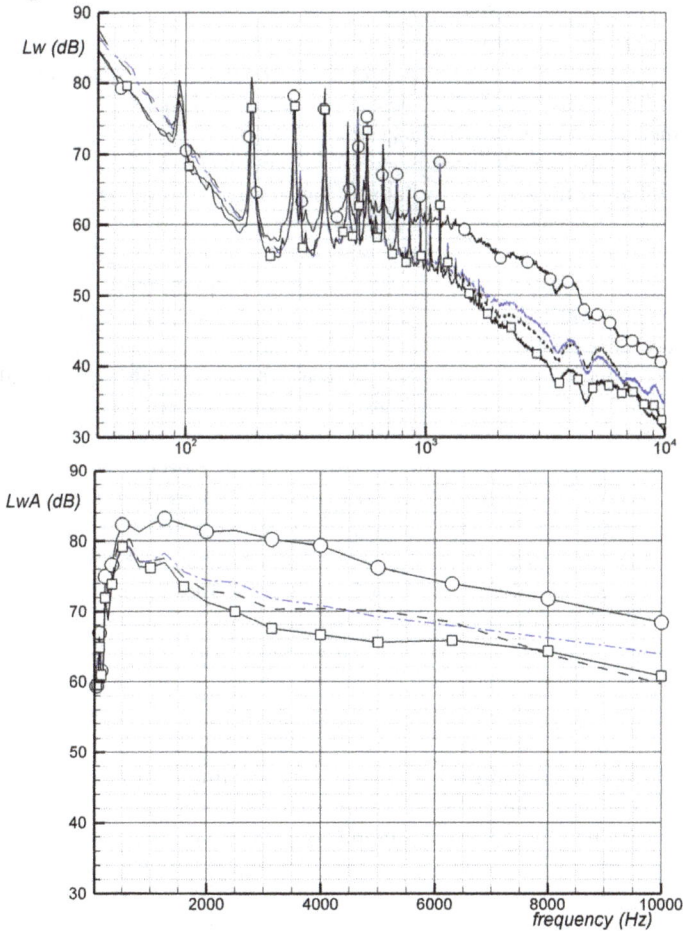

Fig. 7. Comparison of Lw (top) and LwA (bottom) for the different blade geometries

The authors undertook identification of noise sources in the flow exiting the fan rotor (as fitted with the various proposed end-plates) by using a coherence analysis, in the way that Bianchi *et al.* (2011b) recently developed. The coherence function (γ) involves a normalisation by frequency band that tends to highlight highly coherent events independently from their energy content. For this reason it is an effective mean of investigating near- to far-field acoustics, providing a resolution in noise space in the source regions. The authors plotted the coherence between the near-field and far-field microphone as a function of the radial position r and the non-dimensional frequency $\Im$, defined as *f/BPF*. Figure 8 illustrates, first, the coherence function's spanwise map for the rotor geometries at 30°. Similarly, Figures 9 and 10 present coherence distribution at 60° and 90° respectively, for the considered rotor geometries: *datum*, TF, TFvte, and TFmvb. The authors set the coherence threshold at $\gamma = 0.25$, as given by the anechoic room coherence response (Leggat and Siddon (1978) proscribed this value). The authors compared all four blade configurations (*datum*, TF, TFvte, and TFmvb) under near-design operating conditions.

The position at 30°, Figure 8, was closer to the fan axis and should be the one more influenced by the rotor discharged flow noise. The *datum* rotor map inferred about the blade-correlated noise sources, discriminating those capable of radiating downstream, thus affecting the progressive far-field. Some distinguished peaks of high coherence characterised this rotor along the whole span at $\Im = 6$ and at $\Im = 7$ and other tones which radiated mostly from the blade's inner part at $\Im = 3$ and at $\Im = 4$. The coherent tone at $\Im = 5$ appeared to correlate to the sources radiating from the hub to mid-span, together with the inter-tonal coherence between $\Im = 6 - 7$. Moreover, *datum*'s map also provided evidence of non-tonal coherent phenomena governing the far-field emission mechanisms in the blade tip's proximity and at the rotor hub. The first coherence region, mainly concentrated in the region from mid-span to tip for a range of $\Im = 3 - 5$, indicated a noise source related to the interaction between the tip-clearance vortex and the rotor wake in the region of highest load as a result of the blade design. At the blade root, in a similar frequency interval that is now slightly larger, $\Im=3-7$, the coherence iso-lines concentration indicated a second aerodynamic noise source caused by a hub corner stall interacting with the passage vortex.

It was possible to isolate coherent patterns attributable to specific tip vortex from the TF rotor map. The TF coherence map identified some differences from the *datum* one, which were apparently related to the aeroacoustic gains that Corsini *et al.* (2009b) previously studied. At higher radii, the TF end-plate appeared to cancel the coherence of tones, with the exception of the hub-corner interaction, which was only reduced, resulting in coherence loss in the region where the tip leakage vortex played the major influence on noise. Examination of the interaction noise cores showed that the region up to $\Im = 6$, appeared inefficiently radiated due to the reduced coherence. However, at about the sixth BPF harmonic, the authors found tonal-like coherence peaks, which, according to their radial position and frequency range, could relate to the tip-leakage vortex bursting that Corsini and Sheard (2007) detected in previous studies of this configuration under these operating conditions. The TF coherence map also showed evidence of a change in the non-tonal noise emission in the rotor hub's region. The coherent region's reduced extension, in terms of radii and frequencies, was in accordance with the control of secondary flows that attenuation of the near-surface fluid centrifugation produced.

The TFvte end-plate coherence distribution confirmed the test results of the constant-thickness tip concept. In particular, this configuration demonstrated an ability to eliminate the coherence from the tones (remarkably, on the range of low $\mathfrak{I}$) at the tip and to decrease coherence due to tip-flow interaction noise sources. In addition, at the hub, this end-plate was capable of reducing the emission effectiveness from the secondary flow noise sources. Finally, the variable-thickness end-plate, owing to its design concept, reduced the extent and coherence level in the vortex breakdown radial and frequency ranges as a consequence of control of leakage vortex rotation.

The TFmvb end-plate revealed that the sequence of positive and negative *momentum* resulted in specific pattern that clearly distinguished these configurations from the original TF and TFvte configurations. First, both TFmvb and TFvte enhanced tone coherence, but reduced coherence with respect to the tip leakage interaction sources (in the range $\mathfrak{I}$ =3–5). Second, the application of multiple vortex breakdown concept on the end-plate configuration outperformed all the other configurations in reducing the breakdown noise correlation in the region about $\mathfrak{I}$ = 3. These findings provided evidence that the mechanism of the multiple vortex breakdown configurations consisted of: (i) acting as a mixing enhancer in the tip-leakage region; and (ii) then increasing the degree of scattering of the local noise sources.

Figure 9 compares the coherence of the instantaneous pressure/noise correlation of the four rotors at α = 60°. This measurement confirmed the previous behaviour, as Figure 7 indicates, for all the geometries and the coherence shape of the noise sources did not appear to change. The only difference was that the authors found specific traces of ingested noise (low frequency number range) in all the three rotors from the inlet plenum and in the signature of the rotor-only noise. With regard to this source, the authors found it to influence coherence on the first frequency number (spanning from the hub to the tip), irrespective of the tested tip configuration. Figure 9 then provided evidence of non-tonal coherent phenomena governing the far-field emission mechanisms in the low frequency which radiated exclusively to this angular pattern.

Finally, Figure 10 shows the coherence maps for α = 90°. In the narrowband at frequencies lower than the frequency number associated to the BPF ($\mathfrak{I}$ = 0.25 and 0.66), the authors also identified the rotor-alone noise. According to Cumpsty (1977), the existence of such a low-frequency tone is a noise source directly correlated to the upstream flow distortions that the rotor produces. The rotor presence in the duct usually causes the noise sources, without distinction whether the rotor spins or not. To support the evidence of ingested flow noise, Figure 10 represents coherence spectra near the *datum*. The analysis of Figure 10 suggested a maximum correlation was for the sources at very low frequency, located close to the hub and influencing the whole span. This observation infers that the noise already in the $\mathfrak{I}$ = 60° was an ingested noise. The path of this distortion then interacted with the rotor's bottom part and its tone was partly scattered to the high frequency broadband. Although these sources are present in all the geometries, as their cause is something excited by the blade's dynamic effects, the *datum* rotor showed to have less control of these low frequency noise sources. This is because the end-plates control the flow path along the blade, making possible a purely 2D flow span distribution, due to the flow blockage which the end-plate at the tip operates. In more detail, whilst the TFmvb reduced the tones to only one, but produced along all the blade span, the TF and, even better, TFvte also reduced the coherence of the remaining unique source. The rotor alone noise tone in the TF and TFvte geometries were so concentrated mainly in the tip zone and showed a possible, more effective, control of these end-plates on the flow distribution along the span with the consequence of reducing the noise excitation of the blade's bottom part with the incoming flow.

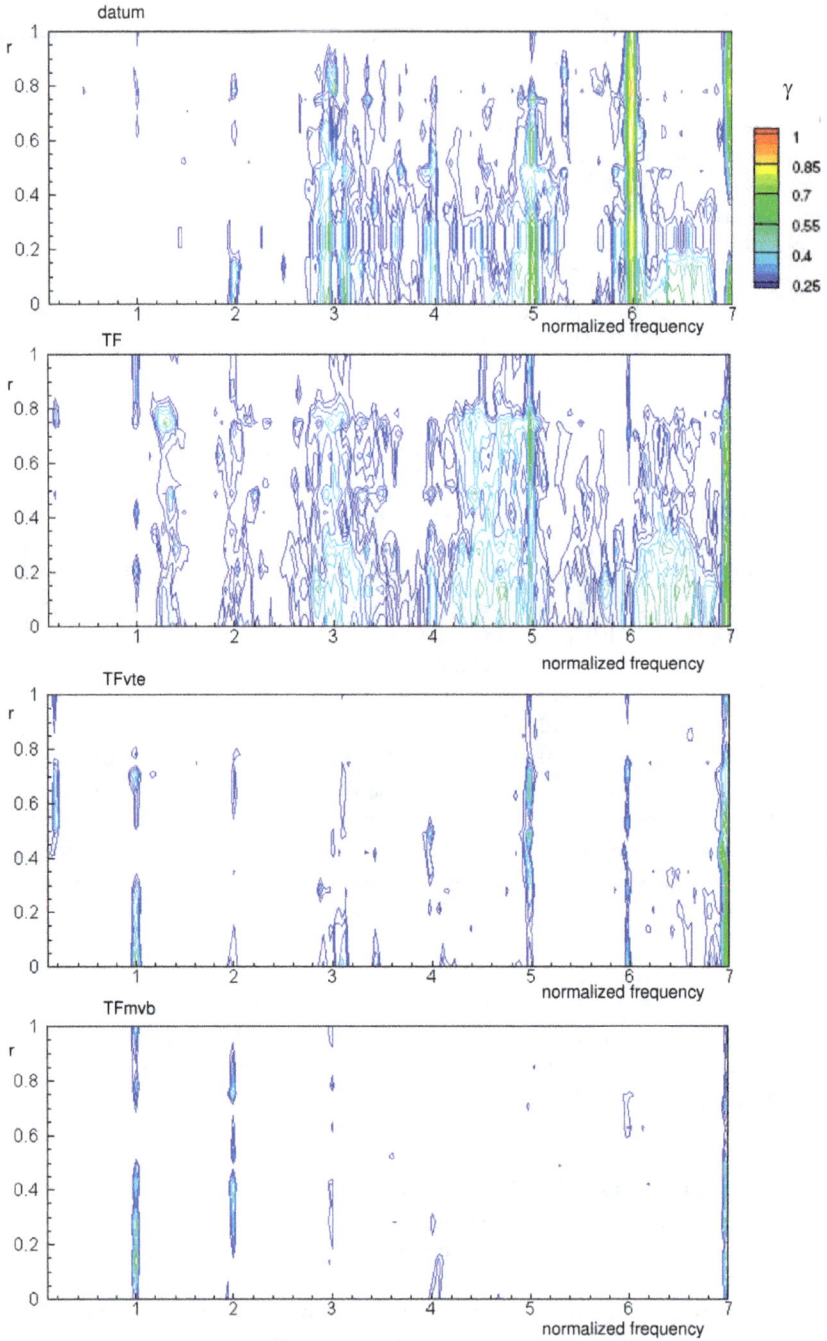

Fig. 8. Spanwise maps of coherence for the different blade geometries at 30°

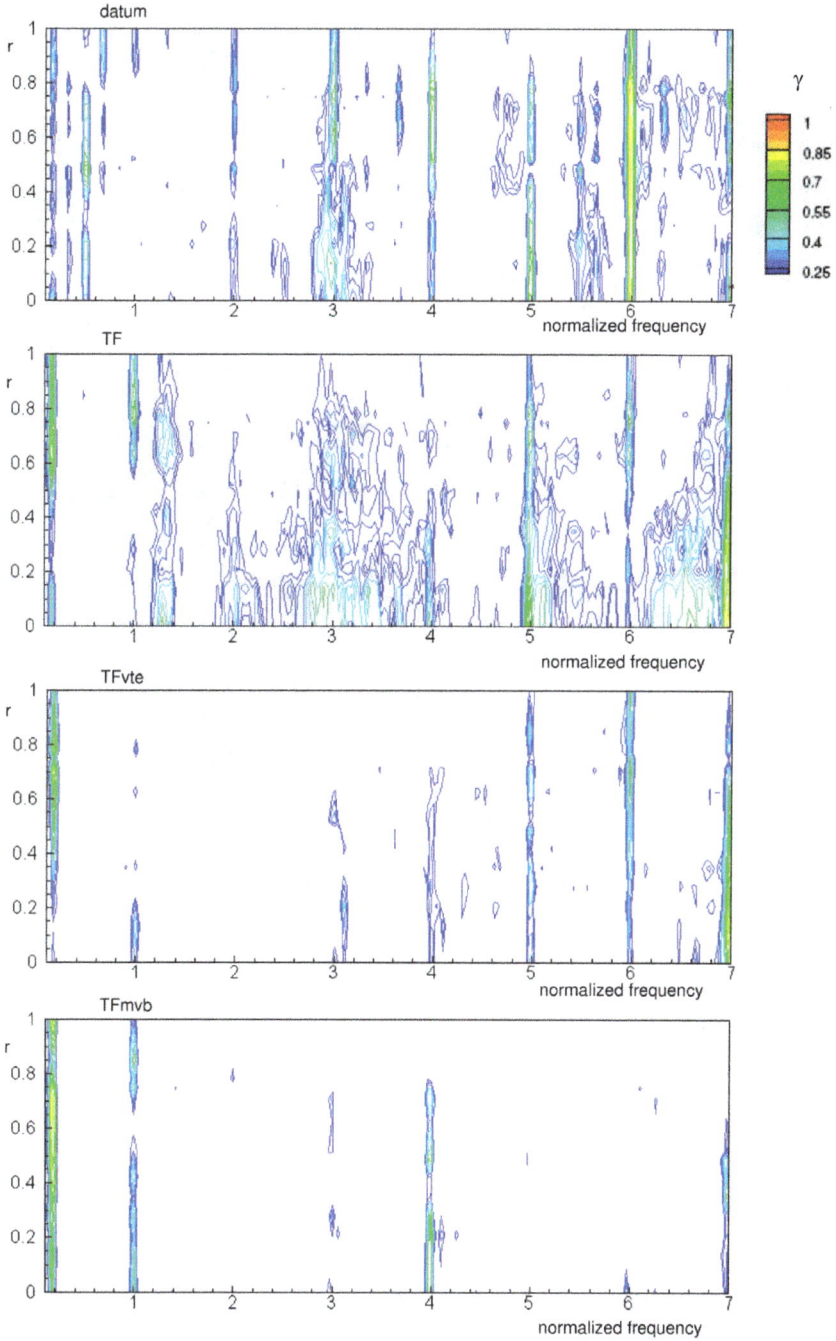

Fig. 9. Spanwise maps of coherence for the different blade geometries at 60°

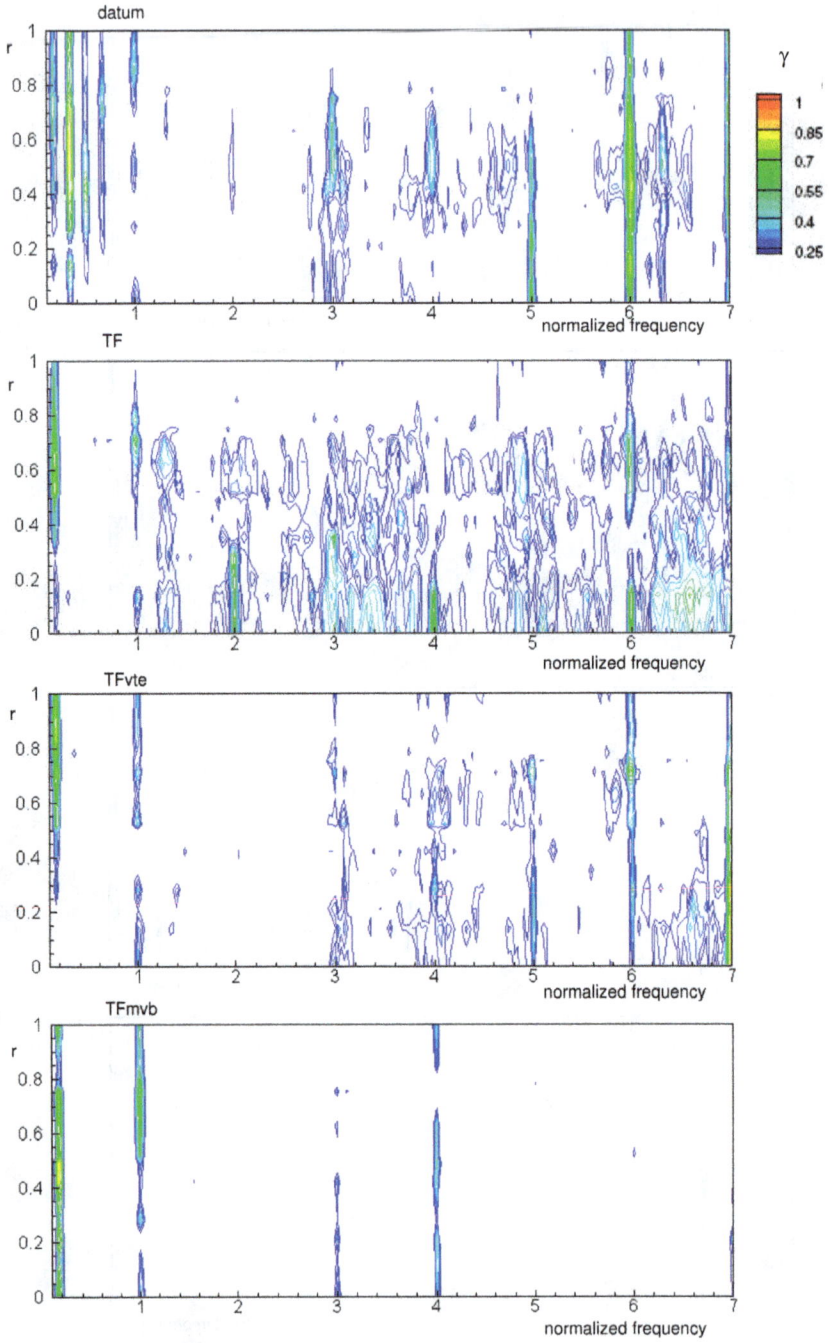

Fig. 10. Spanwise maps of coherence for the different blade geometries at 90°

6. Conclusions

The comparison of different tip features indicated a positive influence on the fan blades' global noise emission to which the authors applied them. The authors found that the end-plate tip geometry modifications reduced acoustic emission most significantly. The study demonstrates that the *datum* blade and the modified fixed-thickness ('TF' and 'TFvte') blades tip both associated with a directional noise emission that had its maximum Lp level in the fan outlet's centre line, thus producing a dipolar noise source. This behaviour has a slight impact on the broadband noise, with the broadband Lp general reduction associated with the TF configuration somewhat greater than that associated with the *datum* fan for the tip vortex breakdown occurrence. With regard to the multiple-vortex-breakdown ('TFmvb') tip concept, the experiments demonstrate that the directivity of the emitted noise changed, thus producing a pattern for this configuration that was more akin to a quadrupole-like source. The strongest component of this quadrupole-like source, which the authors suspect as a significant influence on the TFmvb noise emission, appears to emit in a direction of 90° degrees to the fan centre line, due to the ingested rotor alone noise.

The literature (Shah *et al.*, 2007) reports similar findings, and with this present study the authors can attribute this particular acoustic phenomenon to the tip leakage flow's augmentation of swirl level by means of end-plate shaping that promoted the coherent vortical structure occurence at an increased rotation rate spun at the fan rotor exhaust. These coherent vortical structures interact with the ingested rotor alone noise and thus excite the tones in low frequency in a completely different manner than the rotor with no end-plate. A near-far field "pressure to noise" correlation showed the presence of well defined aerodynamic sources acting to enhance noise enhanced and this, in agreement with the literature, was the cause of a different noise pattern that the modified geometries showed.

The signal correlations also showed the central importance of the low frequency and the influence of the blade tip noise in building the tonal harmonics. The data from the experiments demonstrated that the tip aeroacoustic emissions were sensitive to the rotor alone noise, but also that the end-plate control on the flow was beneficial in noise source reduction.

The results show the control feature of the TF, TFvte and TFmvb geometries on the tip derived noise, with respect of the *datum* blade. The auto-spectra in the broadband range also show a better control of the tip noise exhorted in this frequency range by the TFmvb. ‚Past numerical investigation (Corsini *et al.*, 2009b, 2010) support the authors' asertion regarding the link existing between the broadband noise reduction and the tip flow peculiarities which TFmvb geometry features. The noise radiated by the TF family of impellers also shows a strong influence from a secondary source localised in the bottom part of the rotor. The TFvte and TFmvb blades also have an influence on this noise source, possibly because of the change that they produce on the whole flow circulation along the blade span.

7. Acknowledgment

The authors conducted the present research in the context of contract FW-DMA09-11 between *Fläkt Woods Ltd* and *Dipartimento di Ingegneria Meccanica e Aerospaziale, "Sapienza" University of Rome.*

8. References

BS 848-2.6:2000 (2000), ISO 10302:1996. *Fans for General Purposes. Methods of Noise Testing.*

ISO IEC60651, IEC/EN-61672-1. *Calibration Rules for Hardware in Noise Measurements.*

Akaike, S., Kuroki, S. & Katagiri, M. (1991), "Noise Reduction of Radiator Cooling Fan for Automobile: Three-dimensional Analysis of the Flow between the Blades of the Fan". *Society of Automotive Engineers of Japan,* vol. 22, pp 79–84.

Akturk, A. & Camci, C. (2010), "Axial Flow Fan Tip Leakage Flow Control using Tip Platform Extensions". *Journal of Fluids Engineering,* vol. 132, pp 101–10.

Akturk, A. & Camci, C. (2011a), "Tip Clearance Investigation of a Ducted Fan used in VTOL UAVS. Part 1: Baseline Experiments and Computational Validation". ASME Paper No. GT2011-46356.

Akturk, A. & Camci, C. (2011b), "Tip Clearance Investigation of a Ducted Fan used in VTOL UAVS. Part 2: Novel Treatments via Computational Design and their Experimental Verification". ASME Paper No. GT2011-46359.

Bianchi, S., Corsini, A., Rispoli, F. & Sheard, A.G. (2009a), "Experimental development of a measurement technique to resolve the radial distribution of fan aero-acoustic emissions". Noise Control Engineering Journal, vol. 57(4), pp 360-369.

Bianchi, S., Corsini, A., Rispoli, F. & Sheard, A.G. (2009b), "Detection of Aerodynamic Noise Sources in Low-speed Axial Fan with Tip End-plates". *Proceedings of the IMechE, Part C: Journal of Mechanical Engineering Science,* vol. 223, pp 1379–92.

Bianchi, S., Corsini, A., Rispoli, F. & Sheard, A.G. (2009c), "Experimental Aero-acoustic Studies on Improved Tip Configurations for Passive Control of Noise Signatures in Low-speed Axial Fans". *ASME Journal of Vibration and Acoustics,* vol. 131, pp 1–10.

Bianchi, S., Sheard, A.G., Corsini, A. & Rispoli, F. (2011a), "Far-field Radiation of Aerodynamic Sound Sources In Axial Fans Fitted With Passive Noise Control Features". ASME Journal of Vibration and Acoustics, vol. 133(5), paper 051001 (11 pages).

Blake,W.K. (1986), Mechanics of Flow-Induced Sound and Vibration, Vol. I.Academic Press, London, UK.

Brookfield, J.M. & Waitz, I.A. (2000), "Trailing Edge Blowing for Reduction of Turbomachinery Fan Noise". *AIAA Journal of Propulsion and Power,* vol. 16, pp 57–64.

Brüel & Kjaer (2006), *4954 ¼ inch Prepolarized Free-field Microphone Manual.*

Corsini, A., Rispoli, F., & Santoriello, A., (2005), "A Variational Multi-Scale High-Order Finite Element Formulation for Turbomachinery Flow Computations" , *Comput. Methods Appl. Mech. Eng.,* vol. 194 (45 – 47), pp 4797 – 4823.

Corsini, A. & Sheard, A.G. (2007), "Tip End-plate Concept Based on Leakage Vortex Rotation Number Control". *Journal of Computational and Applied Mechanics,* vol. 8, pp 21–37.

Corsini, A. & Sheard, A.G. (2011), "End-plate Design for Noise-by-flow Control in Axial Fans: Theory and Performance". Under revision in *Journal of Fluids Engineering,* manuscript reference number FE-11-1168.

Corsini, A., Rispoli, F. & Sheard, A.G. (2007), "Development of Improved Blade Tip End-plate Concepts for Low-noise Operation in Industrial Fans". *Journal of Power and Energy,* vol. 221, pp 669–81.

Corsini, A., Rispoli, F. & Sheard, A.G. (2009a), "A Meridional Fan". Patent Application No. WO/2009/090376.

Corsini, A., Rispoli, F. & Sheard, A.G. (2009b), "Aerodynamic Performance of Blade Tip End-plates Designed for Low-noise Operation in Axial Flow Fans". *Journal of Fluids Engineering,* vol. 131, pp 1–13, Paper No. 081101.

Corsini, A., Rispoli, F. & Sheard, A.G. (2010), "Shaping of Tip End-plate to Control Leakage Vortex Swirl in Axial Flow Fans". *Journal of Turbomachinery*, vol. 132, pp 1–9, Paper No. 031005.

Cumpsty, N.A. (1977), A Critical Review of Turbomachinery Noise. *Journal of Fluids Engineering*, vol. 99, pp 278–93.

Escudier, M. (1987), "Confined Vortices in Flow Machinery". *Annual Review of Fluid Mechanics*, vol. 19, pp 27–52.

Escudier, M., & Zehnder, N., (1982), "Vortex flow regimes", *Journal of Fluid Mechanics*, vol. 115, pp 105-121.

European Parliament (2005), *Directive 2005/32/EC Establishing a Framework for the Setting of Ecodesign Requirements for Energy-using Products and Amending Council Directive 92/42/EEC and Directives 96/57/EC and 2000/55/EC of the European Parliament and of the Council.*

Ffowcs Williams, J.E. (1977), "Aeroacoustics". *Annual Review of Fluid Mechanics*, vol. 9, pp 447–68.

Fukano, T., Takamatsu, Y. & Kodama, Y. (1986), "The Effects of Tip Clearance on the Noise of Low Pressure Axial and Mixed Flow Fans". *Journal of Sound and Vibration*, vol. 105, pp 291–308.

Furukawa, M., Inoue, M., Kuroumaru, M., Saiki, K. & Yamada, K., (1999), "The role of tip leakage vortex breakdown in compressor rotor aerodynamics", *Journal of Turbomachinery*, vol. 121, pp 469-480.

Gad-el-Hak, M. (2000), *Flow Control: Passive, Active, and Reactive Flow Management.* Cambridge University Press, Cambridge, UK.

Garg, A.K. & Leibovich, S. (1979), "Spectral Characteristics of Vortex Breakdown Flowfields". *Physics of Fluids*, vol. 22, pp 2053–64.

Herrada, M.A. & Shtern, V. (2003), "Vortex Breakdown Control by Adding Near-axis Swirl and Temperature Gradients". *Physical Review E*, vol. 68, pp 1–8.

Inoue, M. & Kuroumaru, M. (1989), "Structure of Tip Clearance Flow in an Isolated Axial Compressor Rotor". *American Society of Mechanical Engineers, Journal of Turbomachinery*, vol. 111, pp 250–6.

Inoue, M. & Furukawa, M., (2002) "Physics of tip clearance flow in turbomachinery", ASME paper FEDSM2002-31184.

Ito, T., Suematsu, Y. & Hayase, T. (1985), "On the Vortex Breakdown Phenomena in a Swirling Pipe-flow". *Nagoya University, Faculty of Engineering, Memoirs*, vol. 37, pp 117–72.

Jones, M.C., Hourigan, K. & Thompson, M.C. (2001), "The Generation and Suppression of Vortex Breakdown by Upstream Swirl Perturbations". *Proceedings of 14th Australian Fluid Mechanics Conference.* Adelaide, Australia.

Joslin, R.D., Rusell, H.T. & Choudhari, M.M. (2005), "Synergism of Flow and Noise Control Technologies". *Progress in Aerospace Sciences*, vol. 41, pp 363–417.

Kameier, F. & Neise, W. (1997), "Experimental Study of Tip Clearance Losses and Noise in Axial Turbomachines and their Reduction". *American Society of Mechanical Engineers, Journal of Turbomachinery*, vol. 119, pp 460–71.

Karlsson, S. & Holmkvist, T. (1986), "Guide Vane Ring For a Return Flow Passage in Axial Fans and a Method of Protecting It". Patent No. US 4,602,410.

Lakshminarayana, B., Zaccaria, M. & Marathe, B. (1995), "The Structure of Tip Clearance Flow in Axial Flow Compressors". *American Society of Mechanical Engineers, Journal of Turbomachinery*, vol. 117, pp 336–47.

Leggat, L.J. & Siddon, T.E. (1978), "Experimental Study of Aeroacoustic Mechanism of Rotor-vortex Interactions". *Journal of the Acoustical Society of America*, vol. 64, pp 1070–77.

Leibovich, S. (1982), "Wave Propagation, Instability, and Breakdown of Vortices". In Hornung, H.G. & Mueller, E.A. (eds), *Vortex Motion*. Vieweg, Braunschweig, Germany, pp 50–67.

Longet, C.M.L. (2003), "Axial Flow Fan with Noise Reducing Means". US Patent No. 2003/0123987 A1.

Longhouse, R.E. (1978), "Control of Tip-vortex Noise of Axial Flow Fans by Rotating Shrouds". *Journal of Sound and Vibration*, vol. 58, pp 201 - 14.

Lucca-Negro, O., & O'Doherty, T., (2001), "Vortex breakdown: a review", *Progress in Energy and Combustion Science*, vol. 27, pp 431–481.

Miles, J.H. (2006), "Procedure for Separating Noise Sources in Measurements of Turbofan Engine Core Noise" . Report NASA/TM-2006-214352.

Mimura, M. (2003), "Axial Flow Fan". US Patent No. 6,648,598 B2.

Mongeau, L., Thompson, D.E. & McLaughlin, D.K. (1995), "A Method for Characterizing Aerodynamic Sound Sources in Turbomachines" . Journal of Sound and Vibration, vol. 181, pp 369 - 89.

Quinlan, D.A. & Bent, P.H. (1998), "High Frequency Noise Generation in Small Axial Flow Fans". *Journal of Sound and Vibration*, vol. 218, pp 177–204.

Saiyed, N.H., Bridges, J.E. & Mikkelsen, K.L. (2000), "Acoustics and Thrust of Separated-flow Exhaust Nozzles with Mixing Devices for High-bypass-ratio Engines". AIAA Paper No. 2000-1961.

Shah, P.D., Mobed, D., Spakovszky, Z. & Brooks, T.F. (2007), "Aero-Acoustics of Drag Generating Swirling Exhaust Flows". *13th AIAA/CEAS Aeroacoustics Conference (28th AIAA Aeroacoustics Conference)*. Rome, Italy, Paper No. AIAA-2007-3714.

Smith, G.D.J. & Cumpsty, N.A. (1984), "Flow Phenomena in Compressor Casing Treatment". *Journal of Engineering for Gas Turbines and Power*, vol. 106, pp 532–41.

Spall, R.E., Gatski, T.B. & Grosch, C.E. (1987), "A Criterion for Vortex Breakdown". *Physics of Fluids*, vol. 30, pp 3434–40.

Srigrarom, S. & Kurosaka, M. (2000), "Shaping of Delta-wing Platform to Suppress Vortex Breakdown". *AIAA Journal*, vol. 38, pp 183–6.

Storer, J.A. & Cumpsty, N.A. (1991), "Tip Leakage Flow in Axial Compressors". *American Society of Mechanical Engineers, Journal of Turbomachinery*, vol. 113, pp 252–9.

Takata, H. & Tsukuda, Y. (1977), "Stall Margin Improvement by Casing Treatment: Its Mechanism and Effectiveness". *Journal of Engineering for Power*, vol. 99, pp 121–33.

Thomas, R.H., Choudhari, M.M. & Joslin, R.D. (2002), "Flow and Noise Control: Review and Assessment of Future Directions". NASA Report TM-2002-211631.

Thompson, D.W., King, P.I. & Rabe, D.C. (1998), "Experimental and Computational Investigation on Stepped Tip Gap Effects on the Flowfield of a Transonic Axial-flow Compressor Rotor". *Journal of Turbomachinery*, vol. 120, pp 477–86.

Uchida, S., Nakamura, Y. & Ohsawa, M. (1985), "Experiments on the Axisymmetric Vortex Breakdown in a Swirling Air Flow". *Transactions of the Japan Society for Aeronautical and Space Sciences*, vol. 27, pp 206–16.

Uselton, R.B., Cook, L.J. & Wright, T. (2005), "Fan with Reduced Noise Generation". US Patent No. 2005/0147496 A1.

Wright, S.E. (1976), "The Acoustic Spectrum of Axial Flow Machines", *Journal of Sound and Vibration*, vol. 45, pp 165–223.

10

Assessment of Acoustic Quality in Classrooms Based on Measurements, Perception and Noise Control

Paulo Henrique Trombetta Zannin, Daniele Petri Zanardo Zwirtes and
Carolina Reich Marcon Passero
*Federal University of Paraná, LAAICA – Laboratory of Environmental and Industrial
Acoustics and Acoustic Comfort,
Brazil*

1. Introduction

Education plays a fundamental role in the formation of modern society. The importance of education for humans is expressed thus by renowned Brazilian educator Paulo Freire[1]:
"The fountainhead of man's hope is the same as that of his educability: the incompleteness of his being of which he has become aware. It would be a sorry contradiction if, incomplete and aware of this incompleteness, man were not engaged in a permanent process of hopeful search. This process is education." (Freire, p 114)[1]
The long and arduous process of individual and collective education takes place primarily in classrooms. It is here that contact is established between teachers and students and between individual students and their peers. It is here that knowledge is transmitted in its most ancient form, i.e., through oral communication. The quality of this communication, and ultimately, of classroom education itself, is closely linked to the acoustic quality of the classroom. This acoustic quality can be characterized based on the reverberation time, speech transmission index, sound insulation, and the noise levels inside and outside the classroom[2-5]. High noise levels in the classroom impair oral communication, causing students to become tired sooner more often, and this premature fatigue tends to have a negative effect on their cognitive skills[6].
The reason for the existence of acoustic problems in classrooms, according to Seep[7], is not a lack of knowledge about how to solve the problem, but primarily a lack of sensitivity of the professionals involved, both in the field of teaching and that of classroom design, to solve the problem. The problem of acoustic quality in a classroom begins in its design phase and extends all the way to the final quality of the education provided in public and private schools, primary and secondary schools, and in universities.
Many of the aspects that appeared with the evolution of the modern era served to deteriorate the acoustic environment of the classroom. A reflection of our times is the fact that practically every student owns a mobile phone, a digital player, and other electronic devices that tend to render the school environment noisy, hindering its core purpose. Hagen[6] believe that the school environment should promote an atmosphere that encourages everyone's interest in listening and being involved in communication. An acoustically

comfortable environment should be one that provides everyone, individually and collectively, with the proper conditions to develop their skills.

This chapter presents an analysis of the acoustic quality of real classrooms based on *in situ* measurements and computer simulations of acoustic parameters such as Reverberation Time, Speech Transmission Index, Sound Insulation of Façades, and External and Internal Sound Pressure Levels. This chapter also discusses an assessment of the perception of teachers and students about the acoustic quality of the school environment.

Lastly, computer simulations were performed in order to identify, from the standpoint of noise control, what actions would be required to improve the acoustic quality of the evaluated classrooms.

2. Materials and methods

The present work involved an evaluation of the acoustic quality of real classrooms built according to standard designs. For this study, three design standards known as 010, 022 and 023 were selected (see description below). A total of six classrooms, two of each constructive design, were analyzed. To facilitate the identification of these schools, those of design 010 were dubbed C1 and C2, while those of design 022 were identified as C3 and C4, and those of design 023 as C5 and C6. The schools that participated in this evaluation were as follows: Standard 023 schools: 1) Colégio Estadual Walde Rosi Galvão and 2) Escola Estadual Luarlindo dos Reis Borges; Standard 022: 1) Colégio Paulo Freire and 2) Colégio Aníbal Khury Neto; and Standard 010: 1) Colégio Estadual Professor Alfredo Parodi and 2) Colégio Estadual Professora Luiza Ross. Physical aspects of the construction of the schools (choice of land and positioning of the buildings) were evaluated.

The results of this work were obtained by measuring the reverberation time (RT), sound insulation of classroom façades, and background noise (inside and outside the classrooms). In addition to these parameters obtained by measurements, the speech transmission index (STI) was obtained through the computer simulation of the calibrated models of the classrooms.

An investigation was also made of the users' perception of acoustic quality and comfort in the classrooms, based on questionnaires for students and teachers.

Noise control in the classrooms was investigated by means of computer simulations aimed at improving the acoustic quality of the classrooms by amending the insulation of their façades. Other simulations concentrated on observing the influence of the background noise level and reverberation time (RT) on the speech transmission index (STI).

2.1 Evaluation of reverberation time in classrooms

An important parameter affecting the acoustic quality of rooms is the RT. Each type of room (classrooms, theaters, churches) requires a given RT. Therefore, it is crucial that the RT be designed according to the purpose for which the room was conceived.

According to the ISO 3382-1[8] and ISO 3382-2[9], the RT can be measured by the interrupted noise method and by the integrated impulse response method. Measuring the RT by the interrupted noise method, as described in this chapter, consists of exciting the room with a pseudo-random pink noise and calculating the RT from the room's response to this excitation. A common setup to measure the RT by this method comprises: 1) an omnidirectional sound source, 2) a sound power amplifier, 3) a noise generator, 4) omnidirectional microphones, and 5) a sound decay recorder and analyzer.

This chapter describes how the RT was measured with a dual-channel Brüel & Kjaer BK 2260 real-time sound analyzer, a BK 2716 power amplifier, a BK 4296 dodecahedron loudspeaker. The sound thus generated was captured by a microphone connected to the BK 2260 analyzer, which calculated the reverberation time for each frequency of the spectrum of interest. These measurements were then transferred to a computer using Brüel and Kjaer BK 7830 Qualifier software, which calculated the mean reverberation time of each classroom.

2.2 Evaluation of the sound insulation index of façades

The procedures for taking field measurements of sound insulation of façades are set forth in the ISO 140-5[10]. *In situ* measurements require the use of a flat cable to reduce the loss of sound energy through cracks in outside openings – windows or doors. Figure 1, below, shows in detail the use of a flat cable for measuring façade sound insulation.

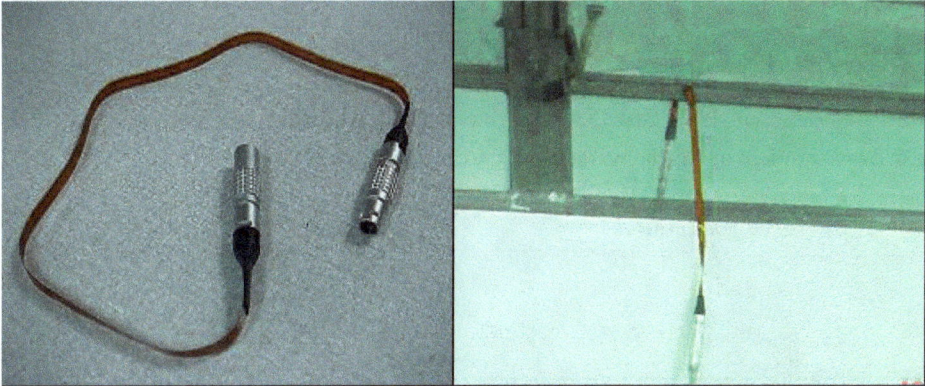

Fig. 1. Measurement of façade sound insulation using a flat cable

The sound insulation of a façade can be measured according to ISO 140-5[10], using a loudspeaker as the outdoor sound source. In this case, the subindex *ls* is used to characterize the loudspeaker in the measurement of the standardized level difference, $D_{ls,2m,nT}$, where $D_{ls,2m,nT,w}$ is the weighted standardized level difference corresponding to this method. The loudspeaker should be tilted at an angle of 45°, according to the ISO 140-5 standard. The loudspeaker in Figure 2 is connected to a noise generator.

Figure 2 illustrates the equipment required for measuring façade insulation using a loudspeaker as the sound source. The figure on the left shows the equipment positioned inside the classroom: a dual-channel sound analyzer, an omnidirectional loudspeaker and a microphone. The figure on the right shows the equipment positioned outside the classroom whose façade insulation is being evaluated: an external loudspeaker tilted at an angle of 45° and a noise generator.

The value of the weighted standardized level difference $D_{ls,2m,nT,w}$, which appears in the upper right-hand corner in Figure 3, can be calculated using the graphic method described in ISO 717-1[11] and the BK 7830 Qualifier software. Since it is a single number, the value of $D_{ls,2m,nT,w}$ is used for comparison with the standardized values of reference to evaluate the performance of the building's façades in terms of sound insulation.

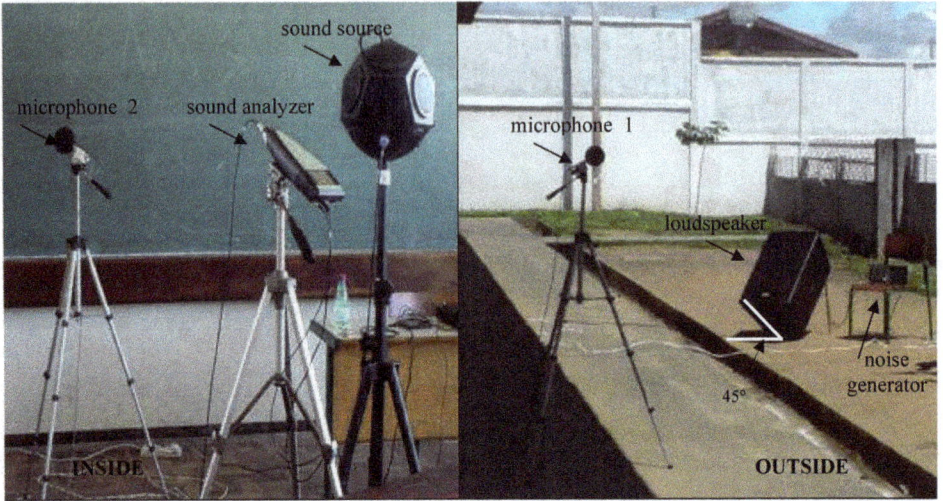

Fig. 2. Measurement of façade insulation according to ISO 140-5[10]

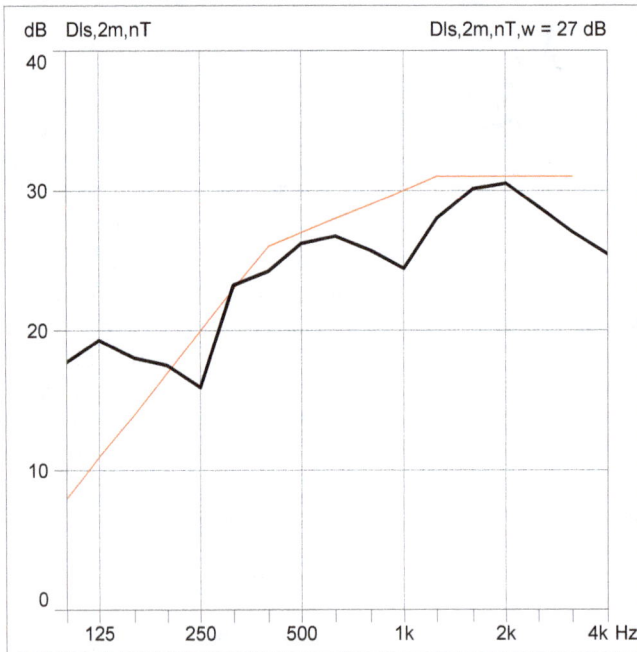

Fig. 3. Measurements of the standardized level difference $D_{ls,2m,nT}$ according to ISO 140-5[10]

The measured data can be expressed in a standard report according to ISO 140-5[10], which presents the value of the weighted standardized level difference $D_{ls,2m,nT,w}$. The standard report should also include the measurements in one-third octave bands of the values of the standardized level difference $D_{ls,2m,nT}$. The standard report is presented in Figure 4.

2.3 Evaluation of background noise in school environment

The sound pressure levels were determined from measurements taken inside the classrooms and in the surroundings of the schools. In both cases, the measurements were taken according to the Brazilian NBR 10151[12] standard, which regulates noise evaluations in inhabited areas for purposes of community comfort.

The purpose of measuring the surroundings was to evaluate the noise produced by the neighborhood (neighbors, street traffic, air traffic, industry, etc.), characterizing the regions where the schools are situated. As for the internal environment, the noise level in the classrooms was evaluated to verify if their acoustic quality favored the development of teaching-learning activities. The influence of noise produced in schoolyards and sports courts on the classrooms was also checked.

The number of samples from each school and the measuring time at each point were selected so as to allow for characterization of the noises of interest. In general, each evaluation involved measurements taken at three points, which resulted in an average value. The measuring time on the streets/roads around the schools was limited to 10 minutes at each point. The noise inside classrooms was assessed based on a 3-minute measurement at each point[4].

The sound pressures were measured with Brüel and Kjaer BK 2237 and BK 2238 sound level meters and the measured values were analyzed using Brüel and Kjaer BK 7820 Evaluator software.

2.4 Assessment of the speech transmission index STI

The STI is an acoustic descriptor that considers the effects of reverberation, background noise and the contribution of the direction of the source to determine speech intelligibility. These elements, which are usually treated individually, are combined in a single index[13,14].

The STI was simulated using Odeon 9.0 software[19]. This software uses the hybrid method to obtain the acoustic parameters. Rindel[15] claims that hybrid methods combine the best characteristics of the image source and ray tracing methods. A comparison of several computer simulation methods indicated that programs that use the hybrid method produce the best results[16].

The STI simulations were performed according to the IEC 60268-16[17] standard. To obtain acoustic parameters through simulations required first making a three-dimensional drawing of the room. Suitable calculation parameters were then inserted (such as the length of the impulse response), the characteristics of the finish surfaces (absorption and scattering coefficients) and the specifications of the sound source and receiver. For the source, the IEC 60268-16 standard establishes that it should be of the pointwise directional type, in order to simulate the characteristics of the human mouth. The noise generated by the source should simulate both the timbre and volume of the human voice.

Standardized Level Difference according to ISO 140-5

Field measurements of airborne sound insulation of facade elements and facades

Client: Date of test: 08/12/2005

Description and identification of the building construction and test arrangement:

Area S of test specimen: 17,00 m²   - - - -   Frequency range according to the
Receiving room volume: 120,00 m³   ———   curve of reference values (ISO 717-1)

Frequency f Hz	Dls,2m,nT 1/3 Octave dB
50	
63	
80	
100	17,7
125	19,3
160	18,0
200	17,5
250	15,9
315	23,2
400	24,2
500	26,2
630	26,7
800	25,7
1000	24,4
1250	28,0
1600	30,1
2000	30,5
2500	28,8
3150	27,0
4000	25,4
5000	

Rating according to ISO 717-1

$D_{ls,2m,nT,w}$ $(C;C_{tr})$ = 27 (-1; -3) dB

Evaluation based on field measurement
results obtained by an engineering
method

No. of test report: Name of test institute:

Date: 08/12/2005   Signature:

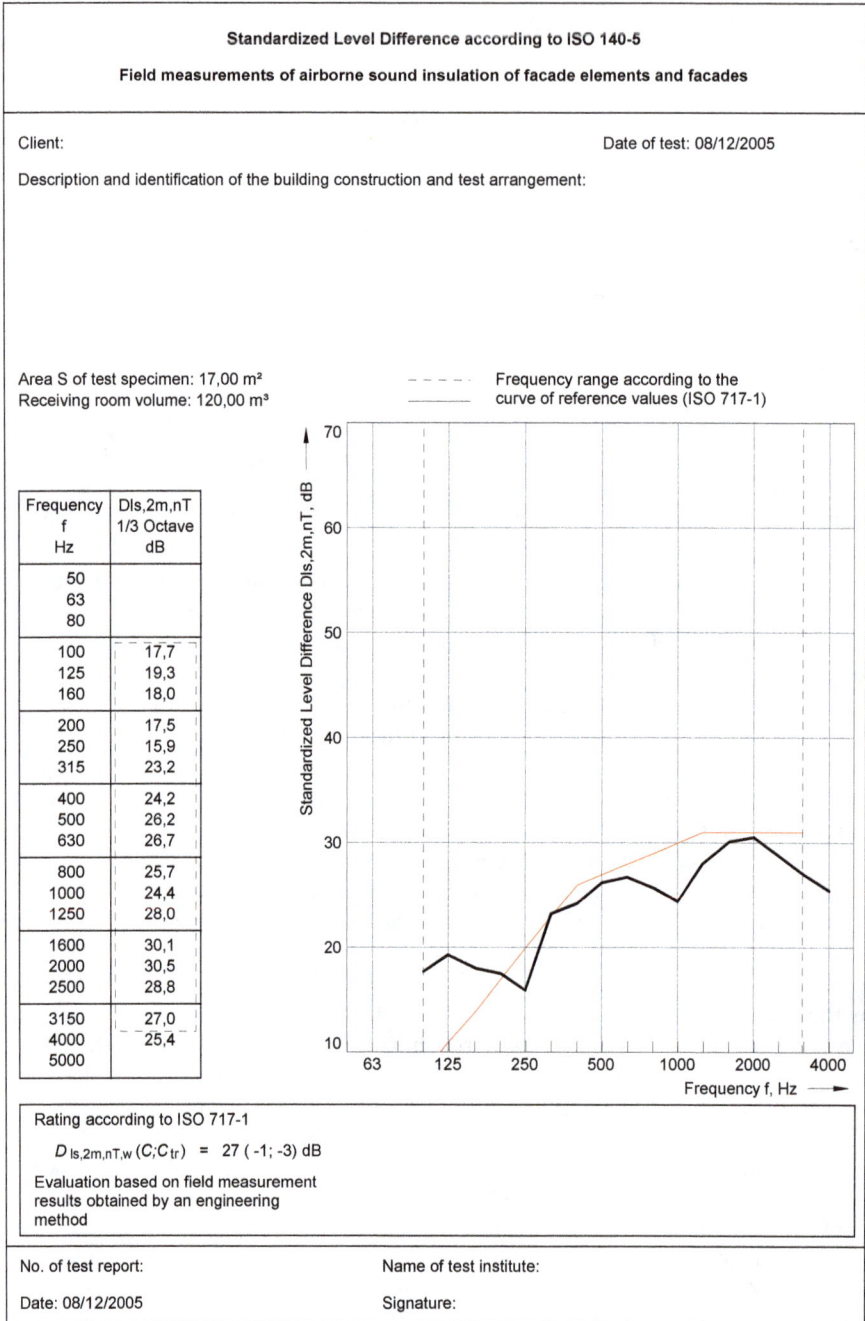

Fig. 4. Measurement report according to ISO 140-5[10] and ISO 717-1[11] – Field measurement of airborne sound insulation of building façades

The three-dimensional models were calibrated based on a comparison of the values of measured and simulated RT. The values of sound pressure level in octave band frequency measured inside a classroom were then inserted in the calibrated model, and the loudspeaker and microphone positions were defined. A 0.50x0.50 m grid was defined for the loudspeakers, and the microphone was placed in the typical position of the teacher and directed towards the students.

2.5 Noise control in classrooms

The simulations of acoustic improvements were performed in two stages: 1) improvement of façade insulation; and 2) improvement of the acoustic conditioning of the classrooms.

The simulations of improved façade insulation were performed using Bastian 2.3 software and the parameter evaluated was the weighted standardized level difference $D_{ls,2m,nT,w}$ (ISO 140-5[10]). The calculations of insulation using Bastian software were based on parts 1 to 3 of the European Standard series EN 12354[18]. The modifications implemented in the simulations considered the alteration of the material and type of window in the classroom façade.

Odeon 9.0[19] software was used to simulate improvements in the acoustic conditioning inside the classrooms, and the parameters evaluated were the RT and the STI. Changes in the finishing material of the rooms' ceilings and in background noise levels were simulated.

2.6 Subjective assessment

To evaluate teachers and students' perception of noise in schools, questionnaires were designed for each group. These questionnaires were based on similar studies conducted by Dockrell et al.[20], Loro[21], Losso[22], Enmarker and Boman[23], and Dockrell and Shield[24].

After validating the questionnaires in a pilot test, they were applied to 71 teachers and 1080 students of the public school system. The questionnaires were applied to the students in the classrooms. The questions were read out loud one by one by the researcher, who allowed sufficient time for the students to write down their answers. As for the teachers, they were given an explanation about the objective of the research and about the design of the questionnaire, after which they answered it individually without the researcher's help.

Out of the total of 1080 questionnaires distributed to the students, 1035 were considered valid. The students' questionnaire contained closed questions and were answered by 5th to 8th grade schoolchildren aged 9 to 18. In the teachers' questionnaire, the answers were given in the form of scores ranging from 0 to 3. The 71 questionnaires distributed to the teachers were all validated.

The data obtained from the students' questionnaire was analyzed statistically following two strategies. The first strategy involved a descriptive analysis using contingency tables, showing the frequency of the individuals' responses as a function of two qualitative variables[34,35]. These tables were the first descriptive instrument for drawing up two hypotheses, whose general formulation is given by: a) hypothesis H0: the two factors are not associated; and b) H1: the two factors are associated[34,35]. The second strategy consisted of using statistical tests of hypotheses that verified the significance of the link between different factors. The R software developed by the R Development Core Team[25] was used to calculate the association tests.

The hypotheses outlined during the first analytical strategy were verified by the Q and Qp statistics, whose approximate probability distribution is the chi-square[34,35]. The decisions about the hypotheses were taken at a 95% level of confidence. In some situations where the expected frequencies in the cells of the contingency table were low, the approximation of the chi-square distribution for the Q and Qp statistics was compromised, so an alternative test was applied[34,35]. In this case, the choice fell on Fisher's Exact Test[34,35].

The analysis of the questionnaire applied to the teachers was similar to that of the students, but the Qs statistics was used for the tests of association. The Qs statistics is used when one of the variables of the contingency table presents an ordinal measure[34,35] (which is the case of the questionnaire applied to the teachers, whose answers were given scores of 0 to 3).

3. Constructive designs of the classrooms

The classrooms in the public schools evaluated here are designed in standard modules that are adjustable to the need for new schools, depending on the forecasted number of students and the type of terrain where they are to be constructed.

The characteristics of the construction designs selected were as follows. 1) Design 010, which consists of independent blocks with a central circulation area and classrooms arranged on both sides of a hall (Figure 5); 2) Design 022, comprising classroom blocks arranged side by side without a hall between them (Figure 6); and 3) Design 023, similar to design 010, composed of independent blocks of classrooms arranged on the two sides of a central hall (Figure 7).

Table 1 presents the characteristics of the classrooms: volume, material of the walls, floor and ceiling and type of window.

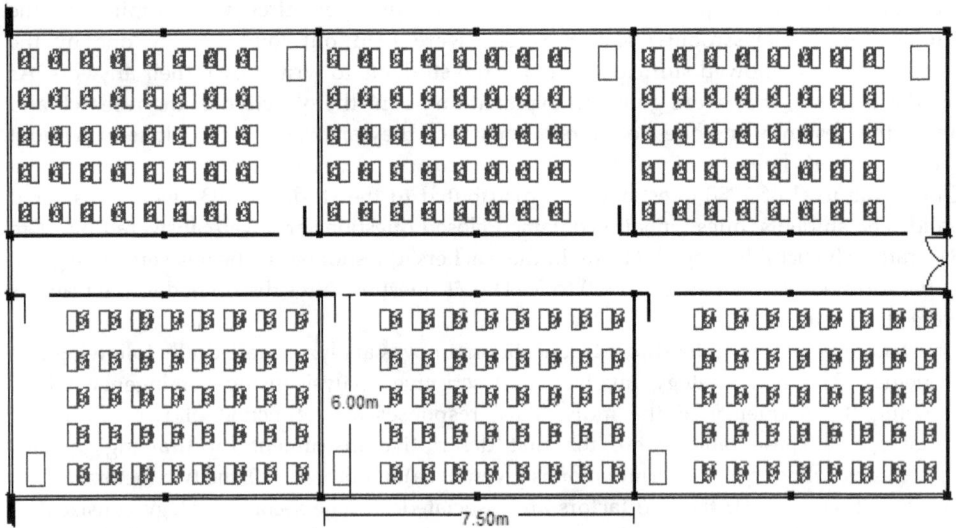

Fig. 5. Classroom construction design 010

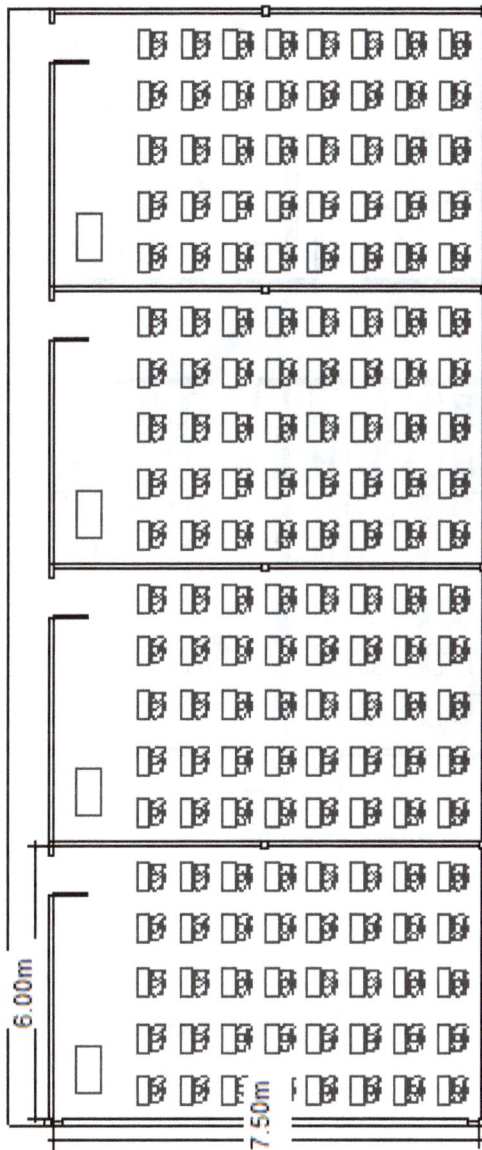

Fig. 6. Classroom construction design 022

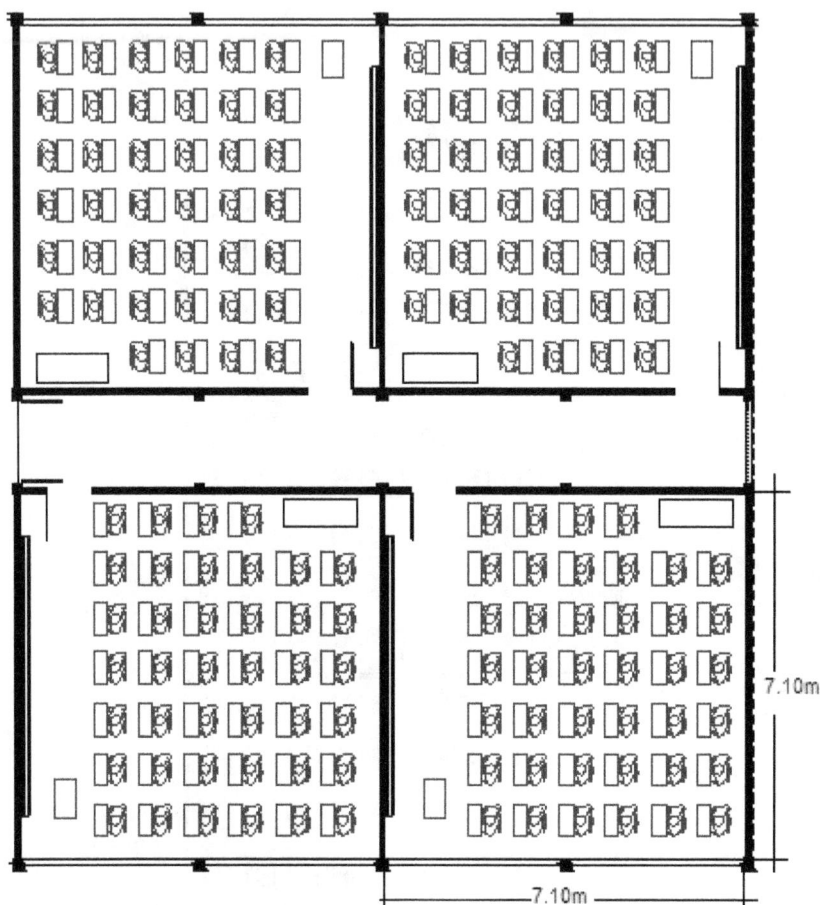

Fig. 7. Classroom construction design 023

Classroom construction designs	Volume (m³)	Wall material	Floor material	Ceiling material	Type of window
010	139	Ordinary brickwork	Parquet	Wood paneled ceiling	Iron window frames with glass panes
022	139	Ordinary brickwork	Parquet	Concrete slab	Iron window frames with glass panes
023	156	Ordinary brickwork	Ceramic tiles	Concrete slab	Iron window frames with glass panes

Table 1. Construction characteristics of the classrooms

4. Results of the measurements and discussion

The schools selected for this study were built during three distinct periods. The schools built to design 010 went up in 1977 (C1) and 1978 (C2). The two schools built according to design 022 were concluded in 1998 (C3 and C4), while those of design 023 were built in 2001 (C5) and 2005 (C6).

4.1 Background noise inside and outside the classrooms

To evaluate the acoustic composition of the classroom environment, measurements were taken of the equivalent sound pressure levels L_{eq} expressed in dB(A). Firstly, the external environment was assessed based on measurements taken on the sidewalks around the schools, at the distances established by the Brazilian standard NBR 10151[26].

The purpose of evaluating the external environment is to check if there is any influence of traffic noise in the classrooms. This evaluation enabled us to investigate the first aspect relating to the location of the schools: the choice of terrain.

According to the Brazilian NBR 10151[26] standard, which establishes sound levels for external environments, the maximum L_{Aeq} admissible for school zones during the daytime is 50 dB. Table 2 presents the average values of the sound levels measured in the proximities of the classrooms[4].

Classroom	Construction Design	L_{eq} dB(A)	Permissible limit for environmental noise Brazilian NBR 10151 standard[26] dB(A)
C1	010	66.3	
C2	010	66.2	
C3	022	68.4	50
C4	022	60.5	
C5	023	59.2	
C6	023	51.8	

Table 2. Traffic noise in the proximities of the classrooms

The values listed in the above table indicate that classroom C5 and C6 of design 023 and classroom C4 (design 022) are located in quieter zones than the other schools. The values measured in all the schools' surroundings were higher than those established by the Brazilian NBR 10151 standard[4,26].

Although the surrounding noise exceeds the limit established by the Brazilian standard, during the field measurements it was found that the traffic noise did not contribute significantly to the composition of background noise in the classrooms. This was confirmed by the sound levels obtained from the other measurements taken inside the schools, as well as by the subjective assessment.

The subjective assessment revealed that, when questioned about the origin of the most disruptive noises in the classroom, 83% of the students considered that the noise coming from inside the classroom itself was worse. Noises generated in the other school environments, such as halls, adjacent classrooms and schoolyards, were cited by 15% of the interviewees. Only 2% of the students mentioned noises coming from outside the school (cars, neighbors, factories, etc.).

The teachers' perception about the origin of the noises that are most disruptive in the classroom coincided with that of the students. Table 3 lists the results of the question asked of the teachers concerning the most disruptive noises in the classroom.

Origin of the noise	Average score
Students in the classroom	2.2
Adjacent classrooms, halls, and schoolyards	1.6
External sources (cars, neighbors, factories, etc.)	0.8

Table 3. Origin of the most disruptive noises, in the teachers' opinion

The teachers' answers in the questionnaire were given in the form of scores varying from 0 to 3 (0 = nothing, 1 = a little, 2 = more or less, 3 = a lot). Table 3 lists the mean scores for each answer. As can be seen, the noises from the schools' surroundings scored lowest among the three choices. Apart from presenting the lowest score, the value of 0.8 indicates that the influence of external sources is negligible.

The analysis of the questionnaires revealed that the most disruptive noises in the classroom come from the school itself, and are completely unrelated to external noises.

As for the acoustic measurements, an example confirming the non-influence of noise from the external surroundings was obtained in classroom C3 (design 022). Although traffic noise in this school exceeds the limit established by the Brazilian standard, our investigations revealed that this noise does not interfere in the classrooms. This statement was confirmed by the noise levels measured during the school vacations, when the school was empty. The noise levels measured in the schoolyard and classroom were, respectively, L_{eq} = 52.3 dB(A) and L_{eq} = 40.4 dB(A). This sound level in the classroom is in line with that recommended by the NBR 10152 standard, which establishes a level of 40 dB(A) for acoustic comfort in classrooms. The noise levels measured here suggest that the condition of acoustic comfort is achieved when the school is empty, i.e., when the noise produced in the surroundings does not impair the acoustic comfort of the classrooms.

The data presented above confirm the correct choice of terrain for the location and construction of the schools. The only exception was classroom C6, where it was found that, although the noise from surrounding traffic is low, another annoying external factor is the proximity of the railway line, Figure 8. The noise levels emitted by the train were measured in the schoolyard close to the classroom blocks. The L_{eq} measured as the train passed by the school, Figure 9, was 71.8 dB(A), with a minimum and maximum of 64.3 dB(A) and 80.8 dB(A), respectively. According to the NBR 10152 standard, the noise level in schoolyards should range from 45 to 55 dB(A). Figure 9 shows the cursor at the frequency of 1000 Hz, for which L_{eq} = 67 dB(A), L_{Max} = 72.2 dB(A) and L_{Min} = 60.4 dB(A).

The passing train generates a very high noise level, contributing significantly to the increase in the background noise inside the classrooms. The school's proximity to the railway line (Figure 8) indicates the inappropriateness of the land for the construction of classroom C6.

Fig. 8. Classroom C6 and the railroad traffic

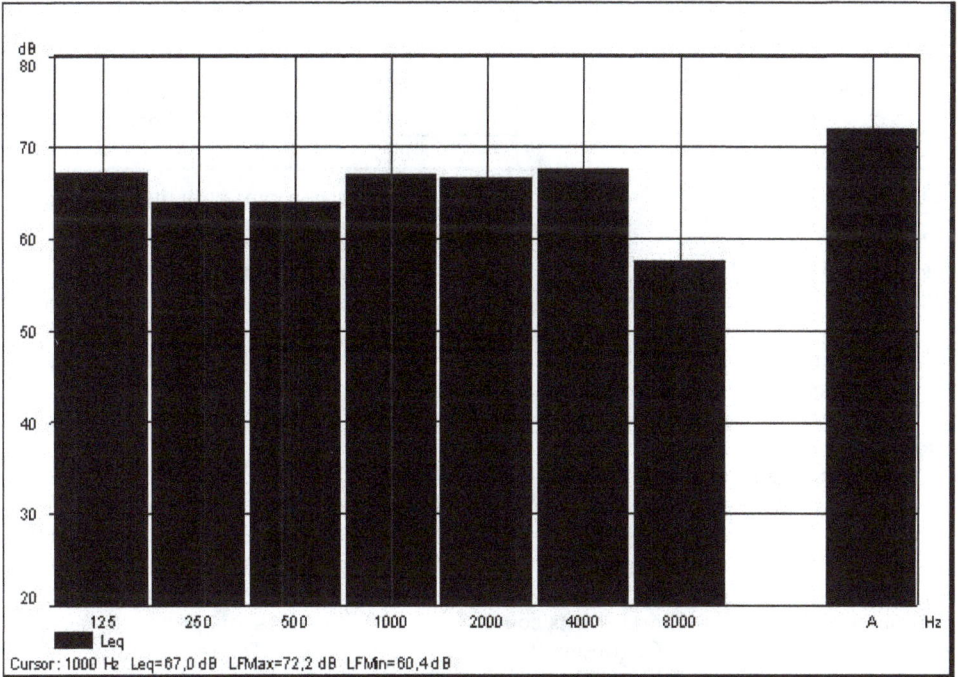

Fig. 9. Frequency analysis of railroad noise in the vicinity of classroom C6

With regard to the internal environment, the background noise in the classroom was investigated. To this end, the background noise originating from classes held in the other classrooms of the same block was measured in an empty classroom. The purpose of these measurements was to ascertain if the noises produced in a classroom affect other classrooms around it.

To determine the daily reality, all the measurements were taken with the windows open, seeking to interfere as little as possible in the schools' daily routines. Some situations could therefore not be evaluated in the same way. This was the case of classroom C3, where it was impossible to evaluate how the noise produced in each classroom affected the other classrooms because the noise coming from the schoolyard exceeded that of adjoining classrooms. Table 4[4] shows the noise levels measured in the empty classroom with the adjoining classrooms engaged in standard activities.

Classroom	Construction Design	L_{eq} dB(A)	Limit for acoustic comfort Brazilian NBR 10152 standard[14] L_{eq} dB(A)
C1	10	59.4	
C2	10	63.2	**40 - 50**
C4	22	51.1	*40 dB(A) is a comfortable noise level in classrooms
C5	23	59.1	**50 dB(A) is an acceptable noise level for classroom purposes
C6	23	60.7	

Table 4. Equivalent sound pressure levels L_{eq} in five empty classrooms with the other rooms engaged

The NBR 10152 standard establishes 40 dB(A) as a comfortable noise level in classrooms, although 50 dB(A) is acceptable for classroom purposes. As can be seen in the above table, the noise levels far exceed the limit determined by this standard. According to the World Health Organization[27], excessive noise levels affect not only the quality of verbal communication but also lead to serious problems in the student's intellectual development, such as slow language learning, difficulties in written and oral language, limitations in reading skills and in the composition of vocabulary.

With regard to the levels listed in Table 4, it can be concluded that the classrooms assessed here have a negative effect on each other, generating high levels of background noise that are incompatible with the values established by the Brazilian standard for acoustic comfort in classrooms. Additional disruptive noise originates from physical education activities. The classrooms of school design standards 010 and 022 showed insufficient distance between the classrooms and schoolyards and sports courts. Figure 10 illustrates this proximity between schoolyards and classrooms C2 (design 010) and C3 (design 022). The figure clearly shows the classroom windows in both schools facing the schoolyard, which contributes to increase the background noise inside the classrooms.

(a) Classroom C2 – design 010   (b) Classroom C3 – design 022

Fig. 10. Proximity between classrooms and schoolyards

Table 5 lists the values measured during physical education activities. These measurements were taken in empty classrooms with open windows[4].

Classroom	Construction Design	Equivalent Sound Pressure Level L_{eq} dB(A)
C1	10	66.7
C2	10	66.0
C3	22	74.6
C4	22	62.5

Table 5. Noise levels in empty classrooms during phys ed activities

The noise levels produced during physical education classes are high. The close proximity of classrooms to the schoolyard where these activities take place is extremely detrimental to the teaching-learning process, not only because of the noise levels that impair speech intelligibility but also due to the students' distraction and diminished concentration resulting from the visual stimuli provided by phys ed activities right outside the classroom windows. The noise levels shown in Table 5 and the photographs in Figure 10 reveal a serious problem in the layout of school spaces, since noisy environments should be far away from the environments that require silence, which is the case of classrooms.

Because the sound levels proved incompatible with the necessary conditions of acoustic quality and comfort in the classroom, the equivalent sound pressure levels were measured during a Portuguese language class (classroom C3) and a mathematics class (classroom C6). The values measured were L_{eq} = 74 dB(A) for the Portuguese language class and L_{eq} = 73.7 dB(A) for the mathematics class and corresponded mainly to the teacher's voice during explanatory classes when the students simply listened[4]. These values are high and demonstrate the vocal effort required of the teachers. This effort is even greater when phys ed activities are being held in the schoolyards, because they raise the noise level inside the classrooms.

The subjective research involving the teachers confirms the measured results. It was found that 21% of the interviewed teachers have had to take a leave of absence from teaching due to noise-related health problems, the main reason being vocal fatigue.

Table 6 shows the results of the teachers' subjective assessment of how noise affects them. The main aspects the teachers listed were the need to raise their tone of voice (2.5), overall fatigue (2), and vocal fatigue (2).

	Average score
Difficulty to concentrate	1.6
Headache	1.5
Irritability	1.9
Overall fatigue	2.0
Buzzing in the ears	1.2
Raising the tone of voice	2.5
Vocal fatigue	2.0

Table 6. Influence of noise in the classroom in the teacher's opinion.

According to Lubman and Sutherland[2], the cost of vocal fatigue of schoolteachers in the United States is US$ 648 million per year. In Brazil there are no estimates of this cost. However, as can be seen in Table 6, the need for teachers to raise their voices is high (2.5), regardless of the schools where they teach (Qs = 6.244, p-value = 0.182).

Noise in the classroom does not affect only teachers, for the findings of the subjective assessment revealed that it disturbs 92% of the students. The activities that suffer the most from noise-related disruption are listening to the teacher's explanations (46%), reading (23%), and doing exams (23%). According to the teachers, noise strongly affects the students' scholastic performance (score = 2.3).

These findings confirm that unfavorable acoustic conditions in the classroom make teaching and learning unnecessarily exhausting for everyone involved in the process[6].

In addition to the noise levels in empty classrooms, the noise level in the halls of the Luiza Ross School (C2) was measured. The measured L_{Aeq} was 72.9 dB with a maximum of 88.4 dB(A) and a minimum of 59.3 dB(A). For school halls, the NBR 10152 standard establishes a noise level of 45 dB(A) for acoustic comfort and 55 dB(A) as acceptable for this purpose. The levels measured in the Luiza Ross School far exceeded the acceptable level.

Another very important environment in the school is the library, where silence is essential and the noise level should be kept below 40 dB(A) (Knudsen and Harris[32]). The library at Luiz Ross (C2) is located in the same block as the classrooms, with windows high up in the wall looking out onto the hall and the entry door facing the outside of the block. The L_{eq} measured in the library was 64.3 dB(A), with a maximum of 75.7 dB(A) and a minimum of 54.7 dB(A). The NBR 10152 standard establishes 45 dB(A) as the noise level for acoustic comfort in libraries, accepting a limit of up to 55 dB(A). The level of 64.3 dB(A) measured in the library far exceeds the upper limit of the standard, impairing aspects inherent to these spaces, such as concentration and reading[6].

4.2 Reverberation time in classrooms

Reverberation time is an extremely important descriptor of the acoustic quality of a room, and several national and/or international standards establish reference values for the RT in rooms. With regard to classrooms, various acoustic standards present reference values for the RT that should be observed in the design of the classroom. Germany, Japan, the United Kingdom, the United States of America, Portugal and France have specific technical standards for evaluating the RT of classrooms.

In Japan, RT values represent the average in 2-octave bands including 500 Hz and 1000 Hz, and RT is measured in the furnished and unoccupied classroom[28]. In the USA, RT is given as the maximum RT for mid-band frequencies of 500 Hz, 1000 Hz and 2000 Hz, and RT is measured in the furnished and unoccupied classroom (ANSI S12.60[29]). In Germany, the DIN 18041[30] standard establishes that RT values represent the average in 2-octave bands including 500 Hz and 1000 Hz, and RT is measured in the furnished and occupied classroom. The German standard DIN 18041:2004 recommends that, in general, the RT of an unoccupied classroom should not be more than 0.2 s above the required value listed in Table 7. In France, RT is calculated as the arithmetic mean, for furnished and unoccupied classrooms, of the values measured at the frequencies of 500 Hz, 1000 Hz and 2000 Hz (WHO[27]), and in Portugal (also furnished and unoccupied classrooms), the RT recommended for classrooms is established as a function of two frequency ranges: 1) 125 Hz $\leq$ f $\leq$ 250 Hz, and 2) 500 Hz $\leq$ f $\leq$ 4000 Hz (WHO[27]). The World Health Organization – WHO recommends the value of 0.6 s for the RT in classrooms (Shield and Dockrell[31]). Table 7 (Zannin et al.[5]) shows the RT recommended in different countries as a function of the volume of the classroom.

Country	Reverberation Time RT, in [s]	Volume V, in [m³]
France	$0.4 < RT \leq 0.8$	$V \leq 250$
	$0.6 < RT \leq 1.2$	$V > 250$
Germany	$RT = 0.5$	$V = 125$
	$RT = 0.6$	$V = 250$
	$RT = 0.7$	$V = 500$
	$RT = 0.8$	$V = 750$
Japan	$RT = 0.6$	$V \sim 200$
	$RT = 0.7$	$V \sim 300$
Portugal	$RT \leq 1.0$ - for 125 Hz $\leq$ f $\leq$ 250 Hz	-
	$0.6 \leq RT \leq 0.8$ - for 500 Hz $\leq$ f $\leq$ 4000 Hz	-
United States of America	$RT = 0.6$	$V \leq 283$
	$RT = 0.7$	$283 < V \leq 566$
WHO	$RT = 0.6$	-

Table 7. Recommended Reverberation Time for Classrooms in different countries

Reverberation time was measured in furnished and unoccupied classrooms, and is listed in Table 8.

Construction Design of the School	Volume of the Classroom [m³]	Maximum Classroom Capacity
010	139	40 students
022	139	40 students
023	156	40 students

Table 8. Characteristics of the classrooms: Volume and Capacity

Comparison Between Reverberation Times Standards 010, 022 e 023

		125	250	500	1000	2000	4000
⊟C1	(010)	1,3	1,1	1,1	1,0	1,0	0,8
■C2	(010)	1,4	0,9	0,7	0,8	0,7	0,6
▲C3	(022)	2,8	2,0	2,0	2,2	2,1	1,6
▲C4	(022)	2,7	1,7	1,8	1,6	1,8	1,4
○C5	(023)	3,3	2,1	1,5	1,4	1,3	1,1
●C6	(023)	3,3	2,7	2,1	2,0	1,9	1,6

Frequency [Hz]

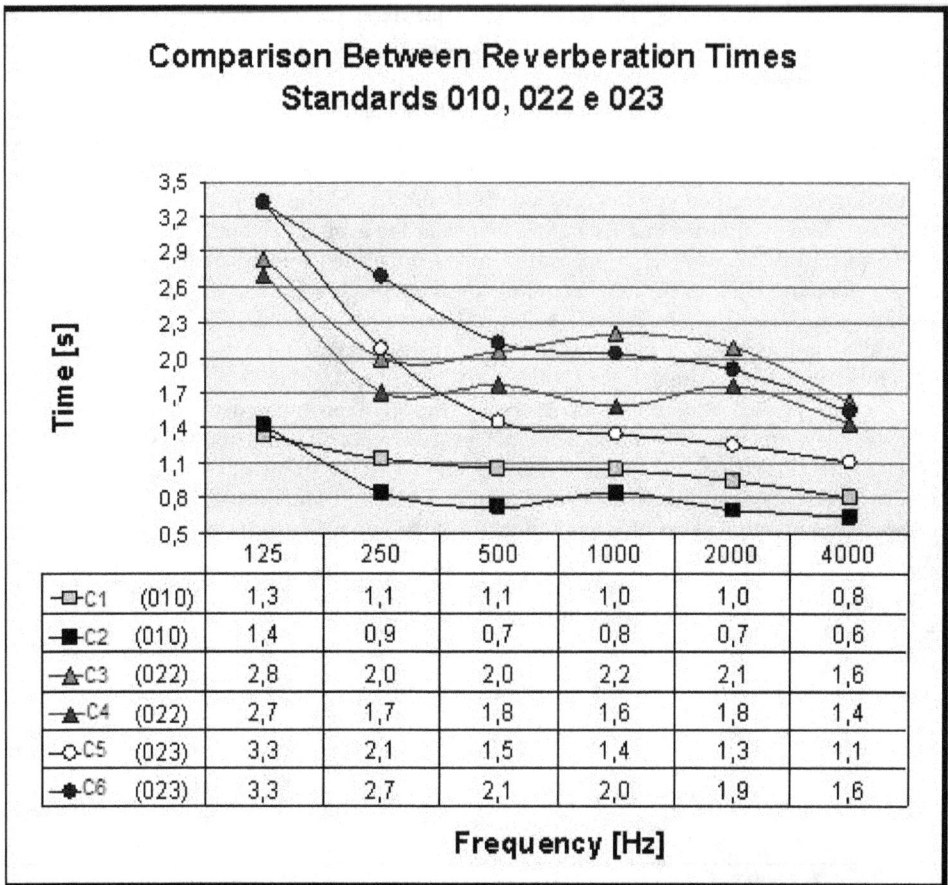

Fig. 11. Reverberation time of the six classrooms measured according to the ISO 3382-2 standard

Note that the RTs of all the classrooms evaluated here exceed the 0.6 s limit established by the ANSI S12.60[29] and Japanese standards. When compared with the French recommendation cited by WHO[27], only classroom C2 (design 010) falls within the established range of 0.4 to 0.8 s, since the mean RT at frequencies of 500, 1000 and 2000 Hz was 0.73 s in this classroom. Only classroom C2 with a mean RT of 0.75 s, at frequencies of 500 and 1000 Hz, complies with the range of values recommended by the DIN 18041[30] standard for furnished unoccupied classrooms.

The differences in the RTs of the constructive designs are due to the different finishing materials employed (Table 1). In the classrooms of design 010, the floors are made of parquet and the ceiling is paneled in wood. The 022 design also has parquet floors, while the 023 design has ceramic tile floors. The ceilings of designs 022 and 023 are not paneled, but simply plastered and painted. The walls of all the constructive designs have painted plaster overlays[4].

The classrooms of design 010 were built about 20 years before those of designs 022 and 023, and their interior finishing (walls, floors and ceilings) and RTs offer better acoustic conditions than do the classrooms in the newer buildings. This is attributed to the low sound absorption coefficients of the interior finishes currently in use.

The RTs measured in all the classrooms showed the lack of acoustic comfort in the classrooms, except for classroom C2. The acoustic deficiency of these spaces impairs communication between students and teachers, since high reverberation times diminish the intelligibility of speech[4,5].

In one of the evaluated classrooms, classroom C3 of construction design 022, the RT measurements were taken considering three different situations of occupancy: a) unoccupied and furnished, b) with 50% occupancy, or 20 students, and c) with 100% occupancy, or 40 students. Figure 12 shows the measured RTs as a function of occupancy.

Figure 12 clearly illustrates the influence of occupancy in the reduction of RT. A comparison of the situation of the unoccupied room and the situation of 100% occupancy indicated that this reduction in RT varied from 0.7s at a frequency of 125 Hz to 1.3s at a frequency of 2000 Hz. Even with full occupancy, the classroom did not reach the RT specified by any of the recommendations listed in Table 7. This demonstrates the urgent need for changes in the interior design of this classroom, using sound absorbing materials whose effect is to reduce the RT, thus contributing to improve its acoustic quality.

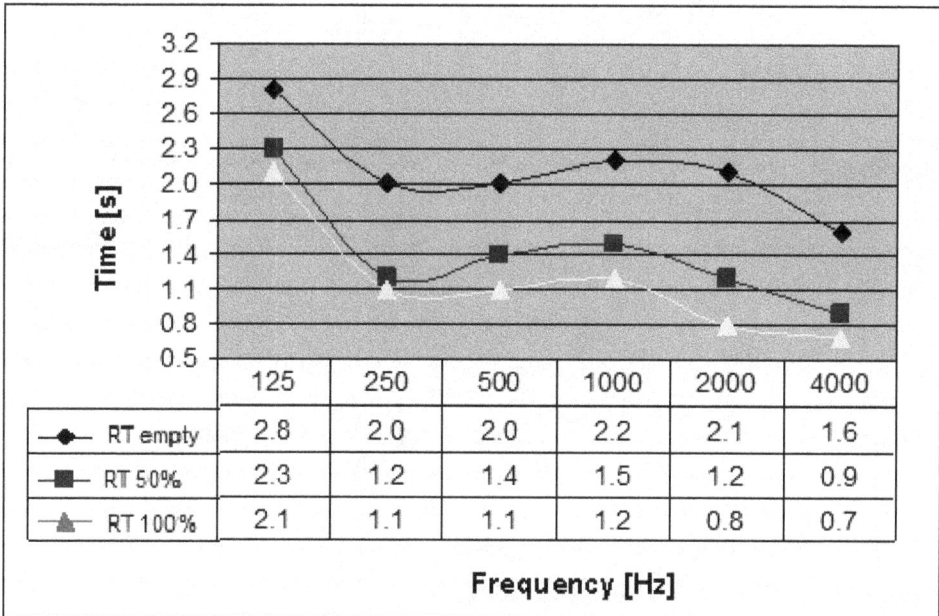

Time [s] / Frequency [Hz]	125	250	500	1000	2000	4000
RT empty	2.8	2.0	2.0	2.2	2.1	1.6
RT 50%	2.3	1.2	1.4	1.5	1.2	0.9
RT 100%	2.1	1.1	1.1	1.2	0.8	0.7

Fig. 12. Influence of classroom occupancy on reverberation time

4.3 Speech transmission index – STI

To obtain objective data on the speech intelligibility in the classrooms, acoustic simulations of the STI were performed using Odeon 9.0 software[19]. The 3D models were calibrated by comparing the measured and simulated RTs.

To simulate the STI, data on the room's background noise must be inserted in the computer model. In these simulations, the background noise was inserted according to the frequency spectrum measured in a classroom (see Figure 13), which is equivalent to a sound pressure level of approximately 60 dB(A).

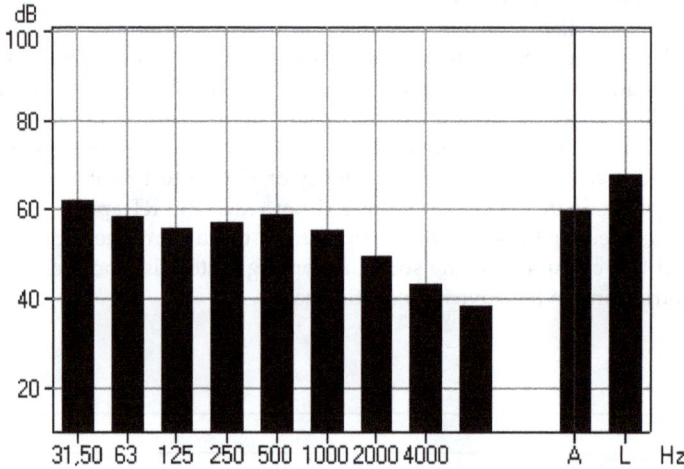

Fig. 13. L_{Leq} graph in octave bands measured in a classroom and used in the simulation of the STI

The maps below present the simulation of the STI in the three construction designs under study.

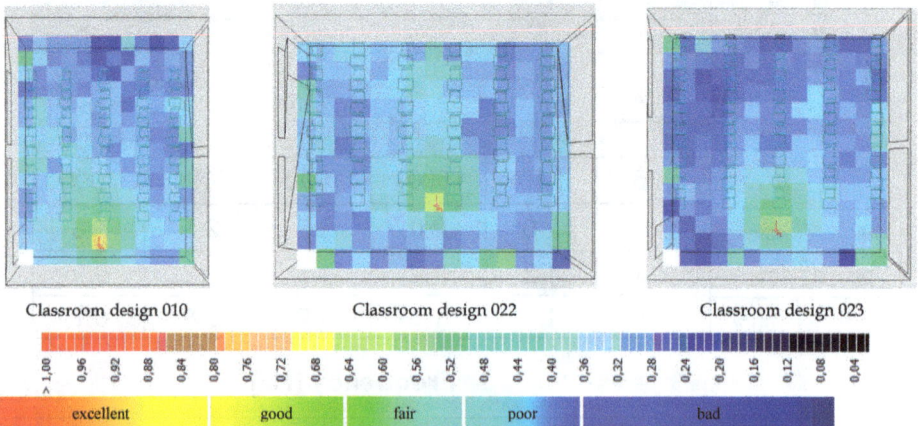

Classroom design 010   Classroom design 022   Classroom design 023

Fig. 14. Simulation of the STI in the construction models under study. The subjective scale equivalent to the objective values of STI are in line with the IEC 60268- 16:2003 standard. The red dot represents the sound source.

As can be seen in the STI maps (Figure 14), the speech intelligibility at most of the simulated points in the classrooms varies from "fair" to "bad," according to the subjective scale of the IEC 60268- 16:2003 standard.

4.4 Façade sound insulation

Sound insulation should be a priority in school environments where the sources of noise cannot be altered, especially in schools affected by high levels of noise from road, air and railroad traffic. Another important factor is the sound insulation between quiet and very noisy spaces, as in the case of the school designs 010 and 022, where phys ed classes are held in schoolyards located very close to the classrooms.

Due to the complexity of the measuring process in terms of the amount of equipment and number of people involved, sound insulation measurements were taken in only one classroom of each design.

The façades of the classrooms blocks of school design 010 are composed of ordinary brickwork overlaid with mortar, inside and outside, and painted. The windows are made of iron frames and ordinary glass panes. Figure 15 shows the façades of this construction design. The evaluations of façade insulation of design 010 were carried out at the Alfredo Parodi School (C1).

The measurements of sound insulation of the classrooms of design 023 were carried out at the Luarlindo dos Reis Borges School (C6). The façades of this design are composed of ordinary brick walls overlaid with ceramic tile, while the inside is overlaid with mortar and painted. The iron frame windows have ordinary glass panes. Figure 16 shows the façade of the Luarlindo dos Reis Borges School during the measuring procedures.

Fig. 15. Classroom Façades of the Luiza Ross and Alfredo Parodi Schools

Fig. 16. Measurement of the sound insulation of the façade according to ISO 140-5[10] (Luarlindo dos Reis Borges School)

The evaluation of sound insulation of the schools of design 022 was carried out at the Anibal Khury Neto School (C3). As Figure 17 indicates, the classroom blocks have different façades on each side, one containing doors and the other windows.

Fig. 17. Façade of a classroom – left: Door and Walls; right: Wall and Windows

Classroom	Construction Design	$D_{ls,2m,nT,w}$ [dB]	Façade Sound Insulation Brazilian NBR 15575 standard[33] [dB]
C1- Alfredo Parodi School	010	21	
C3 - Aníbal Khury Neto School	022	31	25 – 29
C6 – Luarlindo dos Reis Borges School	023	27	

Table 9. Sound insulation of classroom façades

Table 9 indicates that only one of the three classrooms evaluated does not meet the requirements of the Brazilian NBR 15575[33] standard. In fact, one of the classrooms has a higher sound insulation index than recommended by the standard. However, it should be kept in mind that most classrooms have their windows open while they are in use, due to the ambient temperature and the need for air circulation. The sound insulation measurements were taken with the windows and doors closed. Therefore, it is to be expected that the values of sound insulation of the façades are very different from the values measured in ideal conditions. Another point that should be considered is the state of conservation of the buildings, which directly affects their sound insulation. Figure 18 shows a classroom with broken window panes and large spaces between the door and the floor, which are factors that contribute to reduce the sound insulation.

Fig. 18. Classroom com broken window panes and spaces between the door and the floor

4.5 Noise control in classrooms

Based on the objective and subjective evaluations carried out in this study, it can be concluded that there are difficulties in spoken communication between teacher and student inside the classrooms. This is due to the high background noise and inadequate RT, particularly in the classrooms of school design standards 022 and 023, which generate "fair" to "bad" speech intelligibility (Figure 14) according to the subjective scale of the IEC 60268-16:2003 standard.

Because the quality of spoken communication is an extremely important factor for learning in classrooms, acoustic simulations were performed which aimed at improving this quality in the rooms under study. Using Bastian software, changes in the façade construction elements were simulated to obtain the $D_{ls,2m,nT,w}$. In addition, changes in the absorption coefficient of the ceiling finishing material and the background noise were simulated to obtain the RT and STI inside the classrooms, using Odeon software.

4.5.1 Simulations of sound insulation

To study the improvements in sound insulation, simulations were performed by changing the type of window used in the façade. The windows in the evaluated rooms consist of poorly sealed iron window frames and ordinary glass panes. For these simulations, a window contained in the library of the Bastian software[18,36] was used, which is well sealed and double paned – 4 mm + 12 mm + 4 mm (two 4 mm glass panes with 12 mm or air between them).

The table below presents the values of $D_{ls,2m,nT,w}$ measured in the classrooms and simulated using Bastian software.

Standard Classroom	Measured Sound Insulation $D_{ls,2m,nT,w}$ [dB]	Simulated Sound Insulation $D_{ls,2m,nT,w}$ [dB]
010	21	44
022	31	44
023	27	45

Table 10. Sound insulation measured *in situ* and simulated with changed window element

Table 10 shows the improvement in the sound insulation of the classroom façades attained by replacing the existing windows for better sealed windows with double panes.

Any improvement in the sound insulation of façades means a reduction of the background noise inside the classroom, especially in areas with intense external noise. This reduction in background noise translates into improved acoustic quality inside the classroom.

It should be kept in mind that, in the cases studied here, due to the climatic conditions and also in view of the need for air circulation, the classrooms usually have their windows open when in use. The measurements and simulations considered closed windows.

4.5.2 Simulations of reverberation and speech intelligibility

The reverberation time in a room is related to the absorption coefficient of the materials that cover the surfaces of the room. Therefore, the RT simulations were performed by replacing the ceiling finishing material of the rooms for another with a higher absorption coefficient.

Table 11 presents, in octave band frequency, the absorption coefficients of the ceiling finishing materials in the rooms and used in the simulations. In construction design 010, the ceiling is wood paneled, while the rooms of designs 022 and 023 are devoid of ceiling finishing material, consisting simply of concrete slabs. A fiberglass acoustic ceiling board was used in the simulations.

	63Hz	125Hz	250Hz	500Hz	1000Hz	2000Hz	4000Hz	8000Hz
Wood	0.25	0.25	0.30	0.40	0.40	0.55	0.60	0.60
Concrete	0.018	0.018	0.02	0.03	0.03	0.03	0.03	0.03
Fiberglass	0.33	0.33	0.79	0.99	0.91	0.76	0.64	0.64

Table 11. Absorption coefficient of the ceiling materials in the classrooms and used in the acoustic simulations

Figure 19 shows the RT simulated with the existing ceilings in the classrooms and with the fiberglass acoustic ceiling.

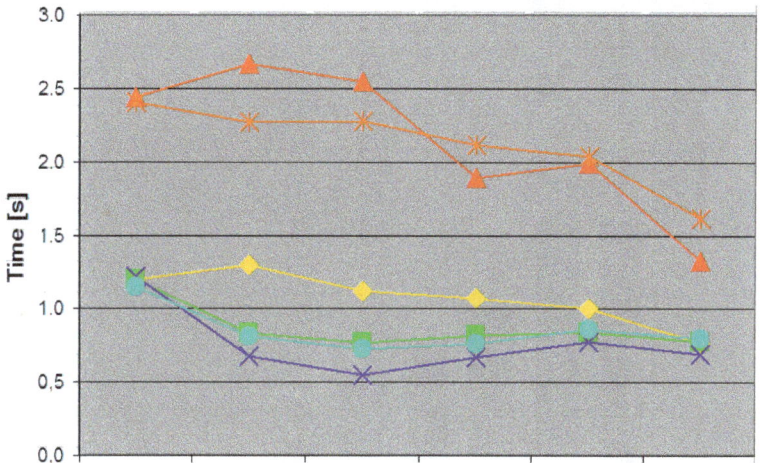

			125	250	500	1000	2000	4000
Design 010	current		1.2	1.3	1.1	1.1	1.0	0.8
	amended		1.2	0.8	0.8	0.8	0.8	0.8
Design 022	current		2.4	2.7	2.6	1.9	2.0	1.3
	amended		1.2	0.7	0.6	0.7	0.8	0.7
Design 023	current		2.4	2.3	2.3	2.1	2.0	1.6
	amended		1.2	0.8	0.7	0.8	0.9	0.8

Fig. 19. Reverberation time simulated in the real conditions of the classroom (current) and with amended ceiling material

As can be seen in Figure 19, there is a significant decrease in the values of RT, especially in the classrooms of design standards 022 and 023, which have no type of ceiling finishing.

If one compares the values of simulated RT (Figure 19) against the values of RT established by various standards (Table 7), one finds that, upon amending the ceiling material, all the rooms are considered suitable according to the French standard. Comparing the simulated RT with the German standard, only the classroom of the 010 model remains inadequate even after replacing the ceiling material, although the RT values are very close to those stipulated by this standard. With regard to the other standards listed in Table 7, the modified rooms presented a mean RT of 0.1s to 0.2s higher than that established by these standards.

In addition to the RT, the STI was simulated with amended ceiling finish. Two sound pressure levels (34 and 60 dBA) were used for the background noise, which were measured in classrooms. The graph in Figure 20 presents a frequency spectrum of the background noise of 34 dB(A).

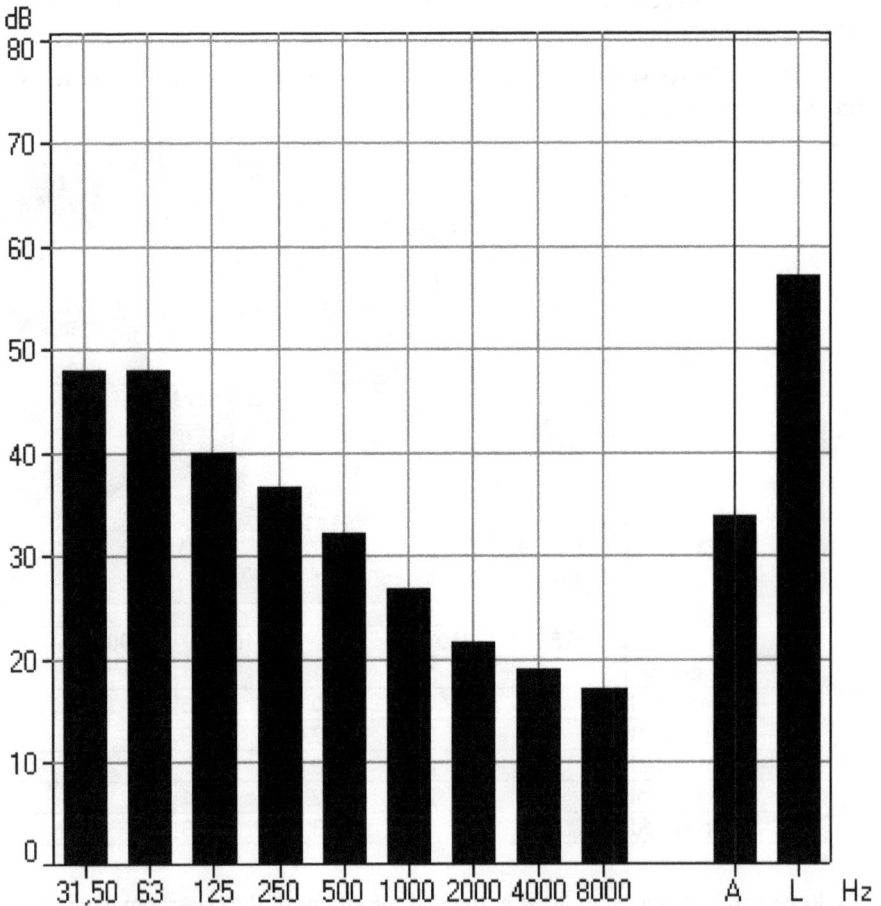

Fig. 20. Non-weighted frequency spectrum of background noise of 34 dB(A)

The figures below present the STI maps of the design standards 010, 022 and 023, for two background noise levels (60 and 34 dBA) and two situations of RT (current situation and situation with amended ceiling material).

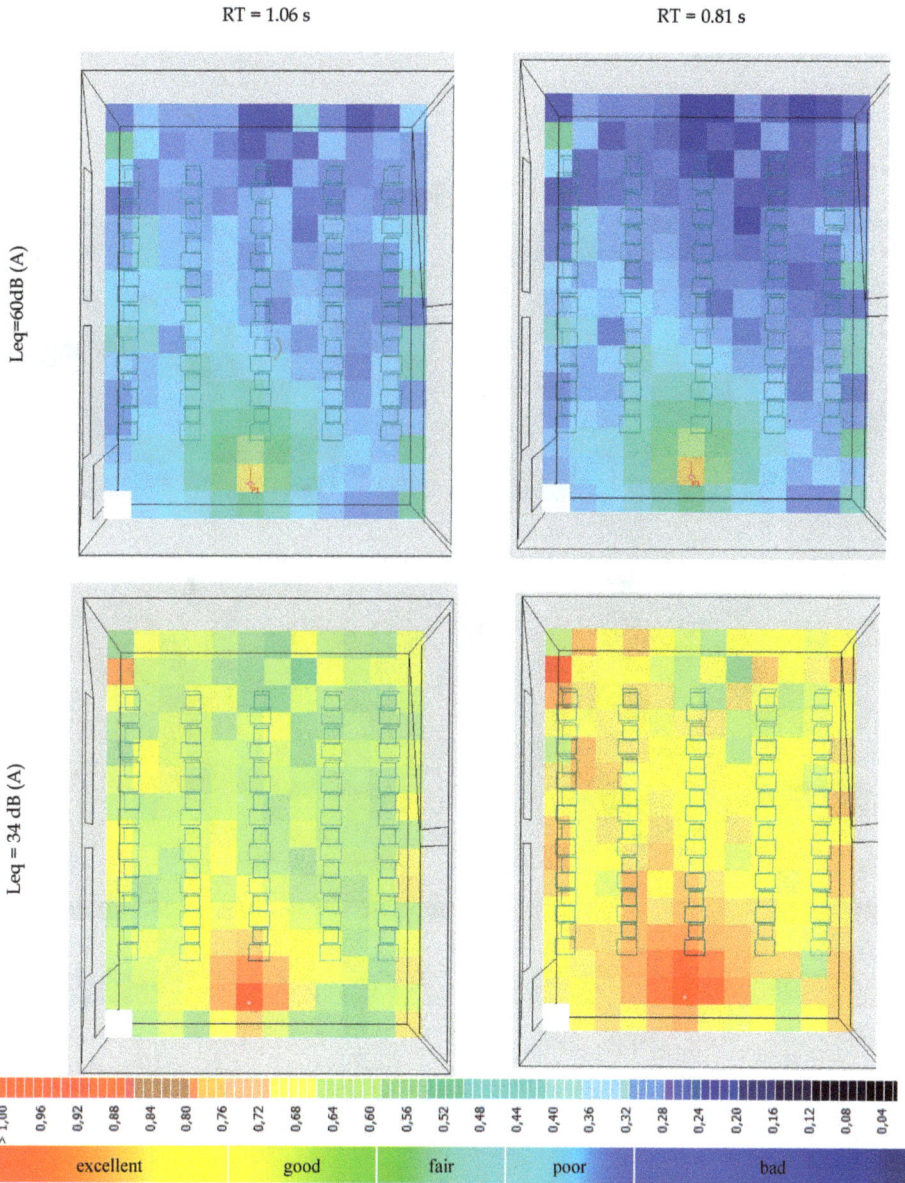

Fig. 21. STI simulations in the classroom of design standard 010. The red dot represents the sound source and the red line indicates its direction (towards the students)

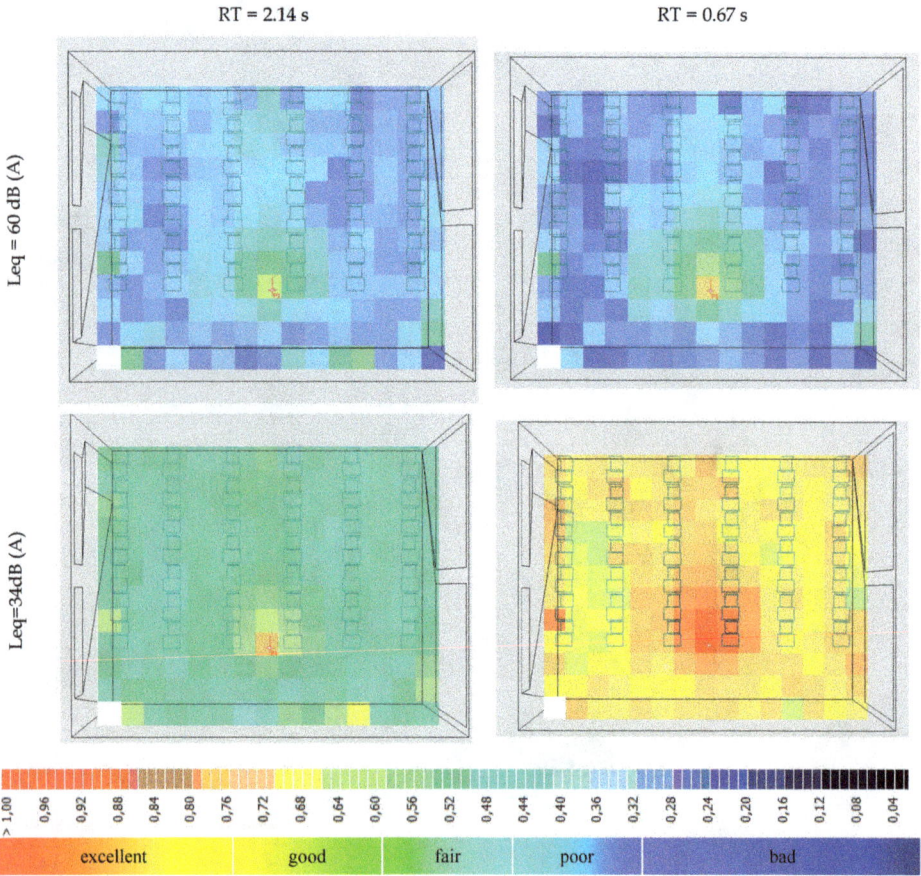

Fig. 22. STI simulations in the classroom of design standard 022. The red dot represents the sound source and the red line indicates its direction (towards the students)

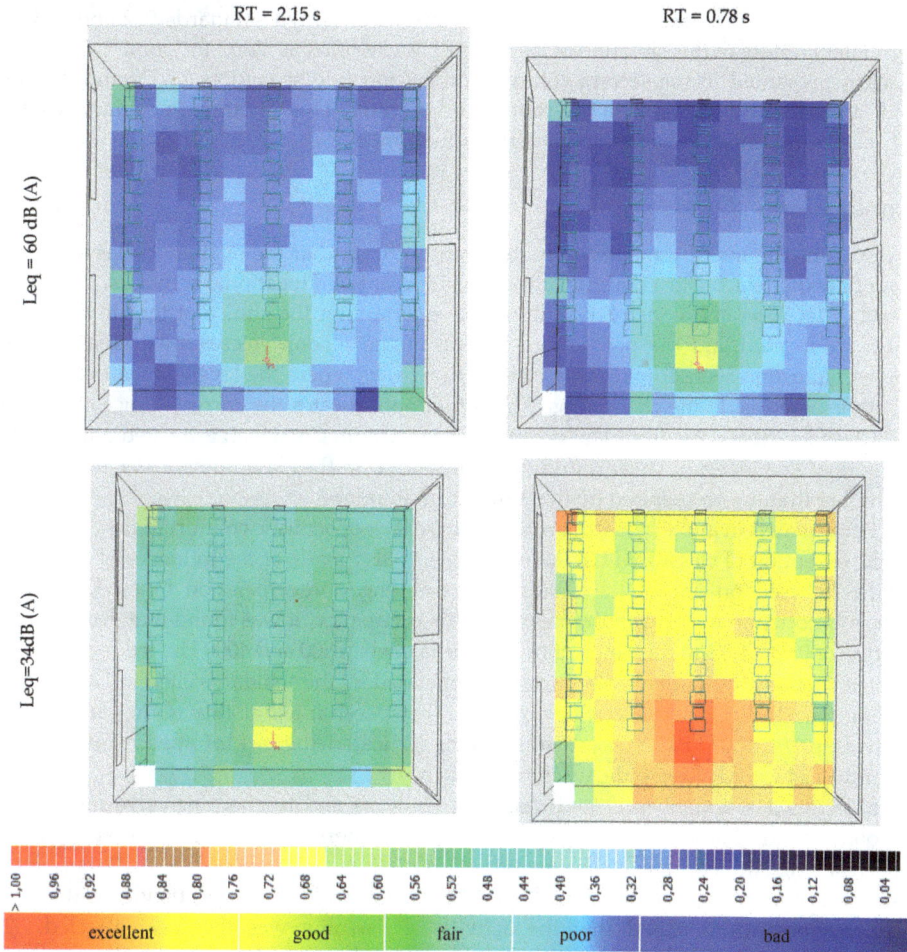

Fig. 23. STI simulations in the classroom of design standard 023. The red dot represents the sound source and the red line indicates its direction (towards the students)

Figures 21, 22 and 23 present the STI simulation maps of the classrooms of design standards 010, 022 and 023. The map in the first column and first line of the figures represents the current situation of the classrooms in terms of RT and SPL.

The maps in the second column and first line represent the simulation with reduction of the RT by replacement of the current ceiling finishing material by a material with higher acoustic absorption, and maintenance of the background noise at 60 dB(A). A comparison of these maps with the current situation indicates that the STI worsened at the points located farther away from the sound source, thus classifying these points as having "bad" to "poor" intelligibility.

The maps on the second line and first column in Figures 21, 22 and 23 represent the STI for the current situation of RT in the classrooms, with reduced background noise. These maps show a significant improvement in speech intelligibility, particularly the room of design standard 010, whose RT was lower than that of the other classrooms. For this room, the map in this situation generated a "good" to "excellent" STI. For the rooms of design standards 022 and 023, the speech intelligibility in this situation of RT and SPL is classified as "fair" to "good."

The maps presented in the second column and second line of Figures 21, 22 and 23 show "good" to "excellent" STI for all the rooms under study. This situation is characterized by low background noise and adequate reverberation time.

5. Conclusions

The results of the measurements presented here indicate that the evaluated classrooms do not offer adequate acoustic comfort for the development of educational activities.

Aspects of location and construction involved ascertaining the choice of terrain where the schools are located and the position, or layout, of the schools' recreational areas. The six schools in question were found to be located in regions where the levels of traffic noise are not bothersome inside the classrooms. This situation indicates the correct choice of land for five of the six schools assessed. The exception is classroom C6, which is located adjacent to a railway line. The levels of background noise measured in the classrooms, halls and libraries were higher than recommended by the NBR 10152 standard.

The RTs measured in all the classrooms showed the lack of acoustic comfort in the classrooms, with the exception of classroom C2. An analysis was also made of the influence of occupancy on the RT, based on one of the rooms with the longest reverberation times. Despite the significant reduction in RT with the room at 100% occupancy, it was found that even with the reduction of 0.9 to 1.3 s in the range of frequencies between 500 and 4000 Hz, the room still did not present values compatible with those recommended by the standards cited in Table 7.

Based on the STI simulations, it was found that the speech intelligibility in the classrooms is classified as "fair" to "bad." These simulations indicated the lack of speech intelligibility even at desks located close to the teacher/speaker.

The subjective assessment indicated that both students and teachers perceive noise in the classroom and are bothered by it. According to the teachers, noise is a factor that negatively affects teaching and learning.

Most of the students stated they could hear the teacher well. However, they considered the classrooms noisy and stated that the activity of listening was the most affected. This statement was confirmed by the results of the questionnaire which the teachers answered.

The results of the measurements and the questionnaires revealed that the noise that impairs classroom activities comes from the school itself, not only from adjacent classrooms, halls and schoolyards but also, and mainly, from inside the classroom itself.

The subjective research involving the teachers confirmed the measured results. It was found that 21% of the interviewed teachers have had to take a leave of absence from teaching due to noise-related health problems, the main reason being vocal fatigue.

The results of the evaluations in the classrooms demonstrated the need for noise control in these environments, in order to align them with the requirements of good oral communication. The simulations revealed that it is possible to improve the façade insulation by substituting ordinary windows (about 5 mm thick) for double paned windows. The amendment of ceiling finishing materials proved efficient, since it resulted in a reduction of the RT to values that are adequate or very close to those established by the standards listed in Table 7. As for speech intelligibility, the STI classification of "good" to "excellent" was only achieved through a combination of reduced background noise and reduced RT inside the classrooms.

6. Acknowledgements

The authors gratefully acknowledge CNPq – Conselho Nacional de Desenvolvimento Científico e Tecnológico (Brazil), DAAD – Deutscher Akademischer Austauschdienst (Germany), and Fundação Araucária (Brazil) for the financial resources that enabled the purchase of all the equipment and software for this research. The authors would also like to thank the directors, teachers and students who contributed to this work by answering the questionnaires.

7. References

[1] Freire P. Pedagogia da indignação: cartas pedagógicas e outros escritos. São Paulo: Editora Unesp, 2000. (in Portuguese).

[2] Lubman D, Sutherland LC. Good Classrooms Acoustic is a god investment. In: International Congress on Acoustics ICA, 2001, Roma, Itália. Anais...

[3] Zannin PHT, Marcon CR. Objective and subjective evaluation of the acoustic comfort in classrooms. Applied Ergonomics, 2007, 38: 675-680.

[4] Zannin PHT, ZwirteS DPZ. Evaluation of the acoustic performance of classrooms in public schools. Applied Acoustics, 2009, 70: 626–635.

[5] Zannin PHT, Passero CRM, Sant'ana DQ, Bunn F, Fiedler PEK, Ferreira AMC. Classroom Acoustics: Measurements, Simulations and Applications. In: Rebecca J. Newley (Org). Classrooms: Management, Effectiveness and Challenges. New York: Nova Science Publishers. (in Press, 2011).

[6] Hagen M, Huber L, Kahlert J. Acoustic school desing. In: Forum Acusticum, 2002, Sevilha. Anais...

[7] Seep B, Glosemeyer R, Hulce E, Linn M, Aytar P. Acústica de salas de aula. Revista de Acústica e Vibrações, 2002, 29: 2-22. (in Portuguese).

[8] International Organization for Standardization. ISO 3382-1: Acoustics — Measurement of room acoustic parameters - Part 1: Performance spaces. Switzerland, 2009.

[9] International Organization for Standardization-ISO. ISO 3382-2: Acoustics — Measurement of room acoustic parameters - Part 2: Reverberation time in ordinary rooms. Switzerland, 2008.

[10] [10] International Organization for Standardization – ISO. ISO 140-5: Acoustic – Measurement of sound insulation in building and of building elements – Part 5: Field measurements of airborne sound insulation of façade elements and façades. Suíça, 1998.

[11] International Organization for Standardization – ISO. ISO 717-1: Acoustics – Rating of sound insulation in buildings and of building elements – Part 1: Airborne sound insulation. Suíça, 1996.

[12] Associação Brasileira de Normas Técnicas. NBR 10152: Níveis de ruído para conforto acústico – procedimento. Rio de Janeiro. 1987. (in Portuguese).

[13] Harris CM. Handbook of acoustical measurements and noise control . USA: Acoustical Society of America, 1998.

[14] Kang J. Numerical modeling of speech intelligibility in dining spaces. Applied Acoustics, 2002, 63: 1315-1333.

[15] Rindel JH. The use of computer modeling in room acoustics. Journal of Vibroengineering, 2000, 3 (4): 219-224.

[16] Bradley DT, Wang LM. Comparison of Measured and Computer-Modeled Objective Parameters for an Existing Coupled Volume Concert Hall. Building Acoustics, 2007, 14 (2): 79–90.

[17] International Electrotechnical Commission. IEC 60268- 16: Sound system equipment-Part 16: Objective rating of speech intelligibility by speech transmission index. Switzerland, 2003.

[18] European Norm – EN 12354-3 Building Acoustics – Estimation of acoustic performance of buildings from the performance of elements. Part 3: Airborne sound insulation against outdoor sound. UK, 2000.

[19] Odeon Software Handbook - Version 9.0.

[20] Dockrell J, Tachmatzidis I, Shield B, Jeffery R. Children's perceptions of noise in schools. In: Proceedings of International Congress on Acoustics, 17, 2001. Roma. Anais...

[21] Loro CLP. Avaliação acústica de salas de aula – Estudo de caso em salas de aula Padrão – 023 da rede pública. Curitiba, 2003. Dissertação (Mestrado), Universidade Federal do Paraná. (in Portuguese).

[22] Losso MAF. Qualidade acústica de edificações escolares em Santa Catarina: Avaliação e Elaboração de diretrizes para projeto e implantação. Florianóplis. 2003, 149 f. Dissertação (Mestrado). Universidade Federal de Santa Catarina. (in Portuguese).

[23] Enmarker I, Boman E. Noise annoyance responses of middle school pupils and teachers. Journal of Environmental Psychology, Nova York, 2004, 24: 527-536.

[24] Dockrell J E, Shield B. Children's perceptions of their acoustic environment at school and at home. Journal of the Acoustical Society of America, USA, 2004, 115 (6):2964–2973.

[25] R Development Core Team. R : A language and environment for statistical computing. R Foundation for Statistical Computing. Vienna, Áustria (2005).

[26] Associação Brasileira de Normas Técnicas. NBR 10151: Acústica – Avaliação do ruído em áreas habitadas, visando o conforto da comunidade - Procedimento. Rio de Janeiro. 2000. (in Portuguese).

[27] World Health Organization – WHO. Noise in schools. Geneva, 2001.

[28] Fukuchi, T.; Ueno, K. Guidelines on acoustic treatments for school buildings proposed by the Architectural Institute of Japan. In: ICA – International Conference on Acoustic, Kyoto, Japan, 2004.

[29] american national standard – ANSI S12.60: Acoustical performance criteria, design requirements, and guidelines for schools. Melville, 2002.

[30] DIN 18041: Acoustic quality in small to medium-sized rooms (Hörsamkeit in kleinen bis mittelgrossen Räumen). Germany, 2004.

[31] Shield BM, Dockrell JE. The effects of noise on children at school: A review. Building Acoustics, 2003, 10 (2): 97-106.

[32] Knudsen VO, Harris, CM. Acoustical Designing in Architecture. 5ed. Acoustical Society of America, 1988.

[33] ABNT - Associação Brasileira de Normas Técnicas - NBR 15575 – Norma Brasileira – Edifícios habitacionais de até cinco pavimentos – Desempenho – Parte 4: Sistemas de vedações verticais. Segunda Edição 12.11.2010. (Brazilian Standard - Residential buildings up to five storied – Performance – Part 4: Internal and External wall systems. (in Portuguese).

[34] Agresti, A. A survey of exact inference for contigency tables. Statistical Science, 1992; 7: 131-153.

[35] Agresti, A. An introduction to categorical data analysis. Nova York: Wiley, 1996.

[36] Bastian - Handbook – Bastian Software 2.3 – The building Acoustics Planning System – User Manual – DataAkustica, 2003.

11

Application of Sound Level for Estimating Rock Properties

Harsha Vardhan[1] and Rajesh Kumar Bayar[2]
[1]National Institute of Technology Karnataka, Surathkal
[2]N.M.A.M. Institute of Technology, Nitte
India

1. Introduction

The process of drilling, in general, always produces sound as a by-product. This sound is generated from the rock-bit interface, regardless of the material the bit is drilling in. The drillability of rock depends on many factors, like bit type and diameter, rotational speed, thrust, flushing and penetration rate. Sound is used as a diagnostic tool for identification of faulty components in the mechanical industry. However, its application in mining industry for estimating rock properties is not much explored. Knowledge of rock properties is essential for mine planning and design. The rock properties such as compressive strength, porosity, density etc. are uncontrollable parameters during the drilling process. The rock properties must be determined at a mine or construction site by testing a sample. There are various techniques for the determination of rock properties in the laboratory and the field. International Society of Rock Mechanics (ISRM) and American Society for Testing and Materials have suggested or standardised the procedure for measuring the rock strength. However, the method is time consuming and expensive. As an alternative, engineers use empirical and theoretical correlations among various physico-mechanical properties of rock to estimate the required engineering properties of rocks.

Most of the works in the application of sound levels are in other branches of engineering (Vardhan et al., 2004, 2005, 2006; Vardhan & Adhikari, 2006). A couple of studies in oil and gas industries have proposed a technique called "Seismic–While–Drilling" for estimating rock formations. For instance, few studies have proposed the use of noise produced by the bit during drilling as a seismic source for surveying the area around a well and also for formation characterization while drilling (Onyia, 1988; Martinez, 1991; Rector & Hardage, 1992; Miranda, 1996; Asanuma & Niitsuma, 1996; Hsu, 1997; Aleotti et al., 1999; Tsuru & Kozawa, 1998; Hand et al., 1999; Fernandez & Pixton, 2005). A recent study (Stuart et al., 2004) has also reported a method of estimating formation properties by analyzing acoustic waves that are emitted from and received by a bottom hole assembly. It needs to be emphasized that "Seismic–While–Drilling" technique is different from the technique of estimating rock properties using sound levels produced during drilling.

For rock engineering purpose, very limited publications are available on this subject. The usefulness of sound level in determining rock or rock mass properties has been shown clearly only in two publications (Roy & Adhikari, 2007; Vardhan & Murthy, 2007) and the need for further work in this area has been suggested. It is anticipated that the sound level with drilling in rocks of different physico-mechanical properties will be different for the same type of drill machine. Keeping this point in mind, the present research work was undertaken.

This chapter reveals some investigations (both laboratory and field) carried out to estimate the rock properties using sound levels produced during drilling.

2. Laboratory investigations

The noise measurement for the same type of drill machine varies from strata to strata. Thus, the variations in the sound level can indicate the type of rock, which can be used to select suitable explosives and blast designs. Rock characterization while drilling is not a new idea. Devices for monitoring the drilling parameters such as thrust, drilling depth and penetration are available and the information obtained are used for blast designs. However, the concept of rock characterization using sound levels is new. Therefore, laboratory investigations were caried out using small portable pneumatic drilling equipment and Computer Numerical Controlled (CNC) machine with carbide drill bit set-up along with noise measuring equipments.

2.1 Laboratory investigations using portable pneumatic drilling equipment
2.1.1 Experimental setup

In the laboratory, all the sound level measurements were conducted on pneumatic drill machine operated by compressed air. The experimental set-up was in a normal cement plastered room of 5 m width, 9 m length and 5 m height. The important specifications of the pneumatic drill used were:
- Weight of the pneumatic drill machine (28 kg)
- Number of blows per minute - 2200
- Type of drill rod - Integrated drill rod with tungsten carbide drill bit
- Recommended optimum air pressure - 589.96 kPa

A lubricator of capacity 0.5 litres and a pressure gauge with a least count of 49 kPa were provided between the compressor and pneumatic drill machine to lubricate the various components and to regulate the air pressure supplied to the drill machine, respectively. A percussive drill setup to drill vertical holes was fabricated to carry out the drilling experiments for sound level measurement on a laboratory scale (Fig. 1). The base plate of the setup consists of two 12.5 mm thick I-sections (flange width - 1 cm and height – 30 cm) which are welded together all along the centre. They are firmly grouted to the concrete floor with the help of four 3.8 cm diameter anchored bolts. Two circular guiding columns of 60 mm diameter, 175 cm long, and 55 cm apart were secured firmly to the base plate. The verticality of the two columns was maintained with the help of a top plate (3.8 cm thick, 13 cm width and 62.5 cm length). On the top of the base plate, 25.4 mm diameter holes were drilled at close intervals on two opposite sides for accommodating different sizes of rock blocks (up to 500 mm cube). The rock block was firmly held on the base plate with the help of two mild steel plates (1 cm thick, 7.5 cm width and 61 cm length) kept on the top of the rock block and four 25.4 mm bolts, placed at the four corners.

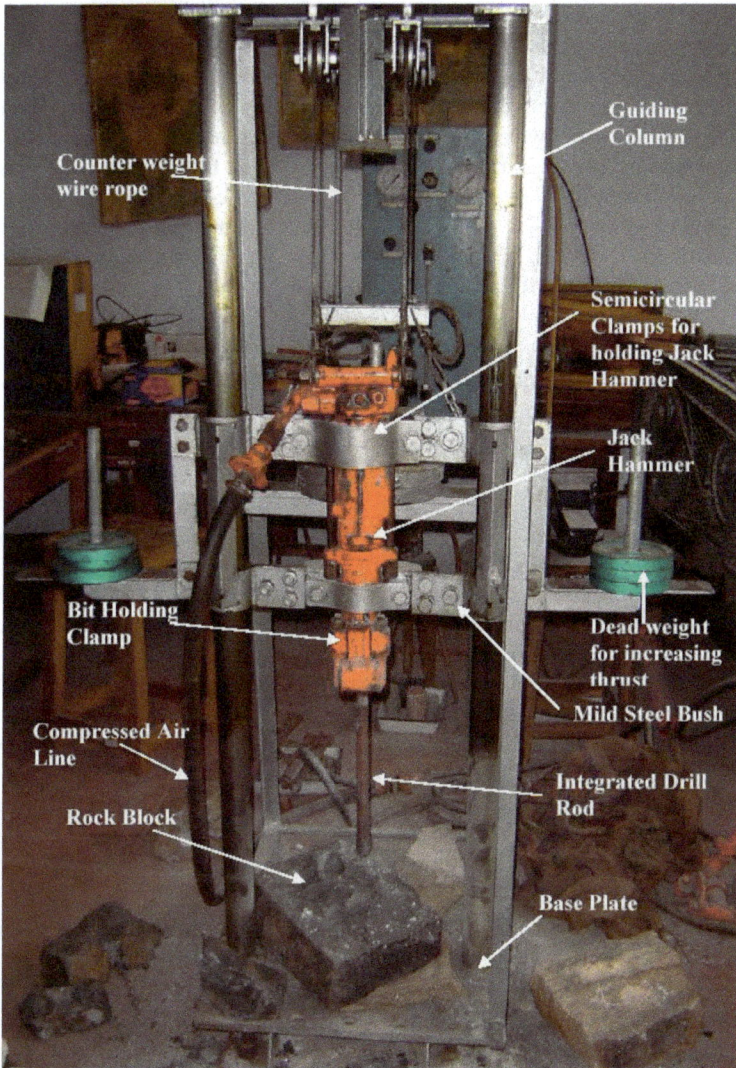

Fig. 1. Pneumatic drill setup for drilling vertical holes in rock blocks

The pneumatic drill was firmly clamped at its top and bottom with the help of four semicircular mild steel clamps, which were in turn bolted firmly to four mild steel bushes for frictionless vertical movement of the unit over the two guiding columns of the setup. In order that the top and bottom clamps work as one unit, they were firmly connected with the help of four vertical mild steel strips (1.3 cm thick, 5 cm width and 50 cm length) on each side of the pneumatic drill. For increasing the vertical thrust, two vertical mild steel strips (1.3 cm thick, 5 cm width and 32 cm length) were bolted to the top and bottom clamps. On this strip, dead weights made up of mild steel were fixed with the help of nut and bolt arrangements. For conducting drilling experiments at low thrust level (less than the dead weight of the drill machine assembly), a counter weight assembly was fabricated. For this purpose a steel wire rope (0.65 cm diameter) was clamped to the top of the pneumatic drill unit which in turn passed through the pulley arrangements located at the top plate of the setup. A rigid frame was firmly grouted to the shop floor at a distance of 86 cm from the experimental setup. The steel wire rope from the experimental setup was made to pass over the pulley mounted on the rigid frame. At the other end of the rope, a plate was fixed for holding the counter weights. The dead weight of pneumatic drill machine and accessories for vertical drilling was 637 N. With the help of counter–weight arrangement, it was possible to achieve a desired thrust value as low as 100 N. Similarly, through the arrangement of increasing the thrust level, it was possible to achieve a thrust value as high as 900 N.

2.1.2 Rock samples used in the investigation
Sound level measurement on pneumatic drill set up was carried out for five different rock samples obtained from the field. These rock samples were gabbros, granite, limestone, hematite and shale. The size of the rock blocks was approximately 30 cm x 20 cm x 20 cm.

2.1.3 Instrumentation for noise measurement
Sound pressure levels were measured with a Larson-Davis model 814 integrating averaging sound level meter. The instrument was equipped with a Larson Davis model 2540 condenser microphone mounted on a model PRM904 preamplifier. The microphone and preamplifier assembly were mounted directly on the sound level meter. The acoustical sensitivity of the sound level meter is checked once a year. For all measurements, the sound level meter was handheld. To determine the noise spectrum, the instrument was set to measure A–weighted, time-averaged one-third-octave-band sound pressure levels with nominal midband frequencies from 25 Hz to 20 kHz. The sound level meter was also set to measure A–weighted equivalent continuous sound levels (Leq). For each measurement, the sound level meter was set for an averaging time of 2 minutes.

2.1.4 Determining the compressive strength and abrasivity of rock specimens
The compressive strength of rock samples was determined indirectly using Protodyakonov's apparatus. Protodyakonov index for estimation of compressive strength of rock samples is an indirect and time-consuming method. However, this method was chosen due to limited availability of any particular type of rock samples in the laboratory. Therefore, first sound level measurement using drilling was carried out. Then the same drilled rock block was used for determining compressive strength and abrasivity. It was difficult to prepare samples for determining uniaxial compressive strength from these drilled rock blocks.

Abrasion test measures the ability of rocks to wear the drill bit. This test includes wear when subject to an abrasive material, wear in contact with metal and wear produced by contact between the rocks. For this purpose, Los Angele's abrasion test apparatus was used.

The results of the experimental study for the compressive strength and the abrasivity of the rock samples are given in Table 1. It is seen that, with increase in compressive strength of rock samples, the abrasivity decreases. This is due to increase in the resistance of rocks to wear with increase in the compressive strength.

Block No.	Rock Type	Compressive strength (kg/cm²)	Abrasivity (%)
Block 1	Shale	1051.35	23.70
Block 2	Hematite	1262.33	21.50
Block 3	Limestone	1542.57	20.30
Block 4	Granite	1937.13	17.50
Block 5	Gabros	2252.35	15.50

Table 1. Compressive strength and abrasivity of different rocks

2.1.5 Noise measurement

A set of four test conditions was defined for measurement of sound spectra which is given in Table 2. The measurement of sound spectra was carried out on pink granite. For the test conditions A2, A3 and A4 mentioned in Table 2, the air pressure was constant at 6 kg/cm². For test condition A1, the sound spectrum was measured at the operator's position and without actually operating the drill machine. This background noise was mainly due to the compressor operating near the pneumatic drill setup. Test condition A2 in the table refers to the measurement of sound spectra at the operator's position by opening the exhaust of the drill but without carrying out any drilling operation. Test conditions A3 and A4 refer to measurement of noise spectra during drilling at the operator's position with 100 N and 300 N thrust respectively.

Noise sources measured at operator position	Test condition
Background	A1
Air only	A2
Air + drill with 100 N thrust	A3
Air + drill with 300 N thrust	A4

Table 2. Test conditions for determination of sound spectra

For measuring the variation in sound level while drilling in rocks of different compressive strength and abrasivity, the rock blocks were kept beneath the integrated drill rod of the pneumatic drill. Sound level measurements were carried out for thrust values of 160, 200, 300 and 360 N on each rock block. It is worth mentioning here that the realistic thrust values used by drill operators in the field vary based on the type of rock encountered at a particular site. Typical thrust values in the field may vary from 150 to beyond 500 N. For each thrust mentioned above, the A–weighted equivalent continuous sound level (L_{eq}) was measured by holding the sound level meter at 15 cm distance from the drill bit, drill rod and the exhaust for air pressure values of 5.0, 5.5, 6.0 and 7.0 kg/cm². Similarly, the L_{eq} level was measured at the operator's position for each thrust of 160 to 360 N and air pressures of 5 to 7 kg/cm² as mentioned above. The operator's position refers to the position of the operator's ear which was at a height of 1.7 m from the ground level and 0.75 m from the center of the experimental set-up. During measurement, all the doors and windows of the room were kept open so as to reduce the effect of reflected sound.

For a particular condition, at each microphone location and for the same rock block, the sound level was determined five times in relatively rapid succession. The arithmetic average of the A-weighted sound pressure levels from each set of five measurements was computed to yield an average A-weighted sound level for a particular condition.

2.1.6 Noise assessment of pneumatic drill under various test conditions at operator's position

The noise spectrum at the operator's position for test conditions A1 and A2 are shown in Fig. 2. It is seen that the background sound level at the measurement location due to the operation of the air compressor alone is below 82 dB with the nominal one-third-octave midband frequencies from 25 Hz to 20 kHz. Also, the increase in sound level with midband frequencies above 50 Hz is more than 10 dB for test condition A2 relative to that of test condition A1. Therefore, the sound level in the frequency range of 63 Hz to 20 kHz for test condition A2 is unlikely to be affected by the background noise due to the compressor. However, the sound level for test condition A2 may be affected due to test condition A1 with nominal midband frequencies from 25 to 50 Hz as the difference in sound level in this range of frequency is below 9 dB.

The noise spectrum at the operator's position for test conditions A2, A3 and A4 are shown in Fig. 3. It is seen that from 50 to 100 Hz, the increase in sound level for test condition A3 relative to that of A2 is from 2.8 to 7.2 dB and that of A4 relative to that of A3 is from 3.2 to 5.9 dB. This shows that drilling operation has increased the sound level with midband frequencies from 50 to 100 Hz. The increase in sound level in this frequency range (50 – 100 Hz) is due to impact between the piston and the drill steel and that between the drill steel and the rock. The increase in sound level for test condition A3 relative to that of A2 with midband frequencies from 125 Hz to 2 kHz is in the range of 1.0 to 11.7 dB and that of A4 relative to that of A3 is in the range of 1.6 to 6.0 dB. The noise in this frequency range (125 Hz – 2 kHz) is due to the exhaust of the drill machine. The combination of drilling noise and exhaust noise has resulted in increase of sound level in this frequency range (125 Hz – 2 kHz). There is significant increase in sound level of the order of 6.6 to 14.2 dB from 2.5 to 20 kHz for test condition A3 relative to that of A2 and 4.0 to 7.7 dB for test condition A4 relative to that of A3. This increase in sound level is due to resonance of the steel parts of the drill steel due to rock drilling.

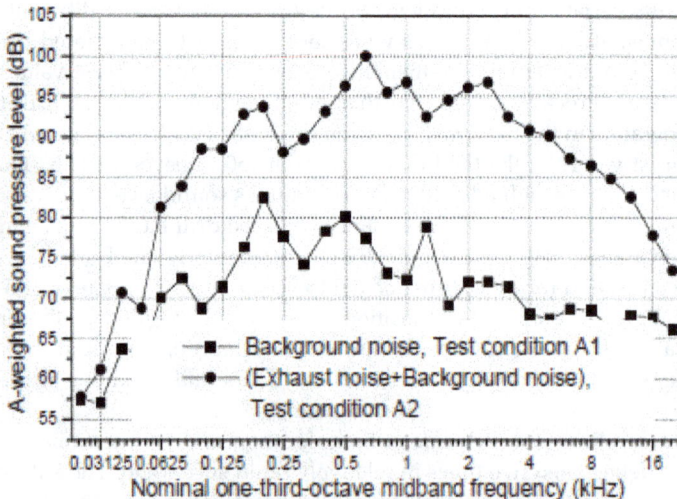

Fig. 2. Effect on L_{eq} levels at the operator's position for test conditions A1 and A2

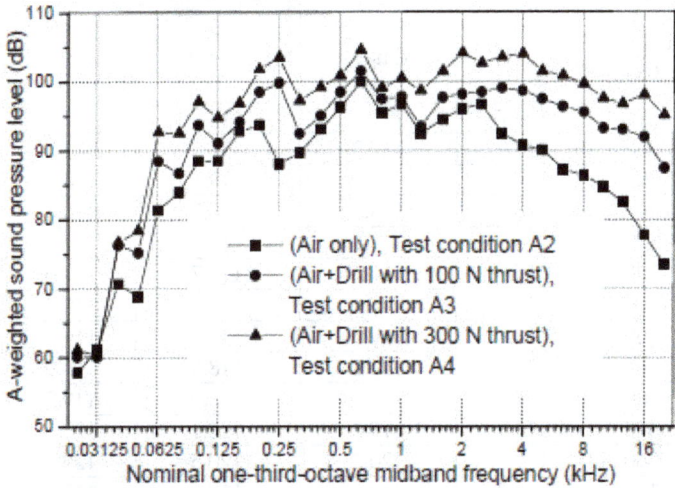

Fig. 3. Effect on L_{eq} levels at the operator's position for test conditions A2, A3 and A4

2.1.7 Effect of rock properties on sound level of pneumatic drill

a. At operators position

The L_{eq} level at the operator's position for different rocks of varying strength at various thrusts and air pressures are given in Table 3. In this table, the compressive strengths of rocks are given in increasing order i.e., shale has the lowest compressive strength and the highest abrasivity whereas gabro has the highest compressive strength and the lowest abrasivity. At an air pressure of 5 kg/cm² and thrust of 160 N, the difference in A–weighted sound level for different rocks was of the order of 0.8 dB, which varied from 0.8 to 1.4 dB with an increase in the thrust from 160 to 360 N. At an air pressure of 5.5 kg/cm², and a thrust of 160 N, the difference in A-weighted sound level for different rocks was 0.9 dB. At this air pressure (5.5 kg/cm²), an increase in the thrust from 160 to 360 N caused an increase in the sound level by 1.6 dB. Similar results were observed at air pressures of 6 and 7 kg/cm² with an increase in the thrust from 160 to 360 N.

The effect of air pressure on sound levels at constant thrust of 160 N for different rock samples at operator's position is shown in Fig. 4. An increase in sound level is observed with increasing air pressure values. With an increase in air pressure by 2 kg/cm², i.e., from 5 to 7 kg/cm² and at a thrust of 160 N, the sound level of block-1 increased by 1.6 dB. Similar results were shown by other rock samples too. The increase in sound level for different rocks (Block-1 to Block-5) with an increase in the air pressure by 2 kg/cm² at a thrust of 160 N is 1.9, 2.1, 2.2 and 2.4 dB respectively.

The effect of compressive strength of rock on sound level at operator's position for a constant thrust of 160 N and for different air pressure values is shown in Fig. 5 The above result shows that an increase in the compressive strength and a decrease in the abrasivity of rocks increase the sound level. It is worth mentioning that, to maintain optimum penetration rate, the thrust and air pressure must be increased in rocks having higher compressive strength and lower abrasivity, which in turn results in higher sound levels.

Air pressure (kg/cm²)	Thrust (N)	Shale	Hematite	Limestone	Granite	Gabros
5	160	116.7	116.9	117.0	117.3	117.5
	200	116.9	117.3	117.3	117.5	117.8
	300	117.8	117.9	118.1	118.3	118.7
	360	118.2	118.3	118.5	118.8	119.6
5.5	160	116.9	117.1	117.2	117.4	117.8
	200	117.3	117.5	117.7	117.9	118.2
	300	118.3	118.9	119.1	119.5	119.7
	360	118.7	119.5	119.8	119.9	120.3
6	160	117.9	118.1	118.6	118.9	119.2
	200	118.4	118.5	118.9	119.3	119.5
	300	119.2	119.8	120.1	120.5	120.7
	360	119.8	120.2	120.5	120.8	121.3
7	160	118.3	118.8	119.1	119.5	119.9
	200	118.6	119.2	119.5	119.7	120.3
	300	119.5	120.3	120.7	121.1	121.7
	360	120.2	120.8	121.1	121.9	122.2

Table 3. L_{eq} level at the operator's position for different rocks at various thrust and air pressures

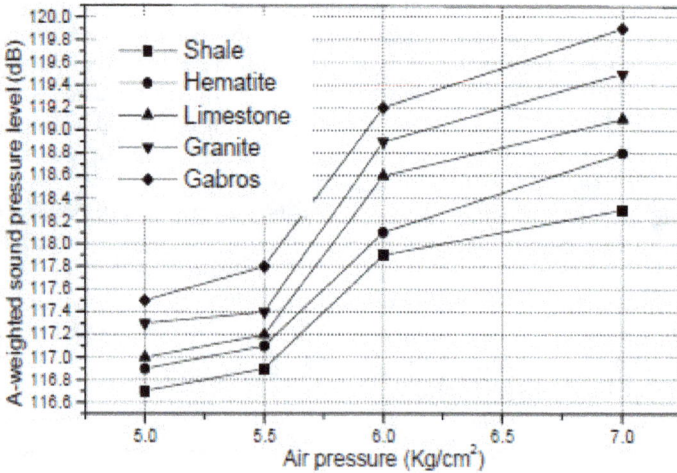

Fig. 4. Effect of air pressure on sound level at the operator's position at a constant thrust of 160 N for different rock blocks

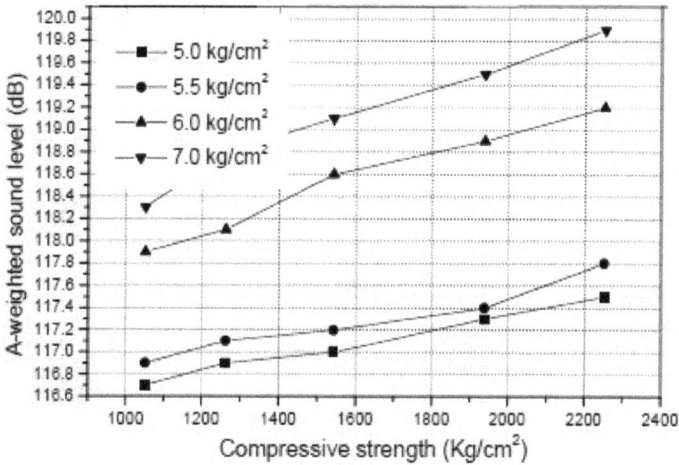

Fig. 5. Effect of compressive strength of rock on sound levels at the operator position for a constant thrust of 160 N and different air pressures

b. At exhaust

The L_{eq} level at exhaust for different rocks of varying strength at various thrusts and air pressures are given in Table 4. A significant increase in the sound level with an increase in the compressive strength and a decrease in the abrasivity is observed for different rocks. For instance, the difference in A-weighted sound level for block-1 and block-5 is 2.2 dB at constant air pressure and thrust of 5 kg/cm² and 160 N respectively. The variation of sound levels in all the five blocks, each with a different compressive strength and abrasivity, at an air pressure of 5 kg/cm² and thrust varying from 160 to 360 N is shown in Fig.6.

It can be seen that, with an increase in the compressive strength and a decrease in the abrasivity of rocks, the L_{eq} level increased near the exhaust at each thrust level for a constant air pressure of 5 kg/cm². Similar results can be seen from Table 4, for air pressures of 5.5, 6.0 and 7.0 kg/cm². At an air pressure of 5 kg/cm², an increase in thrust by 200 N (from 160 to 360 N) caused the sound level difference to vary from 1.4 to 1.8 dB for different rocks at the exhaust. The effect of compressive strength of rock on sound level at exhaust for a constant thrust of 160 N for different air pressure values is shown in Fig. 7. An increase in the air pressure by 2 kg/cm² at a constant thrust of 160 N resulted in an increase in the sound level, varying from 1.2 to 2.4 dB for different rock properties. This shows that, both thrust and air pressure have a significant effect on sound level produced by pneumatic drill at the exhaust.

Air pressure (kg/cm²)	Thrust (N)	Shale	Hematite	Limestone	Granite	Gabros
5	160	118.4	118.7	119.8	120.1	120.6
	200	118.8	119.2	120.6	120.9	121.5
	300	119.3	119.5	121.0	121.6	121.7
	360	119.9	120.5	121.5	121.9	122.2
5.5	160	119.9	120.1	120.2	120.7	120.8
	200	120.2	120.7	120.9	121.2	121.7
	300	120.9	121.3	121.7	121.9	122.3
	360	121.2	121.7	121.8	122.2	122.6
6	160	120.3	120.5	120.8	121.1	121.4
	200	120.6	121.2	121.8	122.2	122.5
	300	121.9	122.5	122.9	123.4	123.8
	360	121.8	122.8	123.2	123.7	124.2
7	160	120.8	120.9	121.2	121.5	121.8
	200	121.3	121.5	121.9	122.4	122.7
	300	122.0	122.7	123.2	123.7	123.9
	360	122.5	123.1	123.7	123.9	124.5

Table 4. L_{eq} level at exhaust for different rocks at various thrust and air pressures

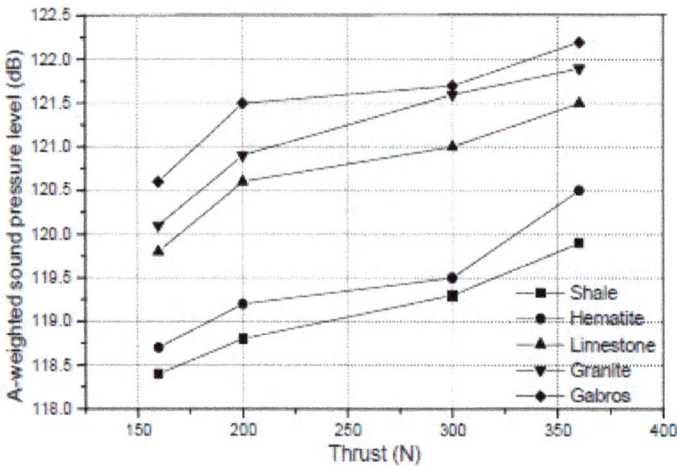

Fig. 6. Effect of thrust on sound level at the exhaust at constant air pressure of 5 kg/cm² for different rock blocks

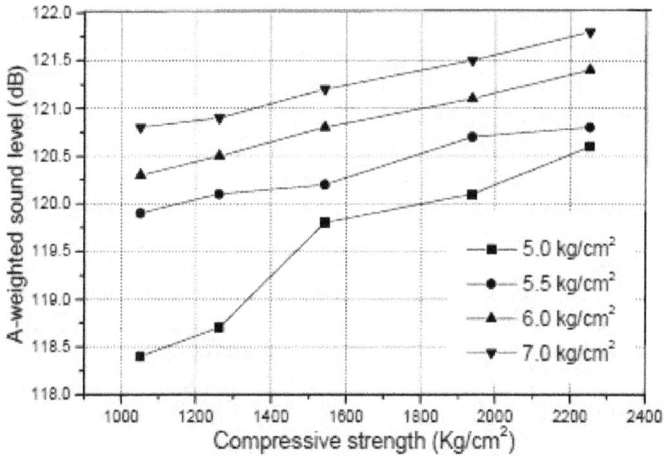

Fig. 7. Effect of compressive strength of rock on sound levels at exhaust at a constant thrust of 160 N and varying air pressure

c. Near drill rod

The L_{eq} level near the drill rod for rocks having varying compressive strength and abrasivity at various thrusts and air pressures is given in Table 5. Maximum increase in the sound level with an increase in the compressive strength and a decrease in the abrasivity was observed near the drill rod compared to that of other positions.

Air pressure (kg/cm²)	Thrust (N)	Shale	Hematite	Limestone	Granite	Gabros
5	160	120.5	121.9	122.3	122.8	123.3
	200	121.2	122.4	123.0	123.4	123.9
	300	122.0	122.7	123.4	124.1	124.2
	360	122.7	123.3	123.7	124.4	125.0
5.5	160	121.1	122.2	122.7	123.1	123.4
	200	121.9	122.8	123.5	123.9	124.1
	300	122.4	123.5	124.2	124.5	124.7
	360	122.9	123.9	124.5	124.8	125.3
6	160	121.7	122.8	123.1	123.5	123.8
	200	122.3	123.1	123.8	124.2	124.5
	300	122.8	123.9	124.6	124.9	125.3
	360	123.2	124.2	124.9	125.3	125.7
7	160	123.1	123.7	123.9	124.2	124.8
	200	123.7	124.2	124.9	125.0	125.5
	300	124.5	125.5	125.2	126.2	126.7
	360	124.9	125.7	125.8	126.7	126.9

Table 5. L_{eq} level near the drill rod for different rocks at various thrusts and air pressures

The sound level difference at an air pressure of 5 kg/cm² with increase in thrust from 160 to 360 N varied from 2.2 to 2.8 dB. At air pressures of 5.5, 6.0 and 7.0 kg/cm², this sound level difference of shale and gabro varied from 2.2 to 2.4 dB, 2.1 to 2.5 dB and 1.7 to 2.2 dB respectively. The above results clearly indicate that the variation in the compressive strength and abrasivity of rock has a significant effect on the sound level near the drill rod and that the sound level near the drill rod increases as the compressive strength increases.

Both the air pressure and thrust were observed to have a significant effect on the sound level produced near the drill rod. For instance, an increase in the air pressure by 2 kg/cm², at a constant thrust of 160 N caused an increase in the sound level varying from 1.4 to 2.6 dB. Similarly, an increase in the sound level with an increase in the thrust of 200 N at an air pressure of 5 kg/cm² varied from 1.4 dB to 2.2 dB for rocks having varying properties.

The effect of the compressive strength of rock on the sound level near the drill rod at a constant thrust of 160 N and varying air pressure is shown in Fig. 8.

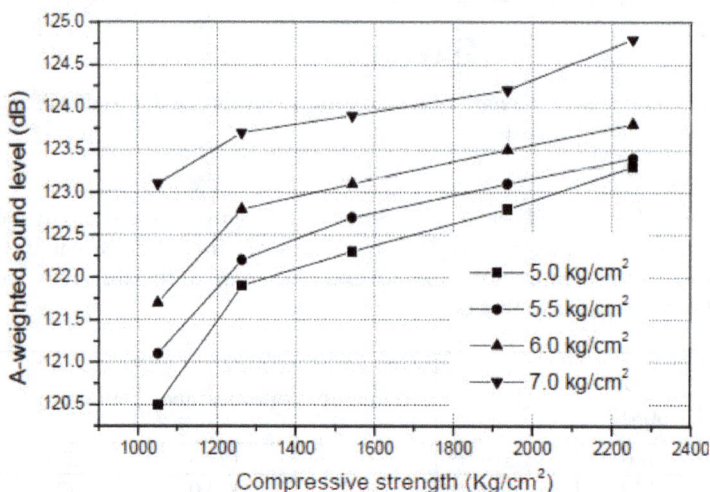

Fig. 8. Effect of compressive strength of rock on sound levels near drill rod for a constant thrust of 160 N and varying air pressure

d. Near the drill bit

The L_{eq} level near the drill bit for rocks having varying compressive strength and abrasivity at various thrusts and air pressures is given in Table 6. In general, an increase in the sound level is observed at each thrust and air pressure with an increase in the compressive strength and a decrease in the abrasivity of the rocks. The difference in the sound level at an air pressure of 5 kg/cm² and with an increase in the thrust from 160 to 360 N varied from 0.9 to 1.9 dB. At air pressures of 5.5, 6.0 and 7.0 kg/cm², this sound level difference in different rocks varied from 1.2 to 2.1 dB. This shows that an increase in the compressive strength and a decrease in the abrasivity of rock increase the sound level significantly.

In this case also, both air pressure and thrust were observed to have a significant effect on the sound level. For example, an increase in the air pressure by 2 kg/cm² at a constant thrust of 160 N indicated an increase in the sound level of 1.7 dB for block-1 and 1.0 dB for block-2 to block-5.

Air pressure (kg/cm²)	Thrust (N)	Shale	Hematite	Limestone	Granite	Gabros
5	160	120.0	121.0	121.2	121.6	121.9
	200	120.8	121.5	121.7	122.0	122.3
	300	121.5	122.0	122.1	122.3	122.5
	360	121.8	122.1	122.3	122.5	122.7
5.5	160	120.8	121.2	121.6	121.8	122.2
	200	121.3	121.7	122.2	122.5	122.7
	300	121.6	122.3	122.7	122.9	122.9
	360	121.9	122.6	122.9	123.3	123.7
6	160	121.5	121.7	122.0	122.4	122.7
	200	121.8	121.9	122.3	122.7	122.9
	300	122.3	122.6	122.9	123.2	123.6
	360	122.7	122.8	123.2	123.7	123.9
7	160	121.7	122.0	122.2	122.5	122.9
	200	121.9	122.4	122.7	122.9	123.1
	300	122.7	123.1	123.6	123.9	124.8
	360	122.9	123.5	123.8	124.0	124.9

Table 6. L_{eq} level near the drill bit for different rocks at various thrust and air pressures

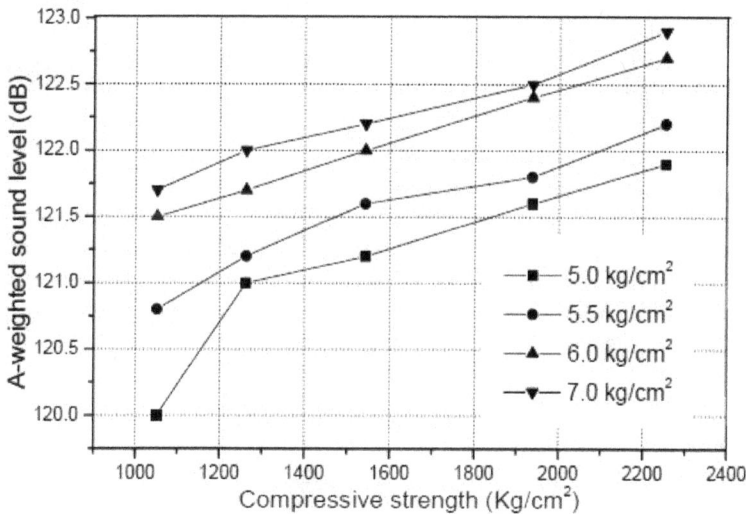

Fig. 9. Effect of compressive strength of rock on sound levels near the drill bit for a constant thrust of 160 N and varying air pressure

The increase in the sound level with an increase in the thrust of 200 N at an air pressure of 5 kg/cm² was 1.8 dB for shale, 1.1 dB for hematite and limestone, 0.9 dB for granite and 0.8 dB for gabro. The effect of compressive strength of rock on the sound level near the drill bit for a constant thrust of 160 N for different air pressure values is shown in Fig. 9.

2.2 Laboratory investigations using CNC machine

Compressor was one of the major sources of noise in the laboratory investigation explained in section 2.1. To overcome this, and also to nullify background noise, another investigation was carried out using Computer Numerical Controlled (CNC) machine with carbide drill bit setup. Further, the main aim of this investigation was to find out the relationship of rock properties with sound level produced during drilling.

2.2.1 Experimental setup

In the laboratory, rock drilling operations were performed on BMV 45 T20, Computer Numerical Controlled (CNC) vertical machining centre. The experimental set-up was in a fibre and glass–paned room of 5 m width, 6 m length and 9 m height. The important specifications of the CNC machine used were:
- Table size 450 mm x 900 mm
- Recommended optimum air pressure – 6 bar.
- Power supply – 415V, 3Phase, 50Hz

Carbide drill bits of shank length 40 mm and diameters of 6, 10, 16 and 20 mm were used for drilling operation. The machine was set to drill 30 mm drillhole length. Since the drilling method affects the sound produced, an attempt was made to standardize the testing procedure. Throughout the drilling process a relatively constant rotation speed (RPM), and penetration rate (mm/min) were provided in order to obtain consistent data.

2.2.2 Rock samples used in the investigation

For this investigation, different igneous rocks were collected from different localities of India taking care of representation of variety of strength. During sample collection, each block was inspected for macroscopic defects so that it would provide test specimens free from fractures and joints. The different igneous rocks used in the investigation and their properties are given in Table 7.

2.2.3 Instrumentation for noise measurement

The instrument used for sound measurement was a Spark 706 from Larson Davis, Inc., USA. The instrument was equipped with a detachable 10.6 mm microphone and 7.6 cm cylindrical mast type preamplifier. The microphone and preamplifier assembly were connected by an integrated 1.0 m cable. A Larson Davis CAL 200 Precision Acoustic Calibrator was used for calibrating the sound level meter. Before taking any measurement, the acoustical sensitivity of the sound level meter was checked using the calibrator.

2.2.4 Determining the rock properties

a. Uniaxial compressive strength

Compressive strength is one of the most important mechanical properties of rock material, used in blast hole design. To determine the UCS of the rock samples, 54 mm diameter NX-size core specimens, having a length-to-diameter ratio of 2.5:1 were prepared as suggested by ISRM. Each block was represented by at least three core specimens. The oven-dried and NX-size core specimens were tested by using a microcontroller compression testing machine. The average results of uniaxial compressive strength values of different rocks are given in Table 7.

Sl. No.	Rock Sample	UCS (MPa)	Dry Density (g/cc)	Tensile Strength (MPa)	A-weighted Equivalent Sound level Leq (dB)	
					Min Leq	Max Leq
1	Koira Grey Granite	77.8	2.773	9.60	112.4	118.1
2	Quartz monzonite	42	2.556	4.72	93.9	97.6
3	Granodiorite	79.6	2.481	9.83	113.1	118.9
4	Peridotite	64.9	2.649	7.82	107.7	112.4
5	Serpentine	37	2.512	4.16	90.7	93.8
6	Syenite	47	2.536	5.28	96.6	100.1
7	Norite	48.7	2.558	5.47	97.5	100.9
8	Granite Porphyry	51.2	2.766	6.17	99.1	102.9
9	Pegmatite	35.4	2.496	3.98	90.0	94.2
10	Charnockite	66.8	2.699	8.05	109.0	114.1
11	Diorite Porphyry	57.2	2.615	6.89	104.0	108.3
12	Grey Granite	46.4	2.571	5.21	95.8	99.6
13	Dolerite	66.1	2.665	7.96	107.9	112.8
14	Gabbro	102.2	3.168	12.62	118.2	121.3

Table 7. Rock properties and range of A-weighted equivalent sound level values obtained during drilling of igneous rocks.

b. Dry density

Density is a measure of mass per unit of volume. Density of rock material varies, and often related to the porosity of the rock. It is sometimes defined by unit weight and specific gravity. The density of each core sample was measured after the removal of moisture from it. The moisture was removed by placing the samples in an electric oven at about 80^0 C for one hour and they were dried at room conditions. The density data of dry samples was obtained from the measurements of bulk volume and mass of each core using the following formula.

$$\rho(g \,/\, cc) = \frac{Mass\,of\,sample}{Volume\,of\,sample}$$

Each test was repeated five times and the average values were recorded. The average results of dry densities of different rocks are given in Table 7.

c. Tensile strength

Rock material generally has a low tensile strength. The low tensile strength is due to the existence of micro cracks in the rock. The existence of micro cracks may also be the cause of rock failing suddenly in tension with a small strain. Tensile strength of rock is obtained from Brazilian test. To determine the Brazilian tensile strength of the rock samples, 54 mm diameter NX-size core specimens, having a length less than 27mm were prepared as suggested by

ISRM. The cylindrical surfaces were made free from any irregularities across the thickness. End faces were made flat to within 0.25 mm and parallel to within 0.25°. The specimen was wrapped around its periphery with one layer of the masking tape and loaded into the Brazil tensile test apparatus across its diameter. Loading was applied continuously at a constant rate such that failure occured within 15-30 seconds. Ten specimens of the same sample were be tested. The average results of Brazilian tensile strength of different rocks are given in Table 7.

2.2.5 Noise measurement

Test samples for rotary drilling, having a dimension of 20 cm x 20 cm x 20 cm were prepared by sawing from block samples. During drilling, to overcome the vibration of rock block, it was firmly held by vise which is kept on the table of the machine. Sound level measurements were carried out for rotation speeds of 150, 200, 250 and 300 RPM and penetration rates of 2, 3, 4 and 5 mm/min on each rock block.

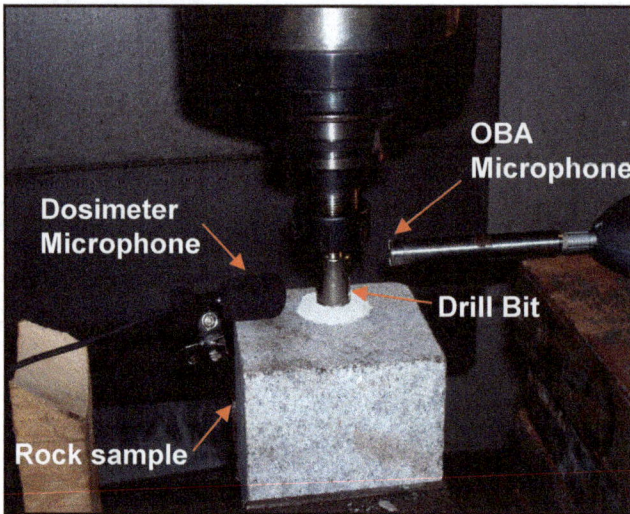

Fig. 10. Position of microphone from drill setup.

For each combination of drill bit diameter, drill bit speed and penetration rate, a total of 64 sets of test conditions were arrived at (drill bit diameter of 6, 10, 16 and 20 mm; drill bit speed of 150, 200, 250 and 300 RPM; penetration rate of 2, 3, 4 and 5 mm/min). A-weighted equivalent continuous sound level (Leq) was recorded for all 64 different drill holes of 30 mm depth on each rock block. For all measurements, the sound level meter was kept at a distance of 1.5 cm from the periphery of the drill bit (Fig. 10).

For a particular condition and for the same rock block, the sound level was determined five times in relatively rapid succession. It was found that the recorded equivalent sound levels were almost consistent. The arithmetic average of each set of five measurements was computed to yield an average A-weighted equivalent sound level for a particular condition.

For 15 minutes, the sound level was measured at 1.5 cm from the drill bit without drilling. The equivalent sound level of 65.2 dB was recorded without drilling which was mainly due to the noise of the CNC machine.

It may be argued that sound produced from the CNC machine itself may affect the sound level measurement during rock drilling. It is important to mention here that if the sound level difference between two sources is more than 10 dB, then the total sound level will remain the same as that of the higher source. Further, taking the measurement very close to the source will reduce the effect of sound produced from other sources.

2.2.6 Regression modelling and analysis of variance (ANOVA)

The results of the measurements of rock properties (UCS, dry density, tensile strength) and range (maximum and minimum) of A–weighted equivalent sound level recorded during drilling of igneous rocks are given in Table 7. These results were analysed using Multiple Regression and Analysis of Variance (ANOVA) technique. For analysis Minitab 15 software for windows was used.

To obtain applicable and practical predictive qualitative relationships it is necessary to model the physico-mechanical rock properties and the drill process variables. These models will be of great use during the optimization of the process. The experimental results were used to model the various responses using multiple regression method by using a non-linear fit among the responses and the corresponding significant parameters. Multiple regression analysis is practical, relatively easy for use and widely used for modelling and analyzing the experimental results. The performance of the model depends on a large number of factors that act and interact in a complex manner. The mathematical modelling of sound level produced during drilling is influenced by many factors. Therefore a detailed process representation anticipates a second order model. ANOVA was carried out to find which input parameter significantly affects the desired response. To facilitate the experiments and measurement, four important factors are considered in the present study. They are drill bit diameter in mm (A), drill bit speed in RPM (B), penetration rate in mm/min (C) and equivalent sound level produced during drilling in dB (D). The responses considered are UCS, Dry Density and Tensile Strength. The mathematical models for the physico- mechanical properties with parameters under consideration can be represented by $Y = f(x_1, x_2, x_3, \dots) + \in$, where Y is the response and x_1, x_2, x_3, are the independent process variables and $\in$ is fitting error. A quadratic model of f can be written as $f = a_0 + \sum_{i=1}^{n} a_i x_i + \sum_{i=1}^{n} a_{ij} x_i^2 + \sum_{i<j}^{n} a_{ij} x_i x_j + \in$ where a_i represents the linear effect of x_i, a_{ij} represents the quadratic effect of x_i and a_{ij} reveals the linear interaction between x_i and x_j. Then the response surface contains linear terms, squared terms and cross product terms.

In order to compare all reasonable regression models, a backward elimination procedure was used as the screening procedure. Then the independent variable having the absolute smallest t statistic was selected. If the t statistic was not significant at the selected α (to test the significance, one needs to set a risk level called the alpha level. In most cases, the "rule of thumb" is to set the alpha level at 0.05, i.e., 95% confidence interval) level, the independent variable under consideration was removed from the model and the regression analysis was performed by using a regression model containing all the remaining independent variables. If the t statistic was significant, the model was selected. The procedure was continued by removing one independent variable at a time from the model. The screening was stopped when the independent variable remaining in the model could not be removed from the system.

For UCS, the best model found was

$$UCS = 468.204 + 2.225 \times A + 0.069 \times B + 2.502 \times C - 10.273 \times D + 0.061 \times D^2$$
$$- 0.025 \times A \times D - 0.001 \times B \times D - 0.028 \times C \times D$$

Significance of regression coefficients for estimation of Universal Compressive strength using Minitab 15 software is listed in Table 8a. The final ANOVA table of the reduced quadratic model for UCS is shown in Table 8b. In addition to the degrees of freedom (DF), mean square (MS), t-ratio and p values associated with factors are represented in this table. As seen form Table 8c, the selected model explains 97.45% of the total variation in the observed UCS tests.

Model terms for UCS	Parameter estimate (coefficients)	t	p
Constant	468.204	22.692	0.000
A	2.225	8.048	0.000
B	0.069	2.671	0.008
C	2.502	1.948	0.032
D	-10.273	-26.767	0.000
D^2	0.061	33.119	0.000
AD	-0.025	-9.465	0.000
BD	-0.001	-3.149	0.002
CD	-0.028	-2.294	0.022

Table 8a. Significance of regression coefficients for estimation of Universal Compressive strength using Minitab 15 software.

For UCS the p-values for all the independent variables are less than 0.05 showing statistical significance. In addition to this the p value of D^2 term and interaction terms related to A, B, C with D are less than 0.05 which establishes the experimental results.
Experimental analysis also shows that for igneous rocks, as the UCS increases the sound level produced during drilling also increases.
Fig. 11 shows the variation between experimentally measured UCS with the UCS calculated from the developed regression model for test data.

Source of variations	Degree of Freedom	Sum of Squares	Mean Squares	F - Value	p - Value
Model	8	221064.60	27633.07	3055.87	0.000
Linear	4	6850.91	1712.73	189.41	0.000
Square	1	9918.64	9918.64	1096.88	0.000
Interaction	3	936.54	312.18	34.52	0.000
Residual Error	631	5705.90	9.04		
Total	639	226770			

Table 8b. Analysis of variance (ANOVA) for the selected quadratic model for estimation of UCS.

R²	Predicted R²	Adjusted R²	Standard error
0.9748	0.9739	0.9745	3.00710

Table 8c. Model summary for dependent variable (UCS)

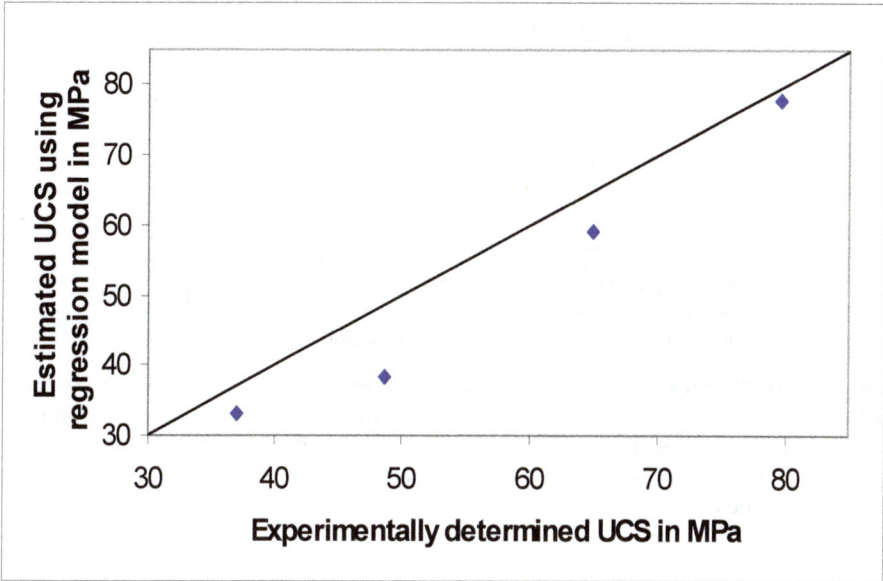

Fig. 11. Estimated UCS using Regression Vs Experimentally determined UCS model of Test data

For Dry Density, the best model found was

$$\rho = 11.0892 + 0.0387 \times A - 0.1813 \times D + 0.0010 \times D^2 - 0.0008 \times A \times D$$

Significance of regression coefficients for estimation of Dry density are listed in Table 9a. The final ANOVA table of the reduced quadratic model for Dry density is shown in Table 9b. In addition to the degrees of freedom (DF), mean square (MS), t-ratio and p values associated with factors are represented in this table. As seen form Table 9c, the selected model explains 77.53% of the total variation in the observed dry density tests.

Model terms for Dry Density	Parameter estimate (coefficients)	t	p
Constant	11.0892	19.120	0.000
A	0.0387	4.814	0.000
D	-0.1813	-16.284	0.000
D²	0.0010	17.966	0.000
AD	-0.0004	-5.205	0.000

Table 9a. Significance of regression coefficients for estimation of dry density using Minitab 15 software.

Source of variations	Degree of Freedom	Sum of Squares	Mean Squares	F - Value	p - Value
Model	4	16.93505	4.23376	552.05	0.000
Linear	2	2.06246	1.03123	134.47	0.000
Square	1	2.47550	2.47550	322.79	0.000
Interaction	1	0.20780	0.20780	27.10	0.000
Residual Error	635	4.86991	0.00767		
Total	639	21.8050			

Table 9b. Analysis of variance (ANOVA) for the selected quadratic model for estimation of dry density.

R^2	Predicted R^2	Adjusted R^2	Standard error
0.7767	0.7729	0.7753	0.0875737

Table 9c. Model summary for dependent variable (dry density)

For Dry density the p-values for independent variables A and D are less than 0.05 showing statistical significance. In addition to this the p value of D^2 term and interaction terms related to A with D are less than 0.05 which establishes the experimental results.

Fig. 12 shows the variation between experimentally measured density with the density calculated from the developed regression model for test data.

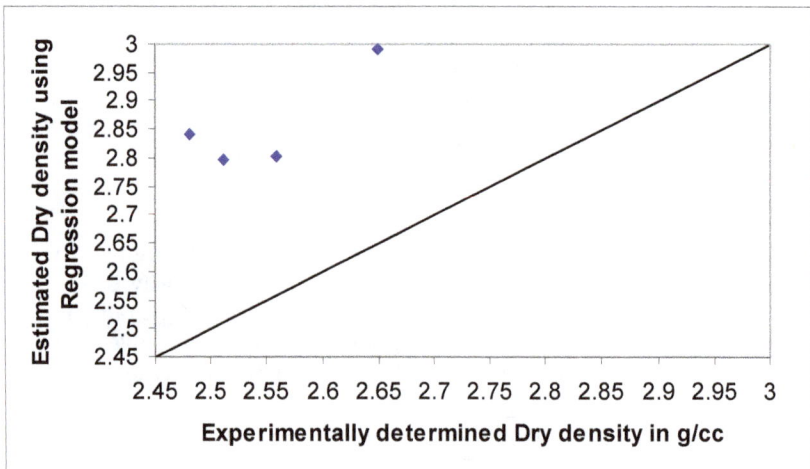

Fig. 12. Estimated dry density using Regression model Vs Experimentally determined dry density of Test data

For Tensile Strength, the best model found was

$$TS = 56.4706 + 0.2730 \times A + 0.0086 \times B + 0.3106 \times C - 1.2657 \times D + 0.0076 \times D^2$$
$$- 0.0031 \times A \times D - 0.0001 \times B \times D - 0.0035 \times C \times D$$

Significance of regression coefficients for estimation of tensile strength is listed in Table 10a. The final ANOVA table of the reduced quadratic model for tensile strength is shown in Table 10b. In addition to the degrees of freedom (DF), mean square (MS), t-ratio and p values associated with factors are represented in this table. As seen form Table 10c, the selected model explains 97.88 % of the total variation in the observed tensile strength tests.

Model terms for Tensile Strength	Parameter estimate (coefficients)	t	p
Constant	56.4706	22.778	0.000
A	0.2730	8.219	0.000
B	0.0086	2.768	0.006
C	0.3106	2.012	0.045
D	-1.2657	-27.448	0.000
D^2	0.0076	34.392	0.000
AD	-0.0031	-9.783	0.000
BD	-0.0001	-3.295	0.001
CD	-0.0035	-2.394	0.017

Table 10a. Significance of regression coefficients for estimation of tensile strength using Minitab 15 software.

For Tensile Strength the p-values for all the independent variables are less than 0.05 showing statistical significance. In addition to this the p value of D^2 term and interaction terms related to A, B and C with D are less than 0.05 which establishes the experimental results.

Source of variations	Degree of Freedom	Sum of Squares	Mean Squares	F - Value	p - Value
Model	8	3855.630	481.954	3691.64	0.000
Linear	4	104.024	26.006	199.20	0.000
Square	1	154.417	154.417	1182.80	0.000
Interaction	3	14.491	4.830	37.00	0.000
Residual Error	631	82.379	0.131		
Total	639	3938.01			

Table 10b. Analysis of variance (ANOVA) for the selected quadratic model for estimation of tensile strength.

R^2	Predicted R^2	Adjusted R^2	Standard error
0.9791	0.9783	0.9788	0.361321

Table 10c. Model summary for dependent variable (tensile strength)

Fig. 13 shows the variation between experimentally measured tensile strength with the tensile strength calculated from the developed regression model for test data.

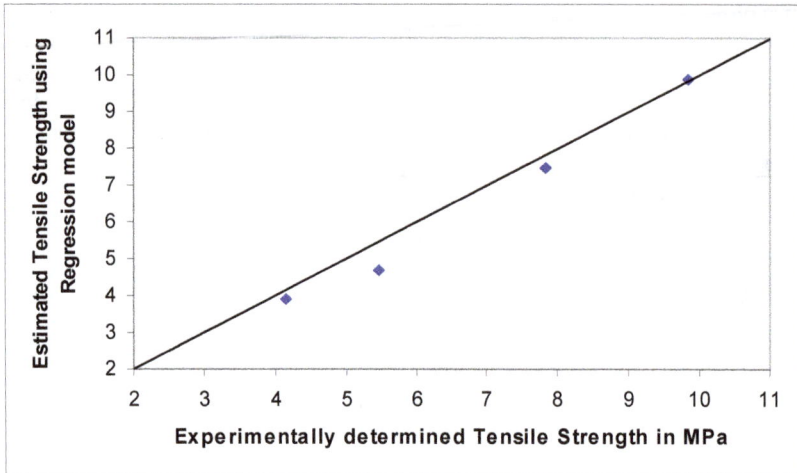

Fig. 13. Estimated tensile strength using Regression model Vs Experimentally determined tensile strength of Test data

3. Field investigations

An attempt was also made to experimentally determine the UCS in the field during drilling blast holes. The Medapalli Open Cast Project (MOCP), belonging to M/S Singareni Colliery Company Limited, situated in the state of Andhra Pradesh in India was used for the field investigations. The rock stratum at the MOCP consists primarily of sandstone, carbonaceous shale, sandy shale, coal, shale, shaly coal, carbonaceous sandstone, and carbonaceous clay. There were a total of five coal seams in that area. Out of these five seams, four coal seams from the top had already been extracted. Borehole data near the investigation area are shown in Fig. 14, which were obtained from the Geology section of the mine. The lithological details from the 4th to 5th seam are also indicated in Fig. 14 (right hand side of the figure), with the depth of each rock formation.

Between the 4th and 5th seam, the strata are classified into Upper Roof (3.0–6.0 m above the top of the coal seam i.e. top of 5th seam), Immediate Roof (0.0–3.0 m above the top of the coal seam), Immediate Floor (0.0–3.0 m below the base of the coal seam), Main Floor (3.0–6.0 m below the base of the coal seam) and Interburden (bounding strata not classified as roof or floor). The details are shown in Fig. 14.

3.1 Noise measuring instruments

The instruments described in section 2.1.3 and 2.2.3 were again used for field investigations. A rotary drill machine was used for drilling blast holes in the mine. The drill bit diameter was 150.0 mm with tungsten carbide button bits. Air was used as the flushing fluid. Compressed air was used as the feed mechanism with a sump pressure of 1.275 MPa and a line pressure of 1.373 MPa.

Both dosimeter and one-third-octave-band analyser were used to record the sound level. For all measurements, both the dosimeter and one-third-octave-band analyser were hand-held at a height of 1.0 m from the ground level and at a distance of 1.5 m and 2.5 m from the blast hole (Fig. 15). Sound levels were recorded for 16 different drill holes. At each second, the equivalent continuous A-weighted sound levels were recorded by the dosimeter. To determine the sound level spectrum, the one-third-octave-band analyser was set to measure A-weighted, time-averaged one-third-octave-band sound levels with nominal mid-band frequencies from 25 Hz to 20 kHz. For each measurement, the one-third-octave-band analyser was set for an averaging time of 2 min. The data recorded during field measurements using the dosimeter and one-third-octave-band analyser were downloaded to the computer for analysis. Some critical observations, such as colour change of flushing dust and the exact time during colour change were also recorded.

For the same drill diameter and type, penetration rate and weight on bit, the sound levels were measured for various drilled holes consisting of strata of different compressive strengths.

For about 3 min, the sound level at about 1.5 m from the drill rod was measured without drilling. The sound level measured without drilling was mainly due to the compressor operating near the drilling machine.

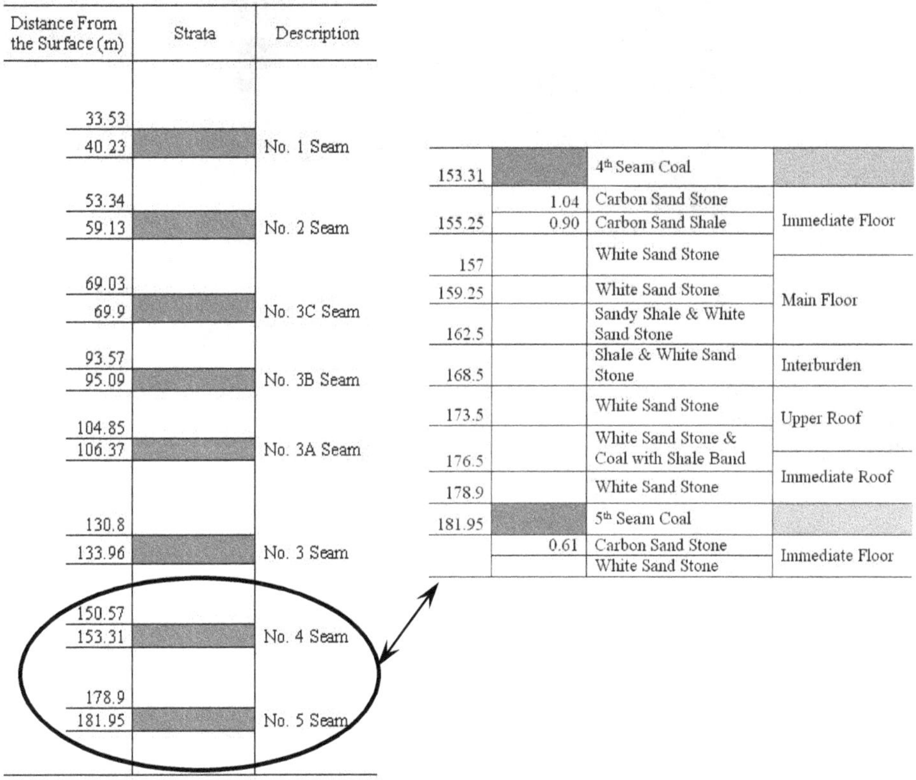

Distance From the Surface (m)	Strata	Description
33.53		
40.23		No. 1 Seam
53.34		
59.13		No. 2 Seam
69.03		
69.9		No. 3C Seam
93.57		
95.09		No. 3B Seam
104.85		
106.37		No. 3A Seam
130.8		
133.96		No. 3 Seam
150.57		
153.31		No. 4 Seam
178.9		
181.95		No. 5 Seam

153.31			4th Seam Coal	
	1.04	Carbon Sand Stone		
155.25	0.90	Carbon Sand Shale	Immediate Floor	
157		White Sand Stone		
159.25		White Sand Stone	Main Floor	
162.5		Sandy Shale & White Sand Stone		
168.5		Shale & White Sand Stone	Interburden	
173.5		White Sand Stone	Upper Roof	
176.5		White Sand Stone & Coal with Shale Band		
178.9		White Sand Stone	Immediate Roof	
181.95		5th Seam Coal		
	0.61	Carbon Sand Stone	Immediate Floor	
		White Sand Stone		

Fig. 14. Lithology of the area (Bore hole data)

3.2 Sound measurement

Field investigation of the sound levels produced during drilling was carried out on the rotary drill machine described in Section 3.1. All the measurements were carried out while drilling blast holes. During field investigation, bit type and diameter, blast hole length, weight on bit, compressed air pressure, net drilling time and rpm of the drill bit were recorded. The penetration rate (m/min) was calculated from the drilled hole length (metres) and the net drilling time (minutes). Blast holes were drilled between the 4th and 5th seams at each classified strata (Fig. 14). Depending on the blast design, the blast hole length was limited to 6.0 m, whereas at other places it was only 3.0 m. For 3.0 m long blast hole, the weight on the bit was 12.0 kg, whereas for the 6.0 m long holes, the weight on the bit was 8.0 kg. The exploratory borehole data were collected from the Geology section of the mine. The UCS, density, tensile strength, young's modulus and impact strength of various strata were collected from the exploratory borehole data near the blast hole drilling as given in Table 11.

Fig. 15. Sound measurement during drilling

3.3 Results of investigation using dosimeter

Using dosimeter, L_{eq} was measured for each second. Drill bit penetration rate in m/sec was calculated. The time taken to drill 3.0 m deep hole was noted down. Then L_{eq} vs drill hole depth was plotted and is as shown in Fig. 16 and Fig. 17.

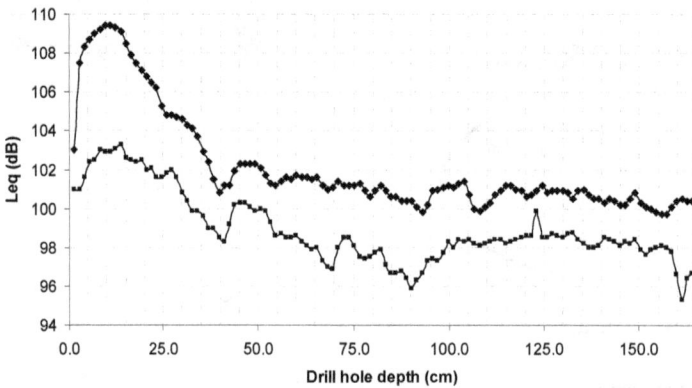

Fig. 16. L_{eq} vs drill hole depth with 8.0 kg weight on drill bit: ♦ Blast hole–1 (UCS 36.49 MPa); ■ Blast hole–12 (UCS 28.35 MPa).

Drill Hole No.	Strata Location	Distance from surface in (m)	Formation	Drill Bit diameter (mm)	Weight on bit (kg)	Drill rod RPM	Observation Distance (m)	Penetration Rate (m/min)	Density (g/cc)	Tensile Strength (MPa)	Compressive Strength (MPa)	Young's Modulus (GPa)	Impact Strength Number
1	Immediate Floor of 4th seam	155.25 to 159.25	White Sand Stone	150	8	85	1.5	0.82	2.28	2.62	36.49	5.10	52.02
2	Interburden between 4&5 seam	162.5 to 168.5	Shale & White Sand Stone	150	12	73	1.5	1.00	2.24	2.36	30.61	3.83	49.42
3	Interburden between 4&5 seam	162.5 to 168.5	Shale & White Sand Stone	150	12	73	2.5	1.00	2.24	2.36	30.61	3.83	49.42
4	Upper Roof of 5th seam	173.5 to 176.5	White Sand Stone & Coal with shale band	150	12	73	4.5	1.00	2.22	1.98	28.84	3.01	47.80
5	Upper Roof of 5th seam	173.5 to 176.5	White Sand Stone & Coal with shale band	150	12	73	3.5	1.00	2.22	1.98	28.84	3.01	47.80
6	Upper Roof of 5th seam	173.5 to 176.5	White Sand Stone & Coal with shale band	150	12	73	1.5	1.00	2.22	1.98	28.84	3.01	47.80
7	Upper Roof of 5th seam	173.5 to 176.5	White Sand Stone & Coal with shale band	150	12	73	2.5	1.00	2.22	1.98	28.84	3.01	47.80
8	Main Floor after 4th seam	157 to 162	Sandy Shale & White Sand Stone	150	8	85	2.5	0.82	2.21	1.81	28.35	2.75	47.32
9	Main Floor after 4th seam	157 to 162	Sandy Shale & White Sand Stone	150	8	85	3.5	0.82	2.21	1.81	28.35	2.75	47.32
10	Main Floor after 4th seam	157 to 162	Sandy Shale & White Sand Stone	150	8	85	4.5	0.82	2.21	1.81	28.35	2.75	47.32
11	Main Floor after 4th seam	157 to 162	Sandy Shale & White Sand Stone	150	8	85	5.5	0.82	2.21	1.81	28.35	2.75	47.32
12	Main Floor after 4th seam	157 to 162	Sandy Shale & White Sand Stone	150	8	85	1.5	0.82	2.21	1.81	28.35	2.75	47.32
13	Immediate Roof of 5th seam	175.9 to 178.9	White Sand Stone	150	12	73	2.5	1.00	2.29	3.14	37.08	6.18	52.36
14	Immediate Roof of 5th seam	175.9 to 178.9	White Sand Stone	150	12	73	1.5	1.00	2.29	3.14	37.08	6.18	52.36
15	Immediate Roof of 5th seam	175.9 to 178.9	White Sand Stone	150	12	73	4.5	1.00	2.29	3.14	37.08	6.18	52.36
16	Immediate Roof of 5th seam	175.9 to 178.9	White Sand Stone	150	12	73	3.5	1.00	2.29	3.14	37.08	6.18	52.36

Table 11. Exploratory borehole data near the blast hole drilling

Investigation with 8.0 kg weight on bit during drilling was also carried out on Blast hole-1 having white sandstone with compressive strength of 36.49 MPa and Blast hole-12 containing sandy shale and white sandstone with compressive strength of 28.35 MPa. From the Fig. 16 it is observed that for the first 45.0 cm depth of drilling, the difference in sound level for Blast hole -1 and Blast hole-12 is as much as 6.7 dB. By neglecting the first 45.0 cm depth, it is observed that for increase in compressive strength by 8.14 MPa (UCS of Blast hole–1 and Blast hole–12), L_{eq} level increases up to 4.0 dB.

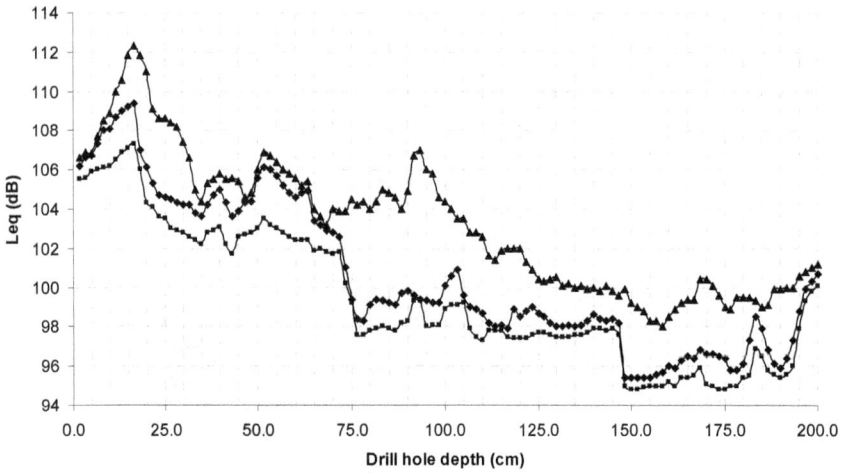

Fig. 17. L_{eq} vs drill hole depth with 12.0 kg weight on drill bit: ♦ Blast hole-2 (UCS 30.61 MPa); ■ Blast hole – 6 (UCS 28.84 MPa); ▲ Blast hole–14 (UCS 37.08 MPa).

Fig. 17 shows results of investigation with 12.0 kg weight on bit during drilling. In this case, Blast hole-2 was shale with white sandstone of compressive strength 30.61 MPa, Blast hole-14 was white sandstone of compressive strength 37.08 MPa whereas Blast hole-6 was white sandstone and coal with shale band of compressive strength 28.84 MPa.

It is observed that for the first 45.0 cm depth of drilling, the increase in sound level for Blast hole-2 compared to that of Blast hole-6 is as much as 2.2 dB. Similarly, the increase in sound level for Blast hole-14 compared to that of Blast hole-2 is as much as 4.9 dB. In addition, the sound level of Blast hole-14 is up to 6.7 dB higher than that of Blast hole-6. By neglecting the first 45.0 cm depth, it is observed that for increase in compressive strength by 1.77 MPa (UCS of Blast hole–2 and Blast hole–6), L_{eq} level increases up to 2.8 dB. For increase in compressive strength by 8.24 MPa (UCS of Blast hole–14 and Blast hole–6) L_{eq} level increases up to 8.0 dB. Similarly, for increase in compressive strength by 6.47 MPa (UCS of Blast hole–14 and Blast hole–2) L_{eq} level increases up to 7.1 dB.

This clearly indicates that as the compressive strength increases, the L_{eq} level produced during drilling also increases. However, this increase in L_{eq} level also depends on the weight on the bit which is indirectly related to the compressor pressure used.

It is also observed that between depths of 75.0 cm to 125.0 cm and 150.0 cm to 175.0 cm, the L_{eq} levels measured at Blast hole-6 and Blast hole-2 were somewhat similar whereas Blast hole-14 had an increase in L_{eq} value of up to 8.0 dB for depths between 75.0 cm and 125.0 cm and up to 5.3 dB for depths between 150.0 cm and 175.0 cm. This is because of the coal present in Blast hole-6 and Blast hole-2 between these depths which was confirmed on observing the coal dust flushing out of the drill holes at these depths.

Table 12 gives the equivalent A–weighted sound levels for Blast holes of different compressive strengths at different measurement distances..

UCS (MPa)	Weight on a bit (kg)	L_{eq} (dB)									
		Blast hole	1.5 m	Blast hole	2.5 m	Blast hole	3.5 m	Blast hole	4.5 m	Blast hole	5.5 m
36.49	8.0	1	102.5	-	-	-	-	-	-	-	-
28.35	8.0	12	98.6	8	96.2	9	95.4	10	94.3	-	-
30.61	12.0	2	101.5	3	98.9	-	-	-	-	-	-
28.84	12.0	6	99.3	7	97.6	5	95.9	4	94.8	11	93.7
37.08	12.0	14	103.2	13	100.9	16	99.6	15	98.5	-	-

Table 12. Comparison of A–weighted equivalent sound level (L_{eq}) for Blast holes of different UCS at different measurement distances for first 2 minutes of drilling.

It was observed that as the measurement distance increases, the equivalent A–weighted sound level decreases. For example, 1.0 meter increase in distance from 1.5 m to 2.5 m, for UCS of 30.61 MPa (Blast hole-2 and Blast hole-3), the sound level decreased by 2.6 dB. Similar results were obtained at strata of different compressive strengths (Blast hole 6 and Blast hole 7, Blast hole 12 and Blast hole 8, Blast hole 14 and Blast hole 13).

3.4 Results of investigation using one-third-octave band analyzer
3.4.1 Comparison of drilling noise with machine noise

The A–weighted sound level spectrum at the measurement location with 8.0 kg weight on bit for Blast hole-1, Blast hole-12 and machine noise is shown in Fig. 18. It is seen that the maximum sound level at measurement location for Blast hole-1 is 96.4 dB, Blast hole-12 is 92.9 dB with nominal one-third–octave midband frequency of 63 Hz.

Similarly, A–weighted sound level at the measurement location with 12.0 kg weight on bit for Blast hole-2, Blast hole-6 and Blast hole-14 is shown in Fig. 19. It is seen that the maximum sound level at measurement location for Blast hole-2 is 100.3 dB, Blast hole-6 is 99.8 dB and Blast hole-14 is 104.1 dB with the nominal one-third–octave midband frequency from 25 Hz to 20 kHz.

In both the cases, the increase in sound level with midband frequencies above 50 Hz is more than 10.0 dB during drilling relative to that of machine noise without drilling. Therefore, the sound level in the frequency range of 63 Hz to 20 kHz, during drilling is unlikely to be affected by the background noise due to the compressor. However, the sound level produced during drilling may be affected due to machine noise with nominal midband frequencies from 25 Hz to 50 Hz as the difference in sound level in this range of frequency is below 10.0 dB.

From Fig. 18, it is seen that from 25 Hz to 50 Hz, the increase in sound level for Blast hole-1 relative to that of machine noise is from 6.7 dB to 9.6 dB and that of Blast hole-12 relative to that of machine noise is from 3.9 dB to 7.4 dB. Similarly, from Fig. 19 it is seen that from 25 Hz to 50 Hz, the increase in sound level, for Blast hole-2 relative to that of machine noise is from 2.7 dB to 8.4 dB, for Blast hole-6 relative to that of machine noise is from 1.2 dB to

7.0 dB and for Blast hole-14 relative to that of machine noise is from 5.3 dB to 9.4 dB. This shows that drilling operation has increased the sound level with midband frequencies from 25 Hz to 50 Hz. The increase in sound level in this frequency range (25 Hz – 50 Hz) may be due to impact between the drill bit and the rock.

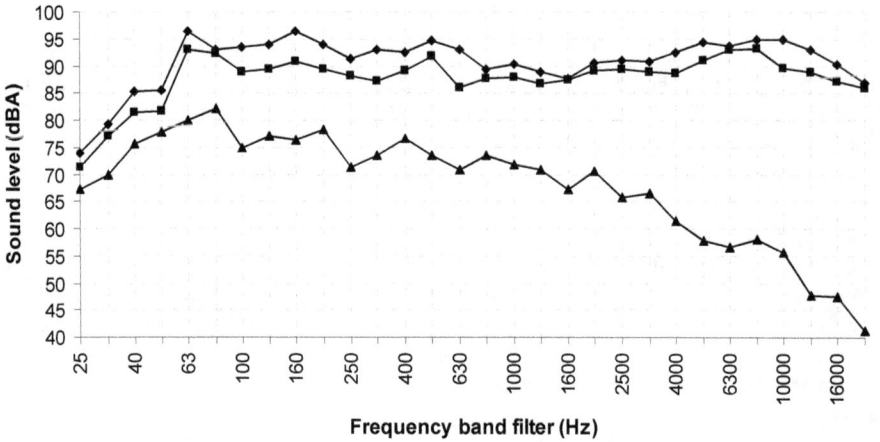

Fig. 18. Sound level vs nominal one-third-octave midband frequency with 8.0 kg weight on drill bit: ♦ Blast hole–1 (UCS 36.49 MPa); ■ Blast hole–12 (UCS 28.35 MPa); ▲ Machine noise (without drilling).

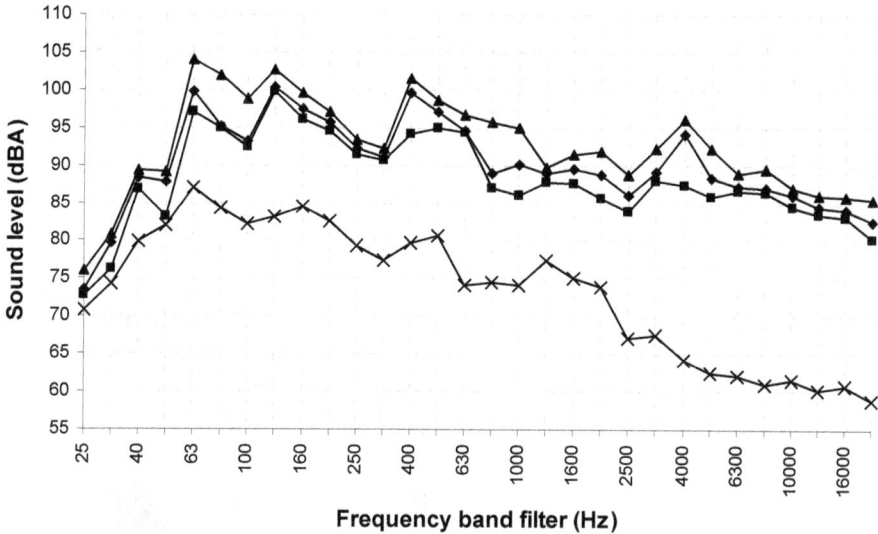

Fig. 19. Sound level vs nominal one-third-octave midband frequency with 12.0 kg weight on drill bit: ♦ Blast hole-2 (UCS 30.61 MPa); ■ Blast hole-6 (UCS 28.84 MPa); ▲ Blast hole-14 (UCS 37.08 MPa); × Machine noise (without drilling).

From Fig. 18, the increase in sound level for Blast hole-1 relative to that of machine noise with midband frequencies from 63 Hz to 2 kHz is from 10.8 dB to 22.1 dB and that of Blast hole-12 relative to that of machine noise is from 10.2 dB to 20.3 dB. Similarly from Fig. 19, the increase in sound level for Blast hole-2 relative to that of machine noise with midband frequencies from 63 Hz to 2 kHz is from 10.9 dB to 20.4 dB, for Blast hole-6 relative to that of machine noise is from 10.1 dB to 20.2 dB and for Blast hole-14 relative to that of machine noise is from 14.4 dB to 22.6 dB.

Also from Fig. 18, it can be observed that there is a significant increase in sound level of the order of 24.3 dB to 45.7 dB from 2.5 kHz to 20 kHz for Blast hole-1 relative to that of machine noise and 22.5 dB to 44.8 dB for Blast hole-12 relative to that of machine noise. Similarly, from Fig. 19, within frequency range of 2.5kHz to 20 kHz, the increase in sound level relative to machine noise for Blast hole-2 is from 18.9 dB 29.9 dB, for Blast hole-6 relative to that of machine noise is from 16.8 dB to 25.5 dB and for Blast hole-14 relative to that of machine noise is from 21.5 dB to 31.9 dB. This increase in sound level is due to resonance of the steel parts of the drill steel due to rock drilling.

3.4.2 Comparison of drilling noise with rock properties

With 8.0 kg weight on bit, the increase in sound level of Blast hole-1 (UCS of 36.49 MPa) compared to that of Blast hole-12 (UCS of 28.35 MPa), with midband frequencies from 25 Hz to 50 Hz, was of the order of 2.0 dB to 3.8 dB. The increase in sound level, with midband frequencies from 63 Hz to 2 kHz, was of the order of 0.3 dB to 6.9 dB. The increase in sound level, with midband frequencies from 2.5 kHz to 20 kHz, was of the order of 0.8 dB to 5.2 dB.

With 12 kg weight on bit, the increase in sound level of Blast hole-14 (UCS of 37.08 MPa) compared to that of Blast hole-2 (UCS of 30.61 MPa), with midband frequencies from 25 Hz to 50 Hz, was of the order of 1.0 dB to 2.6 dB. The increase in sound level, with midband frequencies from 63 Hz to 2 kHz, was of the order of 0.8 dB to 6.9 dB whereas the increase in sound level, with midband frequencies from 2.5 kHz to 20 kHz, was of the order of 1.0 dB to 3.8 dB. The increase in sound level of Blast hole-14 (UCS of 37.08 MPa) compared to Blast hole-6 (UCS of 28.84 MPa), with midband frequencies from 25 Hz to 50 Hz, was of the order of 2.4 dB to 6.0 dB. The increase in sound level, with midband frequencies from 63 Hz to 2 kHz, was of the order of 1.5 dB to 8.9 dB whereas the increase in sound level with midband frequencies from 2.5 kHz to 20 kHz, was of the order of 2.4 dB to 8.7 dB. The increase in sound level for Blast hole-2 (UCS of 30.61 MPa) compared to Blast hole-6 (UCS of 28.84 MPa), with midband frequencies from 25 Hz to 50 Hz, was of the order of 0.7 dB to 4.7 dB. The increase in sound level, with midband frequencies from 63 Hz to 2 kHz, was of the order of 0.2 dB to 5.3 dB whereas the increase in sound level, with midband frequencies from 2.5 kHz to 20 kHz, was of the order of 0.7 dB to 6.7 dB.

4. Conclusions

The laboratory study using portable pneumatic drilling equipment indicated that the sound level near the drill rod is relatively higher than that of the exhaust, the drill bit and the operator' s position for all the rock samples tested. Both the thrust and air pressure were found to have a significant effect on the sound level produced by pneumatic drill at all the measurement locations i.e., at operator' s position, exhaust, drill rod and the drill bit.

The laboratory study using CNC machine was carried out to evaluate the empirical relation between various rock properties and sound level produced during drilling considering the effects of drill bit diameter, drill bit speed and penetration rate. The empirical relationship developed is not aimed at replacing the ISRM suggested testing methods, but rather as a quick and easy method to estimate the physico-mechanical properties of rock. The results of this study could be used to predict the physico-mechancial properties of igneous rocks.

In the field investigation, results of frequency analyser shows that the sound level in the frequency range of 63 Hz to 20 kHz, during drilling is unlikely to be affected by the background noise because above 50 Hz the sound level produced is more than 10 dB during drilling relative to that of machine noise without drilling. However, the sound level produced during drilling maybe affected due to machine noise with nominal midband frequencies from 25 Hz to 50 Hz as the difference in sound level in this range of frequency is below 10 dB.

Results from both laboratory and filed investigations show that there is a possibility to establish relationship between rock properties and sound level produced during drilling. The present investigations lead to further research in this direction.

5. References

[1] Aleotti, L.; Poletto, F.; Miranda, F.; Corubolo, P.; Abramo, F. & Craglietto, A. (1999). Seismic while drilling technology: use and analysis of the drill-bit seismic source in a cross-hole survey. *Geophysical Prospecting*, Vol.47, No.1, pp. 25-39

[2] Asanuma, H. & Niitsuma, H. (1996). Triaxial seismic measurement while drilling and estimation of subsurface structure. *International Journal of Rock Mechanics and Mineral Sciences and Geomechanics Abstracts*, Vol.33, No.7, p 307A

[3] Fernandez, J.V. & Pixton, D.S. (2005). Integrated drilling system using mud actuated down hole hammer as primary engine. Final Technical Report. Report # 34365R05, DOE Award Number: DE-FC26-97FT34365. NOVATEK, Provo, UT

[4] Hand, M.; Rueter, C.; Evans, B.J.; Dodds, K. & Addis, T. (1999). Look-ahead prediction of pore pressure while drilling: Assessment of existing and promising technologies. Document GRI-99/0042, *Gas Research Institute*, Chicago, IL

[5] Hsu, K. (1997). Sonic-while-drilling tool detects over pressured formations. *Oil & Gas Journal*, (August 1997), pp. 59-67

[6] Martinez, R.D. (1991). Formation pressure prediction with seismic data from the Gulf of Mexico. *SPE Formation Evaluation*, Vol.6, No.1, pp. 27-32

[7] Miranda, F. (1996). Impact of the seismic while drilling technique on exploration wells. *International Journal of Rock Mechanics and Mineral Sciences and Geomechanics Abstracts*, Vol.33, No.8, p 360A

[8] Onyia, E.C. (1988). Relationships between formation strength, drilling strength, and electric log properties. SPE 18166, *63rd Annual Technical Conference and Exhibition of the SPE*, Houston, TX, (October 2-5, 1988), pp. 605-618

[9] Rajesh, K.B.; Vardhan, H. & Govindaraj, M. (2010). Estimating rock properties using sound level during drilling: field investigation. *International Journal of Mining and Mineral Engineering*, Vol.2, No.3, pp. 169-184

[10] Rector, J.W. & Hardage, B.A. (1992). Radiation pattern and seismic waves generated by a working roller-cone drill bit. *Geophysics*, Vol.57, No.10, pp.1319-1333

[11] Roy, S. & Adhikari, G.R. (2007). Worker noise exposures from diesel and electric surface coal mining machinery. *Noise Control Engineering Journal*, Vol.55, No.5, pp. 434-437

[12] Stuart, R.K.; Charles, F.P. & Hans, T. (2007). Method for borehole measurement of formation properties. US patent issued on October 30, 2007. (Application No. 10779885 filed on 17-02-2004)

[13] Tsuru, T. & Kozawa, T. (1998). Noise characterization in SWD survey. Society of Exploration Geophysicists of Japan, Tokyo, Butsuri-Tansa, *Geophysical Exploration*, Vol.51, No.1, (February, 1998), pp. 45-54 (English translation of Abstract)

[14] Vardhan, H. & Adhikari, G.R. (2006). Development of noise spectrum based maintenance guideline for reduction of heavy earth moving machinery noise. *Noise Control Engineering Journal*, Vol.54 No.4, pp. 236-244

[15] Vardhan, H. & Murthy, ChSN. (2007). An experimental investigation of jack hammer drill noise with special emphasis on drilling in rocks of different compressive strengths. *Noise Control Engineering Journal*, Vol.55, No.3, pp. 282-293

[16] Vardhan, H.; Karmakar, N.C. & Rao, Y.V. (2006). Assessment of heavy earth-moving machinery noise vis-à-vis routine maintenance, *Noise Control Engineering Journal*, Vol.54, No.2, pp. 64-78

[17] Vardhan, H.; Karmakar, N.C. & Rao, Y.V. (2004) Assessment of machine generated noise with maintenance of various heavy earthmoving machinery. *Journal of the Institution of Engineers* (India), Vol.85, (October 2004), pp. 93-97

[18] Vardhan, H.; Karmakar, N.C. & Rao, Y.V. (2005). Experimental study of sources of noise from heavy earth moving machinery. *Noise Control Engineering Journal*, Vol.53, No.2, pp. 37-42

[19] Vardhan, H.; Adhikari, G.R. & Govindaraj, M. (2009). Estimating rock properties using sound level during drilling, *International Journal of Rock Mechanics and Mining Sciences*, Vol.46, No.3, pp. 604–612

12

From Noise Levels to Sound Quality: The Successful Approach to Improve the Acoustic Comfort

Eleonora Carletti and Francesca Pedrielli
National Research Council of Italy, IMAMOTER Institute
Italy

1. Introduction

This work aims at presenting the experience of the authors in applying the "product sound quality" approach to the noise signals recorded at the operator station of some earth moving machines (EMMs) in order to improve the acoustic comfort for the operator.

For industrial products, the concept of "product sound quality" was defined by Blauert and Jekosch as "*...a descriptor of the adequacy of the sound attached to a product. It results from judgements upon the totality of auditory characteristics of the sound, the judgements being performed with reference to the set of those desired features of the product which are apparent to the users in their actual cognitive, actional and emotional situation*" (Blauert & Jekosch, 1997).

Referring to the operator station of an EMM, health and quality of the workplace are both important aspects to be taken into account. Therefore the reduction of the noise exposure levels and the improvement of the noise quality in terms of low annoyance are both key elements. Unfortunately, these aspects are not automatically correlated. According to the mandatory provisions, the exposure to noise must be assessed by means of physical parameters that have proved to be inaccurate indicators of subjective human response, especially for sounds exceeding 60 dB (Hellman & Zwicker, 1987).

This chapter collects the main results of the research carried out by the authors in the last five years in order to overcome this problem and to identify a methodology that is able to establish the basic criteria for noise control solutions which guarantee the improvement of the operator comfort conditions (Brambilla et al., 2001).

All the results presented below refer to investigations carried out on compact loaders. The particular interest in this kind of machine is due to the fact that it is widely used not only for outdoor work but also in the activities of building construction and renovation. In addition, the compact loader is one of the worst machines as far as the noise emission is concerned. Due to its compactness, indeed, the operator station is located just over the engine compartment which cannot be completely insulated from the outside due to overheating problems. As a consequence, noise and vibration levels at the operator station are extremely high, causing very uncomfortable conditions for workers.

Although the enforcement of the results described in this work is limited to the assessment of annoyance for this kind of product, the philosophy of this approach has a general validity which is to be customised for each different application.

This chapter is divided into three main sections:
- the first section describes the methodologies applied to carry out the binaural recordings at the operator station of these machines;
- the second section presents the psychometric technique chosen for the experiments and the procedures applied to the subjective evaluations of the noise signals;
- the third section collects the most significant experimental investigations and their relevant results.

2. Binaural recordings

Binaural technology offers the best technical solution to pick up and store sound in a way which is compatible to human hearing and then is a fundamental component of an experimental approach oriented to the sound quality evaluation.

2.1 Binaural recording technique

Binaural techniques for aurally-adequate sound recording and playback have evolved greatly in recent decades with the development of digital systems for signal recording and very fast processors for signal processing.

The basic principle of binaural techniques takes into account that the sensation of any sound is processed by the human auditory system from the two sound pressure signals arriving at the right and left ears. It enables us not only to identify noise sources but also to localise them in a 3D space. Starting from these features and performance of the human auditory system, binaural techniques require that sounds are picked up by two microphones and that recordings are performed in such a way that the *"signals can faithfully reproduce all aspects of the auditory experience also with regard to the spatial characteristics of sounds and their direction of origin"* (Møller, 1992).

The playback of the recorded signals is also an important step in the binaural chain as listening and subjective judgements are essential in assessing the quality of sounds. For this reason, the original sound has to be accurately reproduced, counterbalancing the unwanted spectral modifications that occur during recording and playback.

Currently, two methods can be used for binaural recordings:
- the use of artificial heads (faithfully reproducing head, torso and ears of a listener);
- the use of special miniature microphones, positioned at the ear canal of real subjects.

Both methods were used in the experimental investigations.

2.1.1 Artificial heads

All the artificial heads available on the market reproduce an average human head in such a way that sound waves reaching the head follow the same path that they would follow to reach the ears of a real listener placed in the same location. They differ in many details including the shape and the dimensions of nose, orbit, pinna, ear canal and the positions of microphones. These latter can be located at varying positions along the ear canal, from its entrance to the eardrum.

All of these systems are highly reliable with regard to the aurally-adequate sound recording as well as the playback. To faithfully reproduce all aspects of human auditory perception, these systems include specific transfer functions which take into account the different effects on the sound due to the body, the head and the outer ear.

Referring to the investigations at the operator station of compact loaders, as the space was insufficient to place an artificial head beside the worker, the head-torso manikin was placed at the operator station. The use of this system was obviously limited to the case of noise measurements taken in stationary conditions.

2.1.2 Real head recording system
The use of very lightweight devices consisting of miniature microphones or probes positioned at the entrance of the ear canal of a real subject is certainly a feasible alternative to that of artificial heads (Møller, 1992). The advantage of this approach is that it does not require any correction referring to the ear canal and the resulting recordings contain complete spatial information. The chain of reproduction of the signals, however, must be carefully developed and specific equalization curves are to be introduced to remove all the changes induced by the various components of the chain.
Referring to the investigations at the operator station of compact loaders, this technique was extensively used for all the noise measurements carried out in dynamic conditions.

2.2 Noise measurements referring to loaders
A sample of 41 compact loaders belonging to six different families (A, B, C, D, E, F) regarding manufacturer, dimension and engine mechanical power, was involved in the experiments. All the binaural recordings were made at the operator station of these machines, both in stationary and real working conditions, in order to be able to reproduce different operations of loaders. Measurements were taken in open areas, generally used for EMM testing, where stockpiles of different materials could be found.
Referring to stationary conditions, binaural recordings were performed using the Cortex System MK1 which consists of a dummy manikin, a torso simulator and a digital DAT recorder HHB PDR 100. This system was placed at the operator station and measurements were carried out while the tested machine was in stationary idle condition, with the engine running at a fixed speed. These measurements involved five different loaders of the same family (F1 to F5).
Referring to the dynamic conditions, binaural recordings were obtained by means of two miniature pre-polarised condenser microphones placed at the entrance of the operator's ear canals (binaural microphones B&K 4101) while the tested machine was performing the typical work cycle for a loader, which includes two main operations: the loading of material from a stockpile and the unloading of it in a defined position. These measurements involved all the remaining machines. Besides the noise signals, also the tachometer signal was recorded in order to relate at each time the frequencies of the noise spectrum to the rotational frequencies of the different components of the machine.
Twenty-one loaders of five different families (A1 to A5, B1 to B5, C1 to C5, D1 to D3, E1 to E3) repeated this work cycle with two kinds of materials: gravel and loam. Fifteen other machines of three different families (A6 to A10, B6 to B10, C6 to C10) performed the same work cycle without any material (simulated cycle).
In total, 62 different binaural noise signals were available for the different investigations. Figure 1 shows the noise measurement setup both for stationary (left) and dynamic (right) conditions.

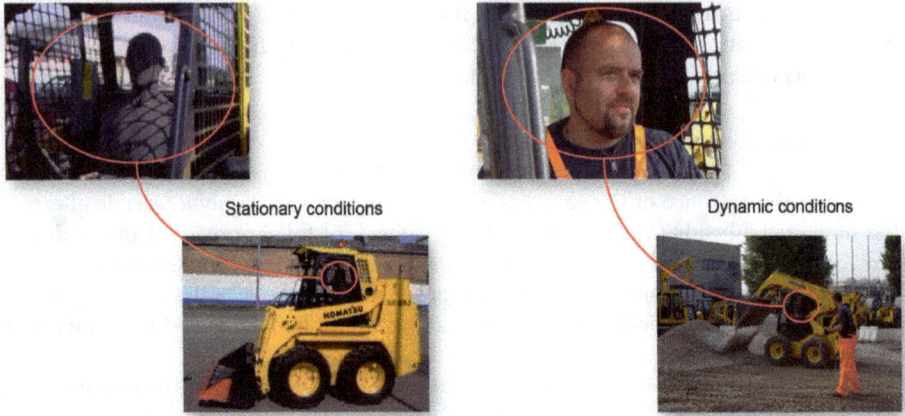

Fig. 1. Binaural recordings

3. Subjective listening tests

Listening test results show a higher variance than that usually encountered in results obtained using instrumental measurements (Blauert, 1994). However, this high uncertainty can be greatly limited by choosing the most appropriate psychometric technique depending on the signals to be characterised and on the listening jury (Fiebig & Genuit, 2010).

3.1 Psychometric techniques

The several psychometric techniques can be broadly divided into two groups: the *absolute* procedures and the *relative* procedures (Van der Auweraer & Wyckaert, 1993).

In listening tests following the *absolute* procedures, the subject has to listen to a sound and judge it referring to one or more of its attributes. In listening tests following the *relative* procedures, on the contrary, the assessment by the subject results from the comparison between at least two different sounds.

In general, the *absolute* classification of a set of sound stimuli or their arrangement in an ordered list according to some criterion, is a process which sometimes can be inappropriate as it involves a series of psychological factors which are uncontrollable. In addition, when the sound stimuli chosen for listening tests are very similar with respect to certain attributes, tests according to absolute procedures can be very difficult, especially if they involve non-expert subjects. In these cases it is advisable to use a relative procedure (Bodden et al., 1998).

A further classification of psychometric procedures is based on the distinction between *non adaptive* and *adaptive* procedures (Gelfand, 1990).

The *non adaptive* procedures include "classic" methods such as:

- the *sequential* procedures (method of limits, method of adjustment), for which the listening level of the sound stimulus is varied step by step or continuously, but always with an ascending or descending sequence (from the lowest to the highest level or vice versa);
- the *non sequential* procedures (method of constant stimuli), for which the listening level of the sound stimulus changes according to a predefined random sequence.

The *adaptive* procedures include methods such as the Békésy's Tracking Method, the Up-Down (Staircase) Method and the PEST procedures (Gelfand, 1990). In these procedures it is necessary to adjust the listening level of a sound stimulus on the basis of the answer given by the subject referring to the sound stimulus previously heard.

3.2 Rating scales
In listening tests the choice of the rating scale on which the subjects express their opinion is a key element to avoid ambiguity in the responses. According to a classification proposed by Stevens in 1951, the rating scales can be divided into: *nominal* scale, *ordinal* scale and *interval* scale (Stevens, 1951).
In *nominal* scales, variables take values represented by names or categories. These values cannot be put in order or treated algebraically. The only relationship that can be established between the various results is that of equality or diversity.
In *ordinal* scales, variables take values which can be sorted by some criterion. Relationships of "greater", "equal", "less" can be established between the different results but without any possibility to establish the distance between classes.
In *interval* scales, each variable is represented by a quantitative value. Therefore, either the different positions on the scale or the distances between the values are significant. In particular, the amplitudes of the intervals between equidistant positions on the scale represent equal differences in the measured phenomenon.
Depending on the type of scale chosen for tests, the most appropriate methods for statistical analysis have to be identified. For *nominal* and *ordinal* scales non-parametric statistical methods are preferred (Spearman correlation coefficient, Kendall's correlation coefficient), while for the *interval* scale the parametric statistical methods are more suitable.

3.3 Procedures applied to the subjective investigations referring to loaders
The binaural noise signals recorded at the operator station of the compact loaders under test were all very similar with respect to the perception of annoyance. A direct estimation of them with respect to this attribute following an *absolute* psychometric procedure would have been very hard, especially for non expert subjects. On the contrary, a *relative* comparison between sounds made the task much easier for the subject and made the detection of the difference among the sound stimuli easier to be assessed. So, the listening tests performed in the several investigations were all carried out according to the relative procedure of paired comparison (Kendall & Babington Smith, 1940).

3.3.1 The listening sequence of sound stimuli
According to the paired comparison procedure, each sound stimulus is directly compared to the others and the subject is asked to give his opinion after listening to each pair. In the several experiments, the *classic* and the *modified* versions of this procedure were both applied: in the *classic* version, the subject had to choose the sound he preferred in each pair and no ties were permitted; in the *modified* version, on the contrary, the subject was permitted to judge the sounds in the pair equally (David, 1988).
In any case, the main advantage of this procedure was that the subject was asked to judge only two stimuli at a time and this helped his concentration and reduced the probability for him of inconsistent judgments.

The number of possible pairs from a group of n sound stimuli is given by the number of combinations of two elements taken in this group, namely by :

$$\binom{n}{2} = \frac{n!}{(n-2)!\,2!} = \frac{n(n-1)}{2} \tag{1}$$

In the listening tests, the two sounds of each pair had always the same duration so that the judgement given by the subjects was not influenced by a different listening period. In addition, the two sound stimuli were always separated by a pause so that the subject could distinguish each of them and was not confused by their similarity. The duration of this pause, however, was not so long as to impair the memory of the first sound heard by the subject.

In order to avoid any sequence effect, all the pairs were arranged in a random sequence according to the well established *digram-balanced Latin Square* design (Wagenaar, 1969). In such a way the first pair to be judged and the order of the pairs in the sequence were different for each subject.

3.3.2 The listening session

All the listening tests were performed in the laboratory, under stable, controlled boundary conditions, with the great advantage of high reproducibility of the test results.

The sound stimuli were presented to the subjects through a high-quality electrostatic headphones (STAX Signature SR-404), with a flat response in the 40-40000 Hz frequency range, after being modified to take into account the transfer function of the headphones and the specific sound card. Listening through headphones reduces the ability of a correct spatial sound localization but for sounds recorded in earth moving machines this effect is not so important as the frequency content is concentrated in the medium-low frequency range that is not directional.

Each listening session started with a *learning* phase during which the person responsible for the experiment gave the subject verbal instructions needed to understand the procedure for the test. This phase was a critical point of the listening session. An interaction between subject and the experimenter was necessary in order to clarify possible doubts before performing the test. This interaction, however, should not be excessive in order not to influence or interfere with the judgements given by the subject.

At the end of this phase, the test started. The subject, after listening to each pair of sound stimuli, was allowed to listen to the pair again as much as necessary. When ready, he gave his rating according to the rules of each specific investigation.

Each listening session ended when the subject had judged all the pairs of sound stimuli in the sequence.

The same procedure was then repeated for all the subjects of the jury involved in that specific test.

3.3.3 The jury of subjects

In listening tests each subject of the jury acts as a measurement device to measure his own perception; so, he must have normal hearing. A further important aspect to take into consideration in the choice of a listening jury is the experience of the subjects involved in the tests. This is related to their familiarity with listening tests and/or with the sounds under examination (Brambilla et al., 1992).

In the several investigations aimed at improving the acoustic comfort at the operator station of compact loaders, the operators were only seldom involved in the listening tests because of the difficulties in finding normal hearing persons. The jury generally included students and/or researchers who were not familiar with these kinds of noise signals but had some knowledge of acoustics and sometimes also prior experience in listening tests.

Nevertheless, the choice of subjects not familiar with EMM sounds did not limit the reliability of the results. An investigation carried out by the authors to verify whether the experience on the use of these machines could provide additional value in the subjective ratings compared to those by non expert subjects, showed similar results between these two groups (Carletti et al., 2002).

The number of subjects involved in the tests varied form test to test but it was always adequate to ensure the statistical significance of the results.

3.3.4 Preference matrices

The ratings given by each subject for all the pairs of sound stimuli in the listening sequence were arranged in a matrix, called the *preference matrix*.

In this matrix, the general element x_{ij} (i = row index and j = column index) represented the judgement expressed by the subject referring to the comparison between stimulus i and stimulus j.

When the specific test was carried out according to the *classic* version, each of these elements could take only the value 0 or the value 1, depending on the preference given by the subject (stimulus j preferred to stimulus i or vice versa).

When the specific test was carried out according to the paired comparison *modified* version, each matrix element could also take the value 0.5 when the two sound stimuli in the pair were equally rated by the subject.

Table 1 shows an example of the preference matrix in the case of a classic paired comparison test involving six sound stimuli (A-F).

	A	B	C	D	E	F
A		0	0	0	0	0
B	1		0	1	0	0
C	1	1		1	1	1
D	1	0	0		0	0
E	1	1	0	1		1
F	1	1	0	1	0	

Table 1. Preference matrix of a subject for a test involving six sound stimuli (A, B, C, D, E, F)

In this matrix the sum by rows gives the preference of each sound stimulus when compared to all the others.

Whichever method was applied, the subjective responses of the entire listening jury were arranged in the *overall preference matrix* which was obtained by adding the scores of the preference matrices of each subject, after the exclusion of the subjects who did not pass the necessary consistency checks explained in the following.

3.3.5 Consistency check of each subject

According to the procedure defined by Kendall and Babington Smith (Kendall & Babington Smith, 1940), in every listening test the *consistency* for each subject and the *agreement* among the subjects have to be evaluated in order to *"guarantee the control of the variance due to the emotional state of the judging individuals"* (Blauert & Jekosch, 1997).

To test for the *consistency* of each subject, two different checks were carried out in the several experiments: a) the check regarding the number of circular triads in the data set; b) the check regarding the judgement given by the subject to the repeated pair of stimuli.

a) Circular triads

In a paired comparison test involving three signals (A, B, C), the judgements given by a subject on comparisons "AB", "BC" and "AC" can be graphically represented using a triangle, usually called *the triad*. Under the hypothesis that a subject chooses A in the first comparison between (A→B) and B in the second comparison (B→C), the choice of C in the third comparison (C→A) does not obey the transitive property and therefore identifies an inconsistency which is graphically represented by a *circular triad*, as shown in figure 2(b).

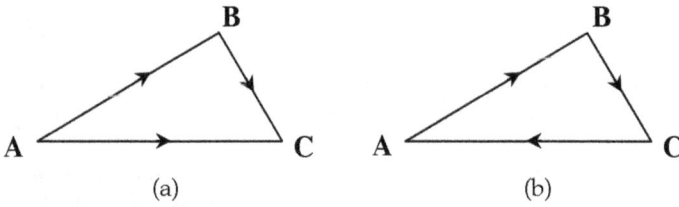

Fig. 2. Not circular triad (a): consistency - Circular triad (b): inconsistency

Referring to the investigations relating to loaders, the paired comparison test often involved 6 sound stimuli. The preference matrix for a classic test may be represented either in tabular form (as previously shown in table 1), or may be represented geometrically as in figure 3. This latter method may help in determining the number of circular triads contained in this polygonal representation.

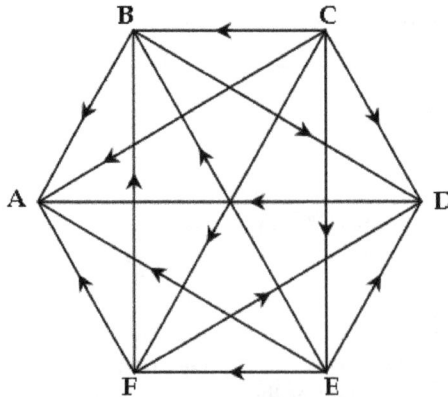

Fig. 3. Geometrical representation of the scheme of preferences of table 1.

In a general test involving n sound stimuli, for each subject the consistency coefficient was calculated on the basis of the number (d) of *circular triads* found in the complete set of judgements, referred to the maximum number of possible *circular triads* for that set of n sounds (d_{max}) (Kendall & Babington Smith, 1940):

$$K = 1 - \frac{d}{d_{max}} \tag{2}$$

The value of this coefficient was then compared to its expected value E(K), calculated under the hypothesis that the observed circular triads were normally distributed (see tables in the Kendall and Babington Smith manuscript). Values of K smaller than E(K) corresponded to a data set where a tendency for inconsistent judgements was observed.

b) Repeated pair

One of the easiest ways to check whether people are consistent with their own answers is to ask them to judge the same pair of sound stimuli twice and compare the results. This check was considered successful when the subject gave concordant answers. However, taking into account the high variability in the subjective perception and the possibility that such an inconsistency could be random or unique, the failure of this test was not considered a sufficient condition to consider the subject unreliable and this check was always complemented by the previous test based on the circular triads.

3.3.6 Agreement among several subjects

To test for the *agreement* among the subjects, the *coefficient of agreement* was calculated, which takes into account the number of concordant judgements between pairs of subjects (Kendall & Babington Smith, 1940). In a test involving n sound stimuli and m subjects, if x_{ij} is the element (i,j) of the preference matrix, the agreement coefficient is defined as:

$$u = \frac{2 \cdot S}{\binom{m}{2} \cdot \binom{n}{2}} - 1 \tag{3}$$

where S is the total number of agreements between pairs of subjects, derived from the following equation:

$$S = \sum_{i,j} \binom{x_{ij}}{2} = \frac{1}{2} \cdot \sum_{i,j} x_{ij}(x_{ij} - 1) \tag{4}$$

The statistical significance of u strictly depends on the probability that its value is exclusively a random value.

In the several experiments, the probability to obtain a specific value of u, as a function both of the distribution of each subject judgments and of the distribution of the judgments given by all the subjects regarding each specific matrix element was obtained by considering the variable (Kendall & Babington Smith, 1940):

$$Z = \frac{4 \cdot S}{m-2} - \frac{m(m-1)(m-3)n(n-1)}{2(m-2)^2} \tag{5}$$

This variable follows the χ^2 distribution, with a number of degrees of freedom given by:

$$\frac{m(m-1)n(n-1)}{2(m-2)^2} \tag{6}$$

When the *modified* paired comparison method was applied, a slight different procedure was used to calculate the *agreement* coefficient u. According to this procedure, each matrix element with value 0.5 has to be excluded from the calculation of the overall judgements and the modified *agreement* coefficient u_m is *defined by*:

$$u_m = \frac{2 \cdot S}{S_{max}} - 1 \quad \text{and} \quad S_{max} = \sum_{i=1}^{X}\binom{k_i}{2} + \binom{m}{2} \cdot \left[\binom{n}{2} - X\right] \tag{7}$$

where X is the number of the preference matrix elements with value 0.5 and S_{max} is given by:

$$S_{max} = \sum_{i=1}^{X}\binom{k_i}{2} + \binom{m}{2} \cdot \left[\binom{n}{2} - X\right] \tag{8}$$

4. The milestones in the investigations referring to loaders

This part of the chapter collects the most significant experimental investigations carried out on compact loaders. The particular interest in this kind of machine is due to the fact that it is widely used. Thanks to its compact size, it goes where bigger machines can not, has a reduced cost, it is easily transportable, agile and productive. Unfortunately, this compactness makes it one of the worst machines as far as the noise emission is concerned, as the operator is very close to the main sources of noise (engine and hydraulics). Consequently, noise and vibration levels at the operator station are extremely high, causing very uncomfortable conditions for workers.

4.1 Noise signals and auditory perception of annoyance

This investigation was performed in order to better understand the relationship between the multidimensional characteristics of the noise signals recorded at the operator position in different working conditions and the relevant auditory perception of annoyance (Carletti et al., 2007).

The tests involved six binaural signals recorded at the operator station of three loaders belonging to the families A, B, and C, while these machines were repeating the same work cycle which included two main operations: the loading of the material from a stockpile and the unloading of it in a specific position. In the following these machines will be indicated as *A1*, *B1*, and *C1* and the different kinds of material as L (loam) and G (gravel).

4.1.1 Objective parameters

Based on the results of a study concerning the sound quality evaluation of wheel loaders (Khan & Dickson, 2002), several acoustic and psychoacoustic parameters were calculated for the left and the right signals, separately. This set included: the overall sound pressure levels L_{eq} and L_{Aeq} (in dB and dBA), the mean values of loudness (in sone), sharpness (in acum), fluctuation strength (in vacil) and roughness (in asper). Referring to the psychoacoustic parameters, they were all calculated according to the models proposed by Fastl and Zwicker (Fastl & Zwicker, 2006).

The results obtained for the six noise signals are summarised in table 2 (columns 3 to 8), while the frequency content of these signals is well described by the sonograms of the sound pressure level shown in figure 4, which refer to the different machines and working conditions.

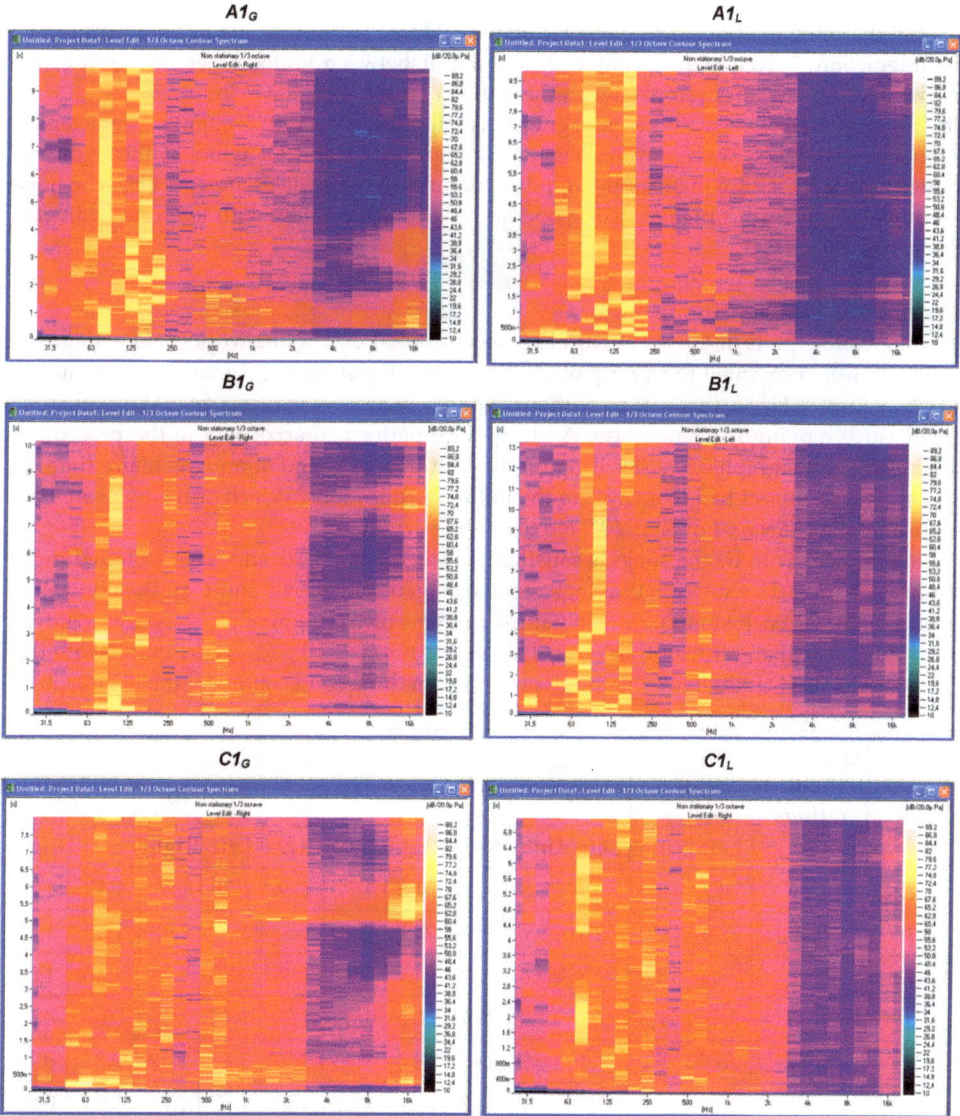

Fig. 4. Sonograms of the sound pressure levels: gravel (left) and loam (right)

Taking into account that during work the engine rotational speed of these machines ranged from 2000 to 2500 rpm, three interesting frequency intervals can be recognised.

The first one, in the 40-400 Hz frequency range, is directly related to the engine noise (engine rotational frequency, firing frequency and higher orders).

The second one, in the 500-3150 Hz frequency range, is related to the noise generated by the engine cooling system and the hydraulic system, this latter which drives arm, boom and bucket. In particular, at frequencies above 1kHz the noise contribution of the hydraulic system becomes the dominant one.

Finally, the third interval, at frequencies above 4 kHz, is related to the noise generated by the interaction between equipment and material or between various metallic parts of the machine (occurring, for example, when the actuators reach their travel limit).

The difference between these sonograms is significant. At the engine characteristic frequencies the noise levels are higher for A1 than for B1 and C1 while at the hydraulic system characteristic frequencies the opposite occurs. At frequencies above 4 kHz, the noise components are always higher during operations with gravel than with loam, regardless of the machine used.

4.1.2 Subjective evaluations

All the possible pairs of the six binaural signals recorded during the work cycle with gravel and loam were presented to a group of 19 normal-hearing subjects (17 males and 2 females), all non expert subjects, that means without experience in listening tests or in the evaluation of the earth moving machine noise. All tests were performed according to the procedures described in paragraph 3.3. After listening to each pair of sound stimuli as many times as necessary, the subject had to answer to the following question: "Which of the two sounds is more annoying? *Sound 1* or *Sound 2*". No ties were permitted.

All the ratings given by the subjects satisfied the consistency tests and were included in the analysis process. The subjective ratings were arranged in matrices and then the annoyance overall score for each stimulus was obtained in terms of the number of cases it was judged more annoying than all the other ones. This value, normalised to the maximum score that the stimulus itself could have obtained, is reported in the second column of table 2.

Machi-nes	Annoyance Subj. ratings (%)	L_{eq} (dB)		L_{Aeq} (dBA)		Loudness (sones)		Sharpness (acum)		Roughness (asper)		Fluctuation strenght (vacil)	
		Left	Right	Left	Right	Left	Right	Left	Right	Left	Right	Left	Right
$C1_G$	0.99	78.7	78.6	74.1	73.8	32.5	31.5	1.77	1.64	1.29	1.31	1.13	1.22
$B1_G$	0.78	77.6	77.0	72.6	71.7	29.8	28.2	1.68	1.58	1.37	1.35	1.13	1.08
$A1_G$	0.43	78.5	78.3	69.5	69.4	25.3	24.7	1.44	1.37	1.72	1.77	1.06	1.03
$C1_L$	0.39	77.6	77.3	72.3	71.7	29.7	28.5	1.49	1.39	1.44	1.39	0.67	0.66
$B1_L$	0.22	76.7	75.9	71.4	69.8	26.7	24.4	1.43	1.37	1.44	1.42	0.81	0.79
$A1_L$	0.19	78.4	78.0	67.5	67.2	22.3	22.0	1.22	1.19	1.57	1.71	0.96	0.85

Table 2. Subjective annoyance ratings and acoustic/psychoacoustic parameters

4.1.3 Results

As shown in table 2, the $C1$ machine handling gravel ($C1_G$) was the most annoying (99%), whilst the $A1$ machine handling loam ($A1_L$) was rated the least annoying (19%).

Referring to the noise emissions of the machines handling different materials, the overall levels indicated in table 2 and the sonograms in figure 4 show that when the machine is working with loam, the overall noise emission is generally lower than that generated when working with gravel, regardless of the machine characteristics. The loam seems to have a damping effect on the different noise components both when the machine is loading the bucket and when it is transporting the material.

Referring to the different machines, $C1$ was always judged more annoying than $B1$, and $B1$ was always judged more annoying than $A1$, for each handled material. The annoyance ratings greater for $C1$ than for $B1$ could be simply related to the highest noise levels. On the contrary, the annoyance ratings greater for $C1$ and $B1$ than for $A1$ could not be explained in terms of the overall noise levels but in terms of the highest noise levels emitted by $B1$ and $C1$ at the characteristic frequencies of the hydraulic system and at frequencies higher than 4 kHz.

The spectral characteristics described above clearly point out the relevance of the noise components at medium and high frequencies in affecting the subjective evaluation of the sound with respect to its annoyance. The subjective ratings of annoyance, however, cannot always be explained taking into account only the energetic characteristics of these signals. As an example, $A1_L$ (19% annoyance rating) has an overall level higher than that of $B1_G$ (78% annoyance rating), even if the first signal is judged significantly less annoying.

This example highlights the absolute necessity to complement this energetic analysis with other considerations involving also the psychoacoustic parameters and their variability over time.

The Pearson correlation coefficients with the annoyance ratings were calculated for all the parameters in table 2. The best correlation was obtained with sharpness ($r = 0.94$) and relatively high values were also found with loudness ($r = 0.87$) and L_{Aeq} ($r = 0.85$). The parameter least correlated with the annoyance ratings was L_{eq}, with $r = 0.40$.

In addition, for each objective parameter the correlation coefficient between left and right signals was also evaluated. These correlations were consistently very high (almost equal to 1) for all the parameters. For this reason, only the signal with the highest correlation coefficient with respect to the subjective judgements was chosen for analysis.

Referring back to $A1_L$ and $B1_G$ noise signals, the different subjective judgements can be found in the sharpness value or, better, in its time history. Figure 5 shows some percentile values of sharpness for the different machines and materials.

The sharpness percentiles of $A1_L$ are significantly lower than those of $B1_G$. In addition, under working condition with gravel, all the noise signals have very high values of percentile sharpness S_5 with respect to their average sharpness S_{50}. As S_5 percentile describes the variability over time of these signals much better than S_{50}, the very high values of this parameter underline that prominent noise events occur very frequently under working conditions with gravel and this leads to a negative subjective evaluation.

The good correlation of S_5 with the annoyance ratings ($r = 0.91$) and the possibility to take into account variability over time of the noise signals makes this parameter very important for acoustic comfort improvement. Moreover, considering that the auditory sensation of this parameter greatly depends on the signal content at medium-high frequencies (Fastl &

Zwicker, 2006), the above results gave a further proof of the relevance on the subjective ratings of annoyance of the noise components generated both by the hydraulic system and the handling of materials.

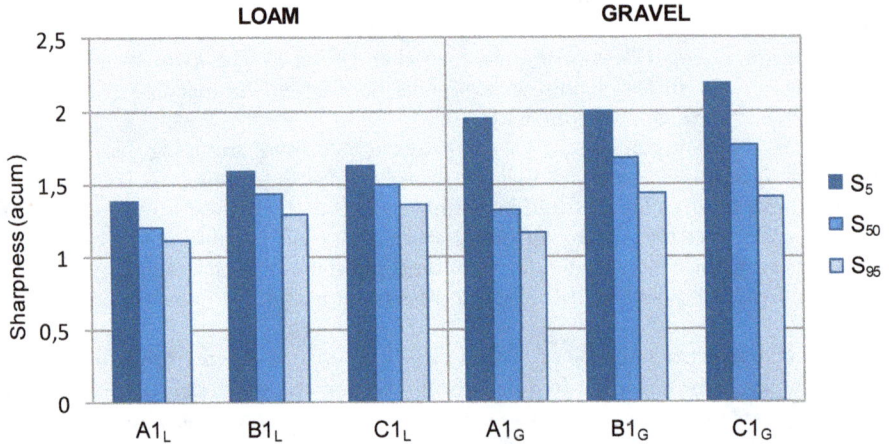

Fig. 5. Sharpness percentile graphs for the different machines and materials handled

A deeper insight of the differences in the subjective judgements for machines handling different materials can be obtained by analysing the time-dependent characteristics of the noise signals in terms of loudness distributions. As an example, figure 6 shows the loudness cumulative distribution for machine *C1*. The same trend, however, could also be obtained for the other machines (*A1* and *B1*).

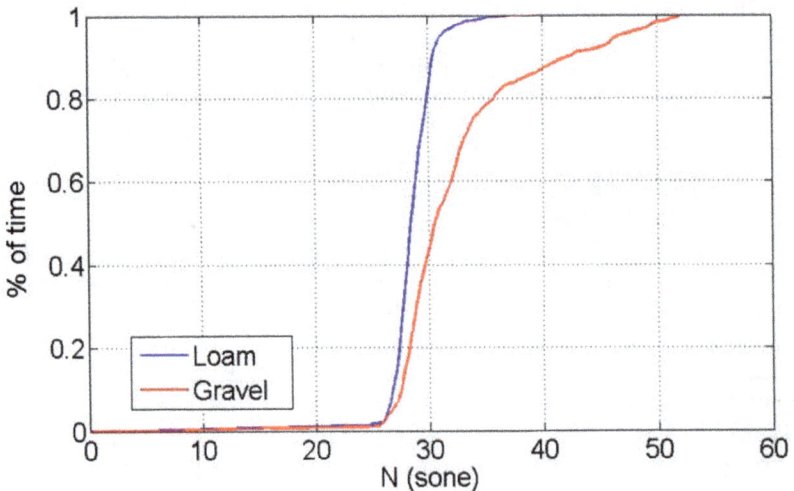

Fig. 6. Loudness cumulative distribution for machine *C1*

This cumulative distribution shows the percentage of time for which a given loudness value (in sone) is not exceeded. The blue line identifies the loudness distribution for the machine working with loam while the red one identifies the loudness distribution for the machine working with gravel. The percentile loudness N_5 can be read at the ordinate point 0.95, N_{10} at the ordinate point 0.90 and N_{95} at the ordinate point 0.05. These two curves have very different gradient, depending on the working conditions (loam or gravel). The curve with lower gradients (gravel) shows that the loudness values are more evenly distributed over time. On the other hand, the noise recordings of the machine working with gravel have always been judged more annoying than those of the machine working with loam. These results seem to confirm some conclusions of previous studies which illustrated how changes of loudness during the considered time frame may be very important in the judgement of annoyance (Genuit, 2006).

4.1.4 Final remarks on the relevance of this investigation
This study provided some fundamental results for the progress of the investigations. Firstly, how to describe the auditory perception of annoyance by means of some objective parameters. Loudness and sharpness are suitable for this purpose and S_5 can be used to better describe the effects on annoyance of the time variability of the noise components at medium-high frequencies.

Secondly, the relevance on the auditory perception of annoyance both of the noise signals overall energy and the frequency distribution. The 400-5000 Hz frequency range, which includes the noise contributions generated by the hydraulic system and the handling of materials, is the most important referring to the annoyance judgements.

Finally, the absolute relevance of the temporal characteristics of these signals in identifying the relationship between machine characteristics/working conditions and auditory perception of annoyance.

4.2 Just noticeable difference in loudness and sharpness
The knowledge of the parameters best correlated to the annoyance sensation is insufficient to develop a methodology able to identify the basic criteria for noise control solutions which guarantee the improvement of the operator comfort conditions. Tiny variations in stimulus magnitude may not lead to a variation in sensation magnitude. In order to detect the step size of the stimulus that leads to a difference in the hearing sensation, the *differential threshold* or *just noticeable difference, JND*, should be known for all the parameters of primary interest (Fastl & Zwicker, 2006). JNDs of amplitude and frequency, as well as duration changes of pure/complex tone or broad band noise, have been investigated for decades. Unfortunately, little is known regarding the JNDs of sound quality metrics in real noises (Sato et al., 2007; You & Jeon, 2008). Regarding this a specific investigation was performed by the authors aimed at evaluating the JNDs for the two psychoacoustic parameters describing at best the auditory perception of noise signals at the operator station of compact loaders with respect to the annoyance subjective ratings (loudness and sharpness) (Pedrielli et al., 2008).

4.2.1 Sound stimuli
This investigation involved a binaural noise signal recorded at the operator station of a compact loader of family F in stationary conditions, with the engine running at 2300 rpm. The recorded signal was post-processed following various steps:

- generation of a sound stimulus with the same signal at both ears (diotic stimulus), in order to help listeners to concentrate only on the difference between the sounds having different loudness or sharpness, without being influenced by interaural differences;
- counterbalance of the spectral modifications that occur during playback, depending on the specific sound card and electrostatic headphones used for the listening tests;
- creation of sound stimuli with different loudness or sharpness values according to the design of experiments typical of the Method of Limits.

For the evaluation of loudness JNDs, the overall sound pressure level of the original sound was varied in order to change the total loudness value by interval steps of +0.3 sone and -0.3 sone. The sharpness value among these stimuli was kept constant.

Apart from the original sound, 9 sounds with higher loudness values and 9 with lower loudness values were created. The specific loudness of all these sound stimuli is reported in the left side of figure 7 where the thick line represents the stimulus used as reference in the listening tests.

For the evaluation of sharpness JNDs, the original sound was filtered in order to change the sharpness value by interval steps of +0.02 acum and -0.02 acum. This effect was achieved with a 1/3 octave band filter with a negative gain in the 40-80 Hz range and a positive gain in the 4-20 kHz range. The maximum difference in loudness among the stimuli with different sharpness values was less than 0.1 sone. As found in a similar study (You & Jeon, 2008), although concerning a different sound source, such a difference should not influence the responses of subjects with respect to the sharpness feature.

Apart from the original sound, 9 sounds with higher sharpness value and 9 with lower sharpness value were created. The 1/3 octave band spectra for the sound pressure level are shown in the right side of figure 7 in order to illustrate the filter effect.

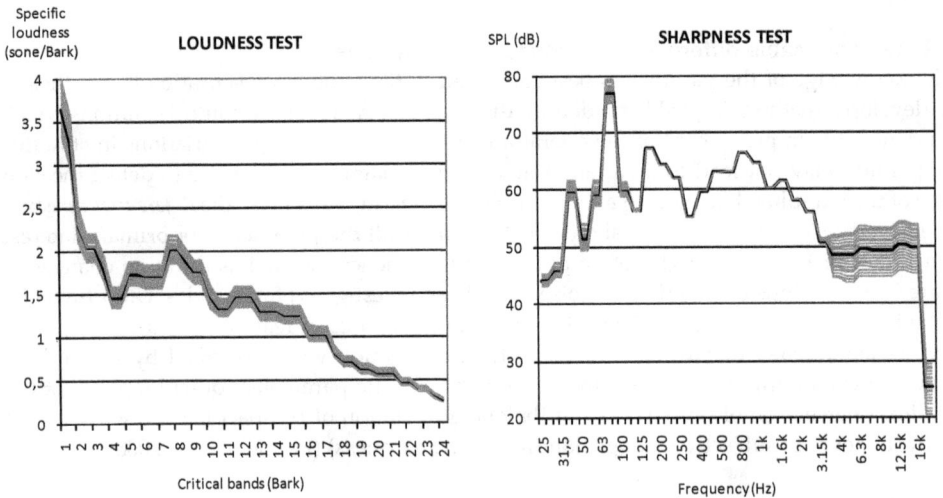

Fig. 7. Specific loudness of the sound stimuli created for the loudness and sharpness JNDs tests

4.2.2 Listening tests

The subjective listening tests were performed following the classical Method of Limits (Gelfand, 1990). According to this method, two stimuli are presented in each trial and the subject is asked whether the second is greater than, less than, or equal to the first with respect to a certain parameter. The first stimulus is held constant (reference stimulus) and the second is varied by the experimenter in specific steps. The procedure is repeated several times in subsequent ascending and descending runs.

In our experiments, a total number of six runs (three ascending alternated to three descending runs) were planned for each loudness and sharpness test.

The entire experiment was divided into three test sessions, different from each other as far as the sound pressure levels of the reference stimulus are concerned. In every test session each subject was asked to perform a test to detect firstly loudness JNDs and then sharpness JNDs. A few minutes' rest was scheduled between the loudness and sharpness tests.

21 subjects (16 males and 5 females) took part in the first and second test sessions, while 16 subjects (12 males and 4 females) took part in the third test session. 50% of the listening jury had prior experience in subjective listening tests, but had never experienced this specific psychophysical procedure (Method of Limits). Moreover, 50% of the listening jury was not familiar with the psychoacoustic parameters for which the evaluations were requested (loudness and sharpness).

Table 3 shows the structure of the experiment, also giving information about the metrics of the reference stimulus in each test.

	Loudness JNDs test	Sharpness JNDs test
1st test session (SPL of the reference stimulus around 80 dB)	Lp = 82.0 dB N = 32.1 sone S = 1.31 acum	Lp = 78.9 dB N = 29.8 sone S = 1.49 acum
2nd test session (SPL of the reference stimulus around 70 dB)	Lp = 73.1 dB N = 18.0 sone S = 1.30 acum	Lp = 69.0 dB N = 15.6 sone S = 1.47 acum
3rd test session (SPL of the reference stimulus around 60 dB)	Lp = 64.9 dB N = 10.3 sone S = 1.27 acum	Lp = 59.1 dB N = 7.74 sone S = 1.42 acum

Table 3. Reference sound stimuli for all the six tests

4.2.3 Results

At the end of the listening tests, the given judgments by each subject were summarised as shown in figure 8.

Referring to the loudness test, the Method of Limits resulted in a range of values in which the second stimulus was louder than the first (reference), a range in which the second was quieter, and a range in which the two sounds appeared to have an equal loudness value. Similar results were found for sharpness test, where "louder" and "quieter" became "higher" and "lower" sharpness, respectively.

The differential threshold (limen) for each subject was estimated once the average upper and lower limens had been defined. The upper limen was halfway between louder/higher and equal judgments, and the lower limen was halfway between quieter/lower and equal judgments. The average limens were obtained by averaging the upper and lower limens across runs. The range between the average upper limen and the average lower limen represents an interval of uncertainty, and the just noticeable difference, or *difference limen*, is generally estimated as half of this uncertainty interval (Gelfand, 1990).

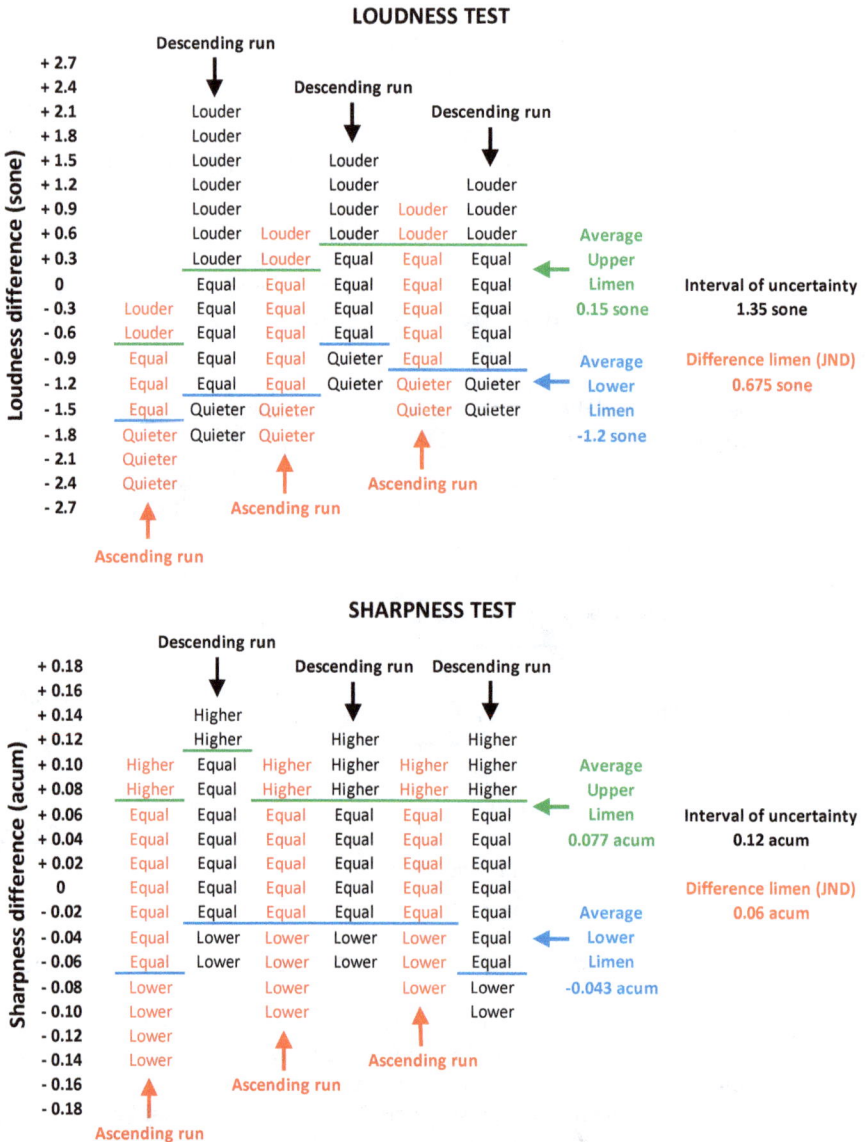

LOUDNESS TEST

Loudness difference (sone)

	Descending run	Descending run	Descending run			
+ 2.7						
+ 2.4						
+ 2.1	Louder					
+ 1.8	Louder					
+ 1.5	Louder	Louder				
+ 1.2	Louder	Louder	Louder			
+ 0.9	Louder	Louder	Louder	Louder		
+ 0.6	Louder	Louder	Louder	Louder	Louder	Average
+ 0.3	Louder	Louder	Equal	Equal	Equal	Upper
0	Equal	Equal	Equal	Equal	Equal	Limen
- 0.3	Louder	Equal	Equal	Equal	Equal	Equal	0.15 sone
- 0.6	Louder	Equal	Equal	Equal	Equal	Equal
- 0.9	Equal	Equal	Equal	Quieter	Equal	Equal	Average
- 1.2	Equal	Equal	Equal	Quieter	Quieter	Quieter	Lower
- 1.5	Equal	Quieter	Quieter		Quieter	Quieter	Limen
- 1.8	Quieter	Quieter	Quieter				-1.2 sone
- 2.1	Quieter					
- 2.4	Quieter			Ascending run		
- 2.7		Ascending run				
	Ascending run					

Interval of uncertainty 1.35 sone

Difference limen (JND) 0.675 sone

SHARPNESS TEST

Sharpness difference (acum)

	Descending run	Descending run	Descending run			
+ 0.18						
+ 0.16						
+ 0.14	Higher					
+ 0.12	Higher		Higher		Higher	
+ 0.10	Higher	Equal	Higher	Higher	Higher	Higher	Average
+ 0.08	Higher	Equal	Higher	Higher	Higher	Higher	Upper
+ 0.06	Equal	Equal	Equal	Equal	Equal	Equal	Limen
+ 0.04	Equal	Equal	Equal	Equal	Equal	Equal	0.077 acum
+ 0.02	Equal	Equal	Equal	Equal	Equal	Equal
0	Equal	Equal	Equal	Equal	Equal	Equal
- 0.02	Equal	Equal	Equal	Equal	Equal	Equal	Average
- 0.04	Equal	Lower	Lower	Lower	Lower	Equal	Lower
- 0.06	Equal	Lower	Lower	Lower	Lower	Equal	Limen
- 0.08	Lower		Lower		Lower	Lower	-0.043 acum
- 0.10	Lower		Lower			Lower
- 0.12	Lower					
- 0.14	Lower			Ascending run		
- 0.16			Ascending run			
- 0.18		Ascending run				
	Ascending run					

Interval of uncertainty 0.12 acum

Difference limen (JND) 0.06 acum

Fig. 8. Judgments given by one subject for the differential thresholds of loudness and sharpness (SPL around 80 dB)

Once the difference limens had been calculated for each subject, some statistical considerations could be outlined for the loudness and sharpness test, separately.

Just noticeable differences in loudness

Table 4 shows the results for the test of just noticeable differences in loudness. In this table, the variation range of the JNDs among the subjects and some percentile values are reported. The loudness value of the reference stimulus of each test is also specified.

	SPL around 80 dB	SPL around 70 dB	SPL around 60 dB
Loudness value	32.1 sone	18.0 sone	10.3 sone
Range	0.4 - 1.2 sone	0.3 - 1.2 sone	0.3 - 0.8 sone
50° percentile	0.7 sone	0.6 sone	0.4 sone
75° percentile	0.8 sone	0.8 sone	0.5 sone
90° percentile	1.0 sone	1.0 sone	0.7 sone

Table 4. Just noticeable differences for loudness tests

The just noticeable difference becomes greater as the overall sound pressure level of the signal increases. This indicates that the greater the level, the more difficult it is for the subject to detect tiny loudness variations in the sounds.

Cumulative distributions rather than unique values of just noticeable differences are more functional and make it possible to choose the just noticeable differences value depending on the specific target.

For this research, the 75° percentile was considered appropriate. An average or median value would not guarantee that the improvement of the operator comfort conditions were extensively appreciated. Consequently, for loaders where the sound pressure levels at the operator position are around 80 dB, the just noticeable difference in loudness is assessed as 0.8 sone.

Just noticeable differences in sharpness

Table 5 shows the results for the test of just noticeable differences in sharpness.

In this table, the variation range of the JNDs among the subjects and some percentile values are reported. The sharpness value of the reference stimulus of each test is also specified even if, as expected, it is almost independent of the sound pressure level variation.

The just noticeable differences show little variations with the presentation level and only for the 90° percentile.

	SPL around 80 dB	SPL around 70 dB	SPL around 60 dB
Sharpness value	1.49 acum	1.47 acum	1.42 acum
Range	0.02 - 0.07 acum	0.01 - 0.08 acum	0.02 - 0.06 acum
50° percentile	0.03 acum	0.03 acum	0.03 acum
75° percentile	0.04 acum	0.04 acum	0.04 acum
90° percentile	0.06 acum	0.04 acum	0.04 acum

Table 5. Just noticeable differences for sharpness tests

Also for this psychoacoustic parameter, the just noticeable difference was defined as the minimum variation in sharpness detected by at least 75% of the jury subjects.

Consequently, at the operator station of earth moving machines, the just noticeable difference in sharpness is assessed as 0.04 acum.

4.2.4 Final remarks on the relevance of this investigation

A specific metrics for loudness and sharpness (the two psychoacoustic parameters describing at best the annoyance auditory perception caused by these noise signals) was developed. In order to describe the step size of these parameters that leads to a difference in the hearing sensation of a group of people, a statistical approach was followed. The 75° percentile was considered appropriate; an average or median value, on the contrary, would not guarantee that the improvement of the operator comfort conditions were extensively appreciated. Focusing on the highest presentation level, 75% of subjects perceived a different sensation when sounds had a loudness difference of at least 0.8 sone and a sharpness difference of 0.04 acum.

These values were chosen as JND of loudness and sharpness to be used in the other investigations.

4.3 Active noise control and sound quality improvement

The effectiveness of the active noise control (ANC) approach to strongly reduce the low frequency noise content has already been shown in many applications involving real and simulated experiments (Fuller, 2002; Hansen, 1997, 2005; Scheuren, 2005). As for the specific field of earth moving machines, only a limited bibliography dealing with the ANC approach is available, despite the significant noise contributions at low frequency. On the other hand, the effectiveness of this approach has been evaluated only in terms of reduction of the overall sound pressure level. Taking into account that the noise level reductions are key elements for worker but they are not always related to improvements in sound quality, a study was carried out aimed at complementing the classical evaluations of such an approach with subjective evaluations of the modifications induced by an ANC system with regard to some noise features important to qualify the comfort and safety conditions (Carletti & Pedrielli, 2009).

4.3.1 The implemented ANC system

All the experiments were carried out on a skid steer loader of family B, equipped with lateral windows and door, in the winter version, as shown in figure 9.

Fig. 9. Skid steer loader used for the implementation of the ANC system

In the EMM industry, where the economic constraints are a key element, noise control solutions with a high economic impact associated with the overall cost of the machine are generally not of interest, even if highly technological. Consequently, a cheap and simple single-input single-output system was adopted, with the further limitation that its implementation inside the cab did not require any significant modification in the standard layout of the cab. On the other hand, this choice could be suitable from a technical point of view as inside EMM cabs the volume of interest is very limited and the ANC system must be effective to create a quiet zone only just around the operator's head.

A commercially-available ANC device, following a single channel adaptive feed-forward scheme, was chosen for the tests. This device (1000 Hz sampling frequency) required a reference signal closely related to the primary noise. This synchronism was simply obtained by picking up the impulses from a reflecting strip fixed on the engine shaft of the machine by an optical probe. In such a way the reference signal was not influenced by the control field and the fundamental frequency of the periodic primary noise could be assessed. Based on the reference signal, the ANC device determined the fundamental frequency of the noise, as well as the harmonics to be cancelled. By means of a series of adaptive filters, the output signal was generated and sent to the secondary source.

In order to minimise the economic impact of this implementation, the two loudspeakers of the Hi-Fi system were used as secondary sources. They were fixed to the vertical rods of the cab, at the same height as the operator's head. The error microphone was placed near the operator's head but in such a position that it did not disturb the operator during his work. A low-cost omnidirectional electret condenser microphone with a flat response in the range 40-400 Hz was used. It measured the resulting sound field due to the primary and secondary sources combined.

The control strategy was based on the minimisation of the mean squared value of the sound pressure at the error microphone position (cost function). For this aim, a gradient descendent algorithm was applied in which each controller coefficient was adjusted at each time step in a way that progressively reduced the cost function (filtered-X LMS algorithm) (Nelson & Elliott, 1993). The functional scheme of this ANC system is shown in figure 10.

Two more microphones (Mc) were placed near the operator's ears (by using an helmet worn by the operator) in order to monitor the acoustic field in the area of interest, in real time.

Many experiments were carried out in order to both check the capability of this system to reduce the overall sound pressure level in the volume around the operator's head and track any changes due to engine speed variations fast enough to maintain the control.

Table 6 shows the modifications brought on by the ANC system for three different values of the engine rotational speed (1500, 1800 and 2350 rpm); the second column shows the reduction of the noise component at the engine firing frequency (Δf) and the following two columns show the reduction of the overall levels, linear (ΔL_{eq}) and A-weighted (ΔL_{Aeq}), respectively.

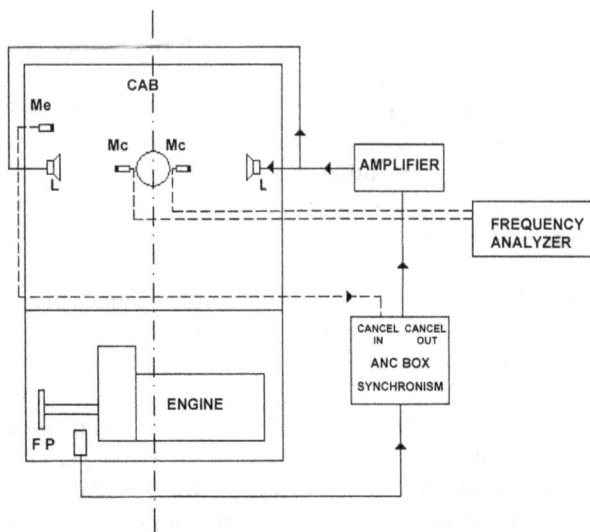

Fig. 10. Layout of the active noise control system. L = loudspeakers, Me = error microphone, Mc = monitoring microphones, FP = photoelectric probe

Rotational speed (rpm)	Δf (dB)	ΔL_{eq} (dB)	ΔL_{Aeq} (dB)
1500	16.8	10.2	2.0
1800	14.8	8.4	1.6
2350	14.9	5.3	0.3

Table 6. Reductions induced by the ANC system at the engine firing frequency (Δf), overall sound pressure level (ΔL_{eq}) and A-weighted overall sound pressure level (ΔL_{Aeq}) for three rpm values

As for the reduction of the overall level, it ranges from 5 to 10 dB and significantly decreases when the engine rotational speed increases: thus the higher the value of rotational speed, the lower the number of tonal components affected by the ANC device. Consequently, a considerable reduction of very few dominant noise components at a low frequency has a small effect on the relevant energetic content of the noise in the frequency range where the system has no influence. This trend is particularly manifest when the effects induced on L_{Aeq} are considered. The reduction of L_{Aeq} is considerably lower than the others (it never exceeds 2 dB) and it turns out to be insignificant at engine speed values higher than 2000 rpm.

From a "physical" point of view, the efficiency of this ANC device decreases when the engine rotational speed increases, the minimum efficiency being reached when the engine speed is at its maximum value (2350 rpm).

4.3.2 Subjective evaluation of the ANC system

Binaural noise recordings were carried out at the operator station of this machine, both with the ANC system activated (C, controlled) and with the ANC system not activated (U, uncontrolled) while the loader was operating in stationary idle conditions with the engine running at 2350 rpm. In such a condition, the ANC had the minimum efficiency and the controlled and uncontrolled noise signals had practically the same energy content at middle-high frequencies but a different distribution of the noise energy at low frequencies, as shown in figure 11. Consequently, subjective tests on controlled and uncontrolled signals would have permitted to check whether this difference, strictly dependent on the ANC action, evoked different subjective reactions despite these two signals had the same L_{Aeq} level.

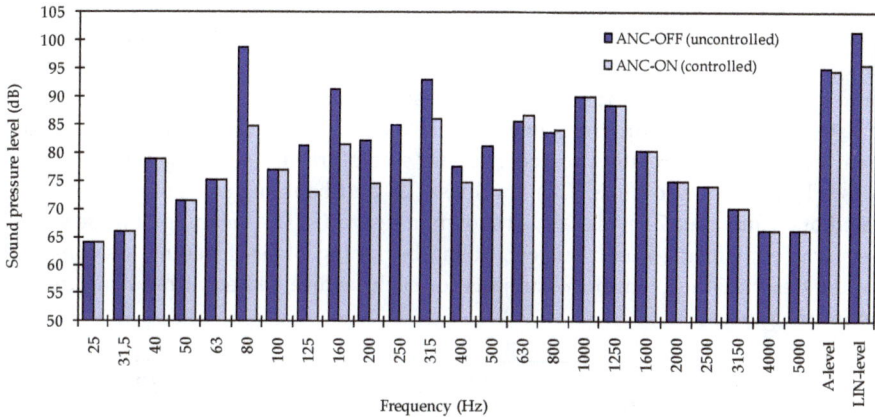

Fig. 11. One-third octave band sound pressure spectra at 2350 rpm with the ANC system on and off

In order to subjectively assess the modifications produced by the ANC system at different levels, both the controlled and uncontrolled sound stimuli were played back at different overall L_{eq} levels, namely 70 dB, 75 dB, and 80 dB. None of these levels actually reproduced the noise at the operator station of the machine (about 20 dB higher). However, these presentation levels were selected mainly to avoid any hazardous hearing effect on the listeners and also because they better highlighted the influence on the auditory perception of specific noise features other than the overall energy content. Table 7 describes the six sound stimuli used in the listening tests.

Sound Stimuli	Description	Overall L_{eq}	Overall L_{Aeq}
U	Original Uncontrolled signal	80 dB	73 dBA
U-5	It has L_{eq} and L_{Aeq} levels 5 dB lower than U	75 dB	68 dBA
U-10	It has L_{eq} and L_{Aeq} levels 10 dB lower than U	70 dB	63 dBA
C	Original Controlled signal	75 dB	73 dBA
C+5	It has L_{eq} and L_{Aeq} levels 5 dB greater than C	80 dB	78 dBA
C-5	It has L_{eq} and L_{Aeq} levels 5 dB lower than C	70 dB	68 dBA

Table 7. Description of the six sound stimuli used in subjective listening tests

As for the subjective evaluation of the modifications produced by the ANC system, particularly interesting was the comparison between uncontrolled and controlled sound stimuli with the same linear or A-weighted overall levels.

Three pairs of sound stimuli had the same linear overall level: U and C+5 (80 dB); U-5 and C (75 dB); U-10 and C-5 (70 dB). In each of these pairs both the reduction due to the active noise control system at the engine firing frequency and its harmonics went with an increase of the noise content at medium-high frequencies, regardless of the overall level.

Only two pairs of sound stimuli had the same A-weighted overall level: U and C (73 dBA); U_{-5} and C_{-5} (68 dBA). In each of these latter pairs the differences are due only to the active noise control system, regardless of the overall level.

The six sound stimuli were arranged in pairs according to the paired comparison procedure and presented to the subjects of the listening jury, tested one at a time. This group of people was formed by eighteen normal-hearing expert operators of earth moving machines, all males aged between twenty-five to fifty years. None of them had previous experience in listening tests but a great experience in using these machines.

After listening to each pair, the subjects were asked to give a rating referring to four different noise features relating to the operator's comfort and working safety conditions: tiredness (T), concentration loss (CL), loudness (L), and booming sensation (B). This rating consisted of a value on a 7-level scale, as shown in figure 12. The meaning of these subjective features was explained to each subject, at the beginning of his listening session. Table 8 details the description given to the subjects for each feature, aimed at reducing the risk of semantic ambiguity.

<table>
<tr><th rowspan="2"></th><th></th><th colspan="3">A Stimulus</th><th></th><th colspan="3">B Stimulus</th></tr>
<tr><th></th><th>Much more than</th><th>More than</th><th>Slightly more than</th><th>Equal to</th><th>Slightly more than</th><th>More than</th><th>Much more than</th></tr>
<tr><td rowspan="4">Features</td><td>Tiring</td><td>A+++</td><td>A++</td><td>A+</td><td>A=B</td><td>B+</td><td>B++</td><td>B+++</td></tr>
<tr><td>Causing concentration loss</td><td>A+++</td><td>A++</td><td>A+</td><td>A=B</td><td>B+</td><td>B++</td><td>B+++</td></tr>
<tr><td>Loud</td><td>A+++</td><td>A++</td><td>A+</td><td>A=B</td><td>B+</td><td>B++</td><td>B+++</td></tr>
<tr><td>Booming</td><td>A+++</td><td>A++</td><td>A+</td><td>A=B</td><td>B+</td><td>B++</td><td>B+++</td></tr>
</table>

Fig. 12. Response scale for each pair of sound stimuli

<table>
<tr><th>Features</th><th>ID</th><th>Description</th></tr>
<tr><td>Tiring</td><td>T</td><td>If the noise is heard for at least two hours non-stop, it may cause either tiredness or mental/physical stress</td></tr>
<tr><td>Causing concentration loss</td><td>CL</td><td>If the noise is heard for at least two hours non-stop, it may cause loss of concentration thus compromising the operator's working tasks</td></tr>
<tr><td>Loud</td><td>L</td><td>A high level in the sound volume</td></tr>
<tr><td>Booming</td><td>B</td><td>A buzzing and echoing sound</td></tr>
</table>

Table 8. Description of the four subjective noise features

4.3.3 Results of the subjective evaluations

For each feature, the subjective ratings of the six stimuli were computed by pooling the marks into two categories: significant difference (marks "+++" and "++" added together) and no significant difference (marks "+" and "=" added together). The ratings given by the entire listening jury for the significant difference of each feature are shown in table 9. These ratings were normalised with respect to the maximum score that each stimulus could have obtained and then expressed as percentage values.

Sound stimuli	Features			
	T	CL	L	B
C_{+5}	85.6 %	81.1 %	88.9 %	44.4 %
U	61.1 %	58.9 %	54.4 %	73.3 %
C	34.4 %	35.6 %	43.3 %	18.9 %
U_{-5}	15.6 %	21.1 %	15.6 %	26.7 %
C_{-5}	4.4 %	4.4 %	6.7 %	1.1 %
U_{-10}	2.2 %	2.2 %	2.2 %	13.3 %

Table 9. Subjective ratings of "significant difference" for the four noise features, in percentage values

The grey area of table 9 shows the subjective ratings of "significant difference" obtained for controlled and uncontrolled signals with the same A-weighted overall sound pressure level: U and C (73 dBA); U_{-5} and C_{-5} (68 dBA). The reductions in the low frequency noise components brought on by the ANC system positively influenced the subjective evaluations in respect of all the noise features when the controlled and uncontrolled signals had significant differences only at low frequencies, no matter what the playback level.

When the subjective ratings of controlled and uncontrolled stimuli with an equal L_{eq} were considered (U and C_{+5} (80 dB); U_{-5} and C (75 dB); U_{-10} and C_{-5} (70 dB)), a different behaviour appeared for the four noise features. As far as the T, CL, and L features are concerned, the subjects always judge the controlled signal worse than the uncontrolled one. This accordance holds at all the different presentation levels, even if the higher the level, the greater the subjective difference between controlled and uncontrolled stimuli. Such results show that the subjective ratings are primarily influenced by the energy content of the noise signal at the medium-high frequencies.

Consequently, the effect of an ANC system in respect of the tiredness, concentration loss and loudness features is negatively judged if the reduction of the low frequency components is accompanied by an increase in the components at high frequencies.

When judging the booming feature, an opposite trend can be noticed: the subjective ratings were always positively influenced by the reduction in the low frequency noise components caused by the ANC system, regardless of the content of the signals at medium-high frequency (stimulus U is more booming than stimulus C_{+5} even if the latter has a higher A-weighted level and then a predominance of the energy content in the medium-high frequency).

4.3.4 Final remarks on the relevance of this investigation
This study showed the feasibility of the ANC approach to improve the sound quality inside loader cabs, provided that the controlled and uncontrolled signals show significant differences only at low frequencies. The sound quality conditions were evaluated by means of subjective evaluations with regards to four different noise features, all related to the operator's comfort and working safety conditions: tiredness (T), concentration loss (CL), loudness (L) and booming sensation (B).

When controlled and uncontrolled signals were forced to have the same overall sound pressure level and then the controlled signal had a higher noise content at the medium-high frequencies, the controlled signal was always judged worse than the uncontrolled one related to concentration loss, tiredness and loudness attributes. Referring to the booming feature, the subjective ratings were always positively influenced by the reduction in the low frequency noise components caused by the ANC system, regardless of the content of the signals at medium-high frequency.

4.4 Annoyance prediction model for loaders

The experience in applying the "product sound quality" approach to the noise signals recorded at the operator station of compact loaders confirmed the effectiveness of this methodology. Besides enlightening the relationship between physical properties of noise signals and auditory perception of some features significant for the acoustic comfort, this approach allowed the authors to identify which noise control criteria could ensure better conditions for the operator.

Unfortunately, this approach requires repeated sessions of jury listening tests which are demanding and time consuming. An annoyance prediction model able to assess the grade of annoyance at the workplace of loaders by using only objective parameters could be an important opportunity for manufacturers and customers. The database of the 62 binaural noise signals recorded at the operator station of several families of compact loaders (see paragraph 2) was therefore used by the authors for this purpose (Carletti et al., 2010a).

4.4.1 Binaural recording and objective characterisations

Based on the results of some other studies (Sato et al., 2007; Kroesen et al., 2008), the following physical parameters were considered relevant for this investigation:

- the overall sound pressure levels: L_{eq}, L_{Aeq}, L_{Ceq}, L_{Peak} ;
- the percentile values of the sound pressure levels (Lp_5, Lp_{10}, Lp_{50}, Lp_{90}, Lp_{95});
- the overall values of loudness N, sharpness S, roughness and fluctuation strength;
- the percentile values of loudness (N_5, N_{10}, N_{50}, N_{90}, N_{95}) and sharpness (S_5, S_{10}, S_{50}, S_{90}, S_{95}).

These parameters were estimated for the complete data set of noise stimuli, for right and left channels separately. Then the stimulus with the highest Pearson correlation coefficient with respect to the subjective annoyance score was considered for subsequent analyses.

4.4.2 Listening tests and subjective annoyance scores

The database of the 62 binaural noise signals was divided into nine different groups. For each noise group the subjective assessment of annoyance was obtained by means of subjective listening tests carried out according to the paired comparison procedure.

80 normal-hearing subjects (60 males and 20 females) aged between 24 and 50 were involved in the various listening tests. None of them was familiar with earth moving machines but all of the subjects had some knowledge in acoustics and some of them had also prior experience in listening tests. In addition, for each of the noise groups, the number of subjects involved in the test was never lower than 15, with the only exception of group 6 test which involved 9 subjects only. The overview of all the binaural noise stimuli belonging to each noise group and the percentage values of the subjective annoyance scores obtained for each of them are shown in table 10.

Group 1 10 binaural signals recorded from 5 loaders of family A during the working cycle with loam (L) and gravel (G)										Group 2 5 binaural signals from 5 loaders of family A during the simulated work cycle (S)				
$A1_L$	$A2_L$	$A3_L$	$A4_L$	$A5_L$	$A1_G$	$A2_G$	$A3_G$	$A4_G$	$A5_G$	$A6_S$	$A7_S$	$A8_S$	$A9_S$	$A10_S$
15.7	71.1	27.9	50.4	21.3	69.3	48.6	50.5	94.6	50.5	66.7	51.7	15.8	27.5	88.3

Group 3 10 binaural signals recorded from 5 loaders of family B during the working cycle with loam (L) and gravel (G)										Group 4 5 binaural signals from 5 loaders of family B during the simulated work cycle (S)				
$B1_L$	$B2_L$	$B3_L$	$B4_L$	$B5_L$	$B1_G$	$B2_G$	$B3_G$	$B4_G$	$B5_G$	$B6_S$	$B7_S$	$B8_S$	$B9_S$	$B10_S$
18.5	18.5	57.0	47.9	13.7	75.9	64.7	86.3	65.4	52.1	55.0	30.0	70.0	65.8	29.2

Group 5 10 binaural signals recorded from 5 loaders of family C during the working cycle with loam (L) and gravel (G)										Group 6 5 binaural signals from 5 loaders of family C during the simulated work cycle				
$C1_L$	$C2_L$	$C3_L$	$C4_L$	$C5_L$	$C1_G$	$C2_G$	$C3_G$	$C4_G$	$C5_G$	$C6_S$	$C7_S$	$C8_S$	$C9_S$	$C10_S$
12.1	26.0	50.0	25.5	44.4	83.4	60.0	68.4	63.9	66.2	33.3	55.8	88.3	36.7	35.8

Group 7 6 binaural signals from 3 loaders of family D during the work cycle with gravel (G) and loam (L)						Group 8 6 binaural signals from 3 loaders of family E during the work cycle with gravel (G) and loam (L)						Group 9 5 binaural signals from 5 loaders of family F recorded in stationary conditions				
$D1_G$	$D2_G$	$D3_G$	$D1_L$	$D2_L$	$D3_L$	$E1_G$	$E2_G$	$E3_G$	$E1_L$	$E2_L$	$E3_L$	$F1$	$F2$	$F3$	$F4$	$F5$
43.2	77.9	98.9	18.9	22.1	38.9	35.6	77.8	93.3	6.7	46.7	40.0	77.9	76.5	70.6	2.9	22.1

Table 10. Groups of noise stimuli and percentage values of subjective annoyance scores

4.4.3 Multiple regression analysis

The first six groups of noise stimuli were used to develop the annoyance prediction model while the last three were kept aside to validate it.

In order to reach the proposed target, multiple regression analysis was chosen as this technique is the most commonly used for analysing multiple dependence between variables and also because the theory is well developed (Kleinbaum et al., 2007).

In this investigation, the Stepwise selection method was firstly applied to each group of noise stimuli in order to identify the smallest set of independent variables which best explained the variation in the subjective annoyance scores (Lindley, 1968). In this respect, the score from subjective listening tests was entered as "dependent variable" and all the objective parameters, considered to be relevant for this investigation, were used as "independent variables". The results obtained for the six groups are shown in table 11.

Noise group	1	2	3	4	5	6
Predictor variables	N	S_{90}, Peak, N_{50}	N_{10}, Peak	Peak, S_5	N_{10}, Peak, N_{50}	N_{95}, S_{95}
R^2	0.63	1.00	0.95	1.00	0.95	1.00
Adjusted R^2	0.58	1.00	0.94	1.00	0.93	1.00

Table 11. Results of the "Stepwise" selection method applied to the six noise groups

In this table, the parameter R^2 is the square value of the correlation coefficient between the subjective scores and the predicted values of the annoyance. It quantifies the suitability of the fit of the model and shows the proportion of variation in the subjective scores which is explained by the set of the identified parameters. In addition, the Adjusted R^2 values, which takes into account the number of variables and the number of observations, were calculated in order to give a most useful measure of the success of the prediction when applied to real world.

For each noise group the variables selected by the Stepwise method account for more than 93% of the variation in the subjective scores, with the only exception of group 1. In addition, the set of the physical parameters which represent loudness, sharpness and peak level are very often included in the model, independently from the specific noise group. On the other hand, all the parameters which reflect the same quantity such as N, N_{10}, N_{50} and N_{95} for loudness, or S_5, S_{90} and S_{95} for sharpness are strongly correlated among each other.

Consequently, in order to identify a common set of predictor variables for each of the six noise groups, further analyses were carried out by substituting some of the parameters shown in table 11 with others reflecting the same acoustic features. The multiple regression analysis was then repeated on the six groups with the "Enter" variable selection method, that is forcing the choice of the set of predictor variables among (Peak, N, S_5), (Peak, N, S_{90}), (Peak, N, S_{95}), (Peak, N_{10}, S_5), ... etc...

The set of predictor variables which led to the highest R^2 values for the correlation between predicted and observed annoyance scores was (Peak, N_{50}, S_5). The multiple regression equations for this set of parameters are shown in table 12.

Predictor variables	Noise group	Multiple Regression Equation	R^2	Adjusted R^2
Peak, N_{50}, S_5	1	$Y = -9.310 + 0.057 \cdot Peak + 0.184 \cdot N_{50} + 0.216 \cdot S_5$	0.79	0.69
	2	$Y = -5.512 + 0.039 \cdot Peak + 0.296 \cdot N_{50} - 3.703 \cdot S_5$	0.99	0.97
	3	$Y = -5.322 + 0.038 \cdot Peak + 0.057 \cdot N_{50} + 0.412 \cdot S_5$	0.89	0.83
	4	$Y = -18.214 + 0.061 \cdot Peak + 0.018 \cdot N_{50} + 9.628 \cdot S_5$	1.00	1.00
	5	$Y = -4.241 + 0.030 \cdot Peak + 0.046 \cdot N_{50} + 0.289 \cdot S_5$	0.96	0.94
	6	$Y = 6.971 - 0.012 \cdot Peak + 0.312 \cdot N_{50} - 11.350 \cdot S_5$	0.89	0.55

Table 12. Results of the "Enter" selection method applied to the six noise groups

For each noise group this set of variables accounts for at least the 89% of the variation in the subjective scores, with the only exception of noise group 1. In addition, the big difference between R^2 and Adjusted R^2 values for group 6 takes into account the limited number of subjects involved in this test.

These results, which might be referred to as compromise solutions are only slightly worse than the best solutions obtained following the "Stepwise" variable selection method.

4.4.4 Predicted annoyance (P.A.)

In order to identify the best annoyance model among the regression equations obtained for the six different noise groups, and listed in table 12, each regression equation was applied to all the other five groups and for each equation the predicted annoyance values were calculated. Then the correlation between these predicted annoyance values and the observed subjective ratings was evaluated for each noise group: the better the correlation the higher the R^2 value. In such a way the best annoyance prediction model was the one that gave the maximum sum of R^2 over all the noise groups except for the one from which that model was issued.

According to this criterion, the regression equation referred to group 3 was the best and was chosen as the prediction model to assess the noise annoyance at the workplace of compact loaders (P.A. = Predicted annoyance) :

$$P.A. = -5.322 + 0.038 \cdot Peak + 0.057 \cdot N_{50} + 0.412 \cdot S_5 \tag{9}$$

4.4.5 Validation of the model

In order to verify whether this prediction model is applicable to noise signals other than those from which the equation was derived, the noise groups 7, 8, and 9 were then involved in the analysis.

Referring to noise signals of groups 7 and 8, the model gave predicted annoyance values that were significantly correlated with the subjective scores (correlation coefficients 0.95 and 0.96). Referring to group 9, it included noise signals recorded in stationary conditions and then with characteristics significantly different from those from which the model was issued (working conditions). These signals had sound pressure levels and loudness values higher than those of all the other signals, approximately 20 dB and 70 sone, respectively. Despite these differences, also this group showed a quite good correlation ($r = 0.85$ corresponding to a significance level of 5.6 percent).

However, considering that subjective listening tests were performed on each group separately, the annoyance scores could not be compared among different groups. For this reason, a further validation was deemed necessary. New subjective listening tests involving all the sound stimuli referred to a certain family of compact loaders (independently from the operating condition of the machine) were carried out according to the experimental procedures described in paragraph 3.3. The results are shown in table 13.

The comparison between the predicted values of annoyance and the subjective annoyance scores obtained by these new listening tests showed again a very good correlation. Figure 13 shows the predicted annoyance (P.A.) values against the observed values for the three groups of signals. The high values of the squared correlation coefficient (R^2) (0.898, 0.909, 0.934, respectively) confirm the suitability of the fit of this model.

Group A														
15 binaural noise signals recorded from 10 loaders of family Aduring the simulated work cycle (S) and during the working cycle with loam (L) and gravel (G)														
$A1_L$	$A2_L$	$A3_L$	$A4_L$	$A5_L$	$A1_G$	$A2_G$	$A3_G$	$A4_G$	$A5_G$	$A6_S$	$A7_S$	$A8_S$	$A9_S$	$A10_S$
15.7	64.9	26.6	46.5	20.7	85.0	66.7	68.4	99.2	68.4	54.3	39.3	3.5	15.1	75.9
Group B														
15 binaural noise signals recorded from 10 loaders of family Bduring the simulated work cycle (S) and during the working cycle with loam (L) and gravel (G)														
$B1_L$	$B2_L$	$B3_L$	$B4_L$	$B5_L$	$B1_G$	$B2_G$	$B3_G$	$B4_G$	$B5_G$	$B6_S$	$B7_S$	$B8_S$	$B9_S$	$B10_S$
34.4	34.4	72.5	63.5	29.6	84.5	73.4	94.8	74.1	60.9	29.7	9.0	42.2	38.7	8.3
Group C														
15 binaural noise signals recorded from 10 loaders of family Cduring the simulated work cycle (S) and during the working cycle with loam (L) and gravel (G)														
$C1_L$	$C2_L$	$C3_L$	$C4_L$	$C5_L$	$C1_G$	$C2_G$	$C3_G$	$C4_G$	$C5_G$	$C6_S$	$C7_S$	$C8_S$	$C9_S$	$C10_S$
30.2	45.0	70.6	44.4	64.7	89.6	64.6	73.5	68.8	71.1	13.6	29.6	52.9	16.0	15.4

Table 13. Subjective annoyance scores (% values) for tests on loaders of family A, B, and C

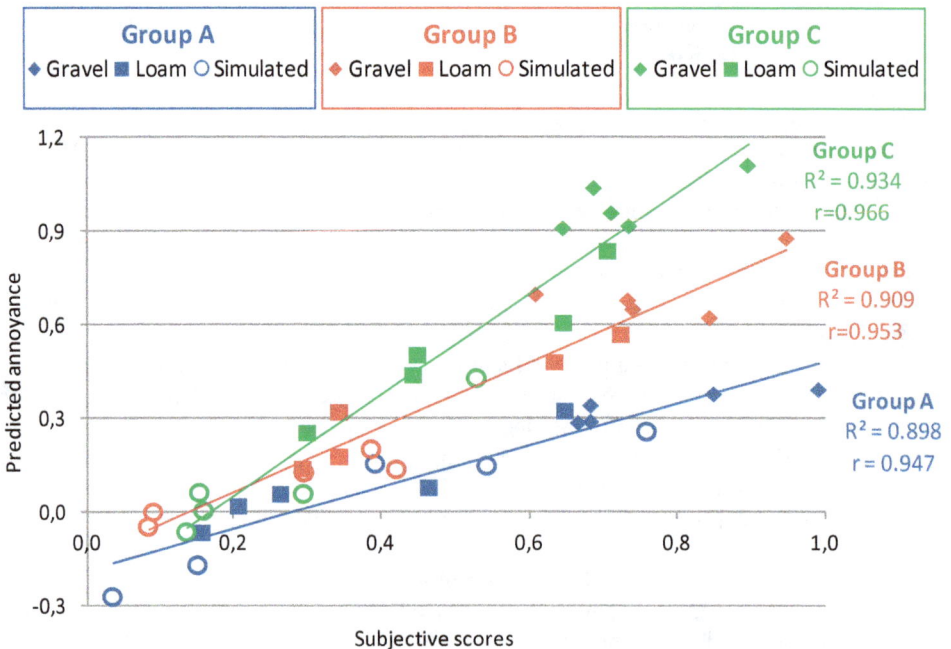

Fig. 13. Comparison between predicted and observed values of annoyance (subjective scores) for the three groups of signals (A, B, C)

4.4.6 Final remarks on the relevance of this investigation

The prediction model was developed on the basis of a huge amount of binaural noise signals recorded at the operator position of several families of loaders. Its regression equation :

$$P.A. = -5.322 + 0.038 \cdot Peak + 0.057 \cdot N_{50} + 0.412 \cdot S_5 \qquad (10)$$

could provide an alternative and simpler way for manufacturers and customers to assess the grade of annoyance at the workplace of any loader. This model, indeed, intrinsically reflects the main results of the sound quality approach but it is obtained by means of objective parameters only.

5. Conclusion

This chapter collects the main results of the research performed by the authors in the last five years in order to identify a methodology that is able to establish the basic criteria for noise control solutions which guarantee the improvement of the operator comfort conditions. All the investigations were carried out on compact loaders and permitted to collect the following main results.

Auditory perception of annoyance (see paragraph 4.1)

This study was aimed at better understanding the relationship between the multidimensional characteristics of the noise signals in different working conditions and the relevant auditory perception of annoyance. It highlighted that sharpness and loudness are suitable for this purpose, that the 400-5000 Hz frequency range - which includes the noise contributions generated by the hydraulic system and the handling of materials - is the most important referring to the annoyance judgements and that the temporal characteristics of the signals play an important role. The sharpness fifth percentile S_5 can be used to better describe the effects on annoyance due to the time variability of the noise components at medium-high frequencies.

Just noticeable difference in loudness and sharpness (see paragraph 4.2)

This study was aimed at evaluating the minimum differences in loudness and sharpness which are subjectively perceived (just noticeable differences, JND). This information is necessary to develop the specific metrics because tiny variations in stimulus magnitude may not lead to a variation in sensation magnitude. It highlighted that the just noticeable difference in loudness becomes greater as the overall sound pressure level of the signal increases while the just noticeable difference in sharpness has very small variations related to the overall level. Referring to sound stimuli with sound pressure levels around 80 dB, 75% of subjects perceived a different hearing sensation when sounds had a loudness difference of at least 0.8 sone and a sharpness difference of 0.04 acum. This step size was chosen as the JND of loudness and sharpness for all the other investigations.

Effectiveness of an active noise control (see paragraph 4.3)

This study was aimed at verifying the feasibility of a simple active noise control (ANC) architecture. The sound quality conditions were evaluated by means of subjective tests with regards to four different noise features, all related to the operator's comfort and working safety conditions: tiredness, concentration loss, loudness and booming sensation. It highlighted that the effect on the subjective responses of a selective reduction, due to the

Permissions

The contributors of this book come from diverse backgrounds, making this book a truly international effort. This book will bring forth new frontiers with its revolutionizing research information and detailed analysis of the nascent developments around the world.

We would like to thank Dr. Siano Daniela for lending his expertise to make the book truly unique. He has played a crucial role in the development of this book. Without his invaluable contribution this book wouldn't have been possible. He has made vital efforts to compile up to date information on the varied aspects of this subject to make this book a valuable addition to the collection of many professionals and students.

This book was conceptualized with the vision of imparting up-to-date information and advanced data in this field. To ensure the same, a matchless editorial board was set up. Every individual on the board went through rigorous rounds of assessment to prove their worth. After which they invested a large part of their time researching and compiling the most relevant data for our readers. Conferences and sessions were held from time to time between the editorial board and the contributing authors to present the data in the most comprehensible form. The editorial team has worked tirelessly to provide valuable and valid information to help people across the globe.

Every chapter published in this book has been scrutinized by our experts. Their significance has been extensively debated. The topics covered herein carry significant findings which will fuel the growth of the discipline. They may even be implemented as practical applications or may be referred to as a beginning point for another development. Chapters in this book were first published by InTech; hereby published with permission under the Creative Commons Attribution License or equivalent.

The editorial board has been involved in producing this book since its inception. They have spent rigorous hours researching and exploring the diverse topics which have resulted in the successful publishing of this book. They have passed on their knowledge of decades through this book. To expedite this challenging task, the publisher supported the team at every step. A small team of assistant editors was also appointed to further simplify the editing procedure and attain best results for the readers.

Our editorial team has been hand-picked from every corner of the world. Their multi-ethnicity adds dynamic inputs to the discussions which result in innovative outcomes. These outcomes are then further discussed with the researchers and contributors who give their valuable feedback and opinion regarding the same. The feedback is then collaborated with the researches and they are edited in a comprehensive manner to aid the understanding of the subject.

Apart from the editorial board, the designing team has also invested a significant amount of their time in understanding the subject and creating the most relevant covers. They scrutinized every image to scout for the most suitable representation of the subject and create an appropriate cover for the book.

The publishing team has been involved in this book since its early stages. They were actively engaged in every process, be it collecting the data, connecting with the contributors or procuring relevant information. The team has been an ardent support to the editorial, designing and production team. Their endless efforts to recruit the best for this project, has resulted in the accomplishment of this book. They are a veteran in the field of academics and their pool of knowledge is as vast as their experience in printing. Their expertise and guidance has proved useful at every step. Their uncompromising quality standards have made this book an exceptional effort. Their encouragement from time to time has been an inspiration for everyone.

The publisher and the editorial board hope that this book will prove to be a valuable piece of knowledge for researchers, students, practitioners and scholars across the globe.

List of Contributors

Alice Elizabeth González
Environmental Engineering Department, IMFIA, Faculty of Engineering, Universidad de la República (UdelaR), Uruguay

Sultan Aldırmaz and Lütfiye Durak–Ata
Department of Electronics and Communications Engineering, Yildiz Technical University, Turkey

Lei Qi and Zhengping Zou
Beihang University, China

Wen-Kung Tseng
National Changhua University of Education, Taiwan, ROC

Guoyue Chen
Akita Prefectural University, Japan

Emiliano Mucchi
Engineering Department, University of Ferrara, Italy

Elena Pierro
DIMeG, Politecnico di Bari, Italy

Antonio Vecchio
ARTEMIS Joint Undertaking, Brussels, Belgium

Wei Cheng
State Key Laboratory for Manufacturing Systems Engineering, Xi'an, PR China
Department of Mechanical Engineering, University of Michigan, Ann Arbor, USA

Zhousuo Zhang and Zhengjia He
State Key Laboratory for Manufacturing Systems Engineering, Xi'an, PR China

Sohei Tsujimura
Institute of Industrial Science, The University of Tokyo, Japan

Takeshi Akita
Tokyo Denki University, Japan

Stefano Bianchi and Alessandro Corsini
Sapienza University of Rome, Italy

Anthony G. Sheard
Fläkt Woods Ltd, UK

Paulo Henrique Trombetta Zannin, Daniele Petri Zanardo Zwirtes and Carolina Reich Marcon Passero
Federal University of Paraná, LAAICA – Laboratory of Environmental and Industrial Acoustics and Acoustic Comfort, Brazil

Harsha Vardhan
National Institute of Technology Karnataka, Surathkal, India

Rajesh Kumar Bayar
N.M.A.M. Institute of Technology, Nitte, India

Eleonora Carletti and Francesca Pedrielli
National Research Council of Italy, IMAMOTER Institute, Italy